Heterojunctions
and
Metal–Semiconductor
Junctions

Heterojunctions and Metal–Semiconductor Junctions

A. G. Milnes and D. L. Feucht

Carnegie-Mellon University
Pittsburgh, Pennsylvania

ACADEMIC PRESS   *New York and London*   *1972*

ACADEMIC PRESS, INC.
111 Fifth Avenue, New York, New York 10003

United Kingdom Edition published by
ACADEMIC PRESS, INC. (LONDON) LTD.
24/28 Oval Road, London NW1 7DD

LIBRARY OF CONGRESS CATALOG CARD NUMBER: 79-127693

PRINTED IN THE UNITED STATES OF AMERICA

To our families

Contents

List of Tables

Semiconductor heterojunction research is an important area of device study which developed from the research of the last decade on semiconductor epitaxy. The barriers introduced into the energy-band diagram by the energy-gap difference of two semiconductors allow a new degree of freedom to the device designer. In GaAs injection lasers, the addition of $Al_xGa_{1-x}As$ confinement barriers has resulted in a major reduction in the 300°K threshold current densities. In heterojunction transistors, interesting performance has been obtained for GaAs/Ge, ZnSe/Ge, ZnSe/GaAs, and Cds/Si structures, and considerable progress is being made with other heterojunction combinations. Electrooptical effects in heterojunctions that are promising include infrared to visible upconversion systems and the window effect in solar cells and phototransistors.

One form of heterojunction considered here, namely, the metal–semiconductor Schottky-barrier junction, has developed into remarkably widespread use in the last few years. For many device applications it has near-ideal $I–V$ characteristics, very low minority carrier storage, and excellent high frequency performance. Schottky-barrier junctions are of value in specialized applications as mixers, detectors, and avalanche photodiodes. The metal–semiconductor junction is also entering the computer and integrated circuit fields.

High-yield photocathodes, cold cathodes, and electron multipliers of the negative-electron affinity type represent yet another family of devices where heterojunctions are involved with considerable success.

Semiconductor–semiconductor heterojunctions and metal–semiconductor heterojunctions are therefore of significant practical importance today and also of considerable scientific interest, with worthwhile problems still to be explored and understood. Many classes of heterojunctions, we believe, will prove to have new and valuable applications.

Although some aspects of heterojunction behavior remain areas for continued scientific and technological study, the main outlines of the subject are clear. These we have attempted to present in this book. There are discussions of the major semiconductor–semiconductor heterojunction ideas and of metal–semiconductor concepts as well as a presentation of the significant experimental findings in each area. Information is also provided on semiconductor epitaxy processes, on etches, and on ohmic contact technology for many important semiconductors. An extensive bibliography is included for the convenience of our readers.

Acknowledgments

In the semiconductor heterojunction chapters we have drawn heavily on the work of our former and present graduate students including J. P. Donnelly, H. J. Hovel, D. S. Howarth, D. K. Jadus, G. O. Ladd, W. G. Oldham, S. S. Perlman, A. R. Riben, R. Sahai, K. J. Sleger, R. N. Sundelin, J. C. Veseley, and J. A. Wyand. Their contributions to the development of the subject are greatly appreciated. We appreciate also discussions with our colleagues, Professors R. L. Longini and A. G. Jordan and with our faculty visitors, Dr. K. Takahashi, Dr. M. S. Tyagi, and Dr. W. D. Baker. At other Universities we acknowledge helpful discussions with Professors R. L. Anderson, M. J. Hampshire, H. K. Henisch, H. Kroemer, and R. H. Rediker. We would also like to express thanks to Drs. M. B. Panish, I. Hayashi, A. D'Asaro, and H. Kressel, and a great many other hetero-junction investigators in industry with whom we have had interaction in the last few years.

In the metal–semiconductor area, we are heavily indebted to many investigators including M. M. Atalla, C. R. Crowell, D. Kahng, D. V. Geppert, C. A. Mead, F. A. Padovani, V. I. Rideout, D. L. Scharfetter, W. G. Spitzer, R. Stratton, and S. M. Sze. Their contributions to the subject have been so substantial that their papers form the source material upon which subsequent writers like ourselves are dependent.

In research funding we gratefully acknowledge the role of the Army Research Office, Durham, North Carolina (Dr. James Murray), the Air Force Cambridge Research Laboratories (Dr. A. C. Yang), and NASA Electronics Research Center, Boston, Massachusetts (Dr. W. C. Dunlap and Mr. D. E. Sawyer).

Our senior semiconductor technician, Mr. H. Reedy, has been an active participant in our heterojunction work and has contributed much to the buildup and effective functioning of our laboratory facilities in this and related areas. Special thanks are due to Betty T. Smith, Lynne Clark, and Effie Lipanovich for extensive editorial, typing, and drafting services.

xiii

List of Principal Symbols

Symbol	Definition	Dimensions
A, a	Area	cm^2
A_{ACT}	Active surface area of a solar cell	cm^2
A_{TOT}	Total surface area of a solar cell	cm^2
A	Richardson's constant	$A\ cm^{-2}\ {}^\circ K^{-2}$
A^*	Richardson's constant with orientation dependence	$A\ cm^{-2}\ {}^\circ K^{-2}$
A_e	Area of emitter	cm^2
a	Radius of Schottky barrier mixer diode	cm
B	Base transport factor	—
B	Magnitude constant: forward tunneling current	$A\ cm$
C	Capacitance	F
C_c	Collector capacitance	F
C_e	Emitter capacitance	F
C_i	Collector depletion region capacitance associated with the emitter area	F
C_s	Collector depletion region capacitance associated with area between base and emitter	F
C_S	Acceptor concentration in base at interface	cm^{-3}
$\bar{c}$	Mean carrier velocity	$cm\ sec^{-1}$
$D_{n,p}$	Diffusion coefficient (electron, hole) as minority carrier	$cm^2\ sec^{-1}$
d	Thickness of a region	cm
d_e	Emitter stripe width	cm
$\mathscr{E}$	Electric field (Chapter 7)	$V\ cm^{-1}$
E_b	Energy barrier for tunneling	eV
E_c	Energy of conduction band edge	eV
ΔE_c	Energy step in conduction band energy diagram of a heterojunction (associated with electron affinity difference)	eV
$E_{Fn,p}$	Fermi level, n and p semiconductor	eV
$E_{g,G}$	Band gap, electrical	eV
E_N	Electronegativity	eV
E_0	Electric field at interface	$V\ cm^{-1}$
E_{00}	Factor defined by Eq. (7.31)	—
E_v	Energy of valence band edge	eV
ΔE_v	Energy step in the valence band energy diagram of a heterojunction	eV

xv

List of Principal Symbols

Symbol	Definition	Dimensions
F	Flux of carriers over a barrier	$cm^{-2}\ sec^{-1}$
F	Force	$J\ cm^{-1}$
f	Frequency	$Hz\ sec^{-1}$
f_{max}	Gain-band width figure of merit [Eq. (3.1)]	—
g_i	Interfacial contact conductance	$mho\ cm^{-1}$
H	Electric field constant $= (2qN_{A2}/\epsilon_2)^{1/2}$	$V^{1/2}\ cm^{-1}$
h_{FE}	Transistor current gain, $I_c/I_e = \beta$	—
h	Planck's constant	$J\ sec$
$\hbar$	$h/2\pi$	$J\ sec$
I	Current	A
I_b	Base current of a transistor (large signal value)	A
I_c	Collector current	A
I_e	Emitter current	A
I_s	Saturation current of a reverse-biased junction	A
J_b	Base current density of a transistor	$A\ cm^{-2}$
J_c	Collector current density of a transistor	$A\ cm^{-2}$
J_{d0}	Diffusion component of reverse saturation current	$A\ cm^{-2}$
J_e	Emitter current density of a transistor	$A\ cm^{-2}$
J_L	Current density resulting from applied illumination	$A\ cm^{-2}$
J_0	Reverse leakage current density of a junction	$A\ cm^{-2}$
J_P	Current density of a photocell for maximum power output	$A\ cm^{-2}$
J_r	Electron capture current density from the emitter depletion region of a heterojunction transistor	$A\ cm^{-2}$
J_{rg0}	Depletion layer recombination–generation current density	$A\ cm^{-2}$
J_S	Recombination current density at the emitter–base interface of a heterojunction transistor	$A\ cm^{-2}$
J_T	Photocurrent density of a solar cell under short-circuit conditions	$A\ cm^{-2}$
J_{th}	Hole current density from the base valence band-to-band gap states in a heterojunction transistor	$A\ cm^{-2}$
$J_{n,p}; j_{n,p}$	Current density, electron, hole	$A\ cm^{-2}$
K	Drift field factor in the base region (p. 65)	—
$K_{n,p}$	Capture rates of holes and electrons into trap levels and interface states	$cm^3\ sec^{-1}$
$K_{1,2}$	Portions of diffusion voltage	—
k	Boltzmann's constant	$eV\ °K^{-1}$
k	Imaginary component of refractive index	—
L	Emitter stripe length	cm

Symbol	Definition	Dimensions
L_{B}	Effective base stripe length	cm
$L_{n,p}$	Diffusion length, electron, hole	cm
$l_{1,2}$	Depletion layer widths	cm
m^*	Electron effective mass	kg
$N_{\mathrm{A,a}}$	Acceptor density	cm^{-3}
N_{be}	Doping impurity density in the base at the emitter edge	cm^{-3}
N_{bc}	Doping impurity density in the base at the collector edge	cm^{-3}
N_{c}	Effective density of energy states at the conduction band edge	cm^{-3}
$N_{\mathrm{D,d}}$	Donor density	cm^{-3}
N_{IS}	Interface state density	cm^{-2}
N_{r}	Density of recombination centers	cm^{-3}
N_{s}	Photon density at solar cell surface	cm^{-2}
N_{t}	Band-gap state density of traps at the energy level E_{t}	cm^{-3}
N_{v}	Effective density of energy states at the valence band edge	cm^{-3}
n	Free electron density in the conduction band	cm^{-3}
n	Refractive index, real part	—
n_{B}	Injected electron concentration in the base at the interface	cm^{-3}
n_{b}	Electron concentration in the base	cm^{-3}
n_{e}	Electron concentration in the emitter	cm^{-3}
P	Tunnel probability	—
P	Polarization dipole produced at a heterojunction interface by strain	
p	Hole density	cm^{-3}
p_{B}	Hole concentration in the base at the interface	cm^{-3}
$p_{\mathrm{b,e}}$	Hole density, base, emitter	cm^{-3}
Q_{s}	Stored charge density in the emitter	C cm^{-3}
Q_{IS}	Total charge per unit area on interface states	C cm^{-2}
q	Charge of an electron, 1.6×10^{-19}	C
R	Interface reflection coefficient (Chapter 4)	—
R	Optical reflection coefficient (Chapter 5)	—
R_0	Differential resistance in the double Schottky diode model	ohm
R_{sb}	Sheet resistance of base diffused layer	ohm/square
R_{sbe}	Sheet resistance of base diffused layer under the emitter	ohm/square
r_{b}	Effective base resistance	ohm
r_{b}'	Defined by $r_{\mathrm{be}} + r_{\mathrm{s}}$	ohm
r_{be}	Base resistance associated with the emitter stripe area	ohm

Symbol	Definition	Dimensions
r_c	Series resistance of the collector bulk	ohm
r_{con}	Base contact resistance	ohm
r_e	Small-signal resistance of the emitter junction	ohm
r_s	Resistance of the base region between the base and emitter stripes	ohm
S	Effective recombination velocity of a surface	cm sec^{-1}
S_n	Effective interface state recombination velocity limited by electron capture	cm sec^{-1}
T	Absolute temperature	°K
T_0	Term in Eq. (7.27)	°K
T_r	Photon transmission coefficient	—
t_r	Transit time	sec
V_a	Voltage applied across a junction	V
$V_{Dn,p}$	Diffusion voltage, n- or p-type side	V
V_{oc}	Open circuit voltage of a photocell with illumination applied	V
V_P	Voltage of a photocell for maximum power output	V
$v\alpha$	Defined by Eq. (7.17)	cm sec^{-1}
v_{ds}	Saturated drift velocity	cm sec^{-1}
v_R	Effective recombination velocity at a Schottky barrier interface	cm sec^{-1}
v_{se}	Saturated limited drift velocity of electrons in the collector	cm sec^{-1}
W_b	Base width	cm
W_c	Collector width	cm
W_D	Collector depletion region width	cm
W_E	Width of semi-insulating emitter region	cm
w	Thickness of epitaxial layer	cm
X	Transmission coefficient for carriers across the junction	—
$x_{n,p}$	Depletion distances	cm
α	Transistor dc gain, J_c/J_e	—
α_1	Optical absorption coefficient	cm^{-1}
β	Transistor current gain, common emitter, J_c/J_b	—
β_T	Dependence of energy gap on temperature	eV °K^{-1}
γ	Emitter injection efficiency	—
$\Delta\phi$	Barrier lowering produced by electric field	eV
$\hat{\delta}_{n,p}$	Separation in energy of the Fermi level and the respective energy band edge	eV
ϵ	Dielectric constant of the semiconductor	F cm^{-1}
ϵ_0	Dielectric constant of vacuum	F cm^{-1}

Symbol	Definition	Dimensions
η	Factor in $\exp(qV/\eta kT)$	—
η	Efficiency of a solar cell	—
μ	Mobility of carriers	$cm^2\ V^{-1}\ sec^{-1}$
μ	Micron	$10^{-4}\ cm$
ν	Frequency of incident light	sec^{-1}
ν	Frequency of oscillations of electrons in the valence band	sec^{-1}
ν	Excess phase factor in a transistor associated with graded base doping	—
ρ	Resistivity	ohm cm
ρ	Reflection coefficient	—
σ	Capture cross section of trap	cm^2
τ_B	Lifetime in the base	sec
τ_b	Transistor base transit time	sec
τ_c	Transistor collector time delay	sec
τ_{cs1}	Collector depletion layer transit time	sec
τ_e	Emitter diode charging time	sec
τ_{ec}	Total time delay of a transistor	sec
$\tau_{n0,\,p0}$	Lifetime of electrons (holes) as minority carriers in a semiconductor	sec
τ_{sl}	Time delay associated with transit time of electrons across collector depletion region	sec
τ_{tr}	Transit time	sec
ϕ_B	Barrier height	eV
ϕ_m	Work function of a metal	eV
ϕ_m	Dipole effect in a heterojunction	eV
ϕ_n	Quasi-Fermi level (Chapter 7)	eV
ϕ_s	Work function of a semiconductor	eV
χ	Electron affinity of a semiconductor	eV
ψ	Electron potential energy	eV

Chapter 1 | Introduction to Semiconductor Heterojunctions

1.1 Heterojunctions

Heterojunctions are contacts with interesting electrical or electrooptical properties between two different materials. The junctions discussed in this work are (a) those between two different semiconductors, such as GaAs and Ge; (b) those between metals and semiconductors, such as the Schottky barrier formed by Au on Si; and (c) those between metals and semiconductors which form ohmic contacts.

Although metal contacts on insulators or very high energy gap semiconductors such as SiO_2 are briefly mentioned (Chapter 6), the properties of metal/insulator/semiconductor structures, such as the MOS transistor, are not considered here.

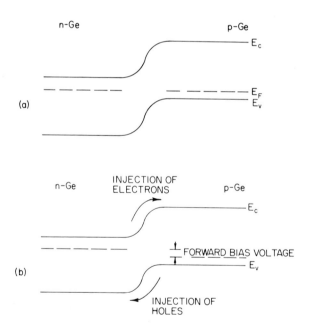

Fig. 1.1. *Energy-band diagram for an n–p Ge homojunction.*
(a) No external applied voltage bias; (b) with forward bias voltage applied (*p*–Ge side made positive).

As a starting point for this discussion of semiconductor heterojunctions, some knowledge of $p-n$ semiconductor homojunctions is assumed. The energy band diagram of Fig. 1.1(a) will therefore be recognized as that of a $n-p$ Ge diode with no voltage bias applied. For comparison Fig. 1.2 represents the probable energy band diagram for a heterojunction of n–GaAs on p–Ge,

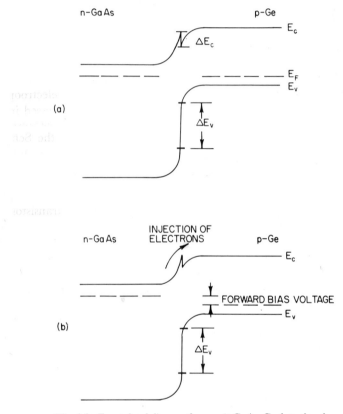

Fig. 1.2. *Energy band diagram for an n–p GaAs–Ge heterojunction.*
(a) No external applied voltage bias; (b) with forward bias voltage applied. Notice that the large barrier in the valence band inhibits injection from the p–Ge.

where (a) shows the zero bias condition and (b) the small forward bias condition. The essential difference to notice between the homojunction and heterojunction diagrams of Figs. 1.1 and 1.2 is the presence of the very large barrier in the valence band of the heterojunction which prevents the injection of holes from p–Ge into n–GaAs. Therefore, p–Ge could be a very heavily doped p^+-type, and the hole flow from p^+ to n should be negligible relative to the electron flow from n to p^+. As we shall show later, this is an important feature of semiconductor heterojunction transistors with wide-gap emitters and heavily doped base regions.

The energy band diagram of a heterojunction is complicated somewhat more than that of a homojunction because of the presence of the energy steps ΔE_c and ΔE_v. These steps arise from energy-gap and work-function differences of the semiconductors. The meaning of such a heterojunction energy diagram is best understood by considering the construction of a particular diagram as an example.

1.2 Construction of a Heterojunction Energy Band Diagram

Consider the energy band diagram for GaAs on Ge, assuming that the bulk properties of the semiconductors are valid right up to the interface where there is an abrupt transition from one material to the other. The energy gaps of the GaAs and Ge will be taken as 1.45 and 0.7 eV. The work function, ϕ_s, of a semiconductor, as for a metal, is defined as the energy required to take an electron from the Fermi level through the bulk surface to the energy level of free space (or vacuum) outside the material. The work function therefore, being dependent on the Fermi level, varies as a function of the doping level. Electron affinity, defined as the energy to take an electron from the conduction band edge to the vacuum level, is more convenient to use since it is a material property that is invariant with normal doping.

Many of the semiconductors of interest in heterojunctions are of the diamond structure (two interpenetrating face-centered cubic forms). This structure is characterized by the cube edge distance, a. The distance between an atom and its four nearest neighbors in the diamond structure is $a\sqrt{3}$. GaAs and Ge have lattice constants at 300°K which are almost identical (within 0.08%) and coefficients of linear expansion with temperature which are also very similar. These materials therefore form potentially very interesting heterojunctions.

To make the band diagram example specific, assume that the GaAs is n-type, doped to 10^{16} donors cm^{-3}, and that the Ge is p-type, doped to 3×10^{16} acceptors cm^{-3}. The relevant material properties are summarized in Table 1.1.

The construction of the energy band diagram proceeds by first drawing the energy diagrams for the two materials separately as in Fig. 1.3(a) with the vacuum level of energy common. Inspection shows that in Fig. 1.3(a), E_{F_p} is lower in electron energy than is E_{F_n}. Therefore, for equalization of the Fermi energy levels as the materials are considered to be brought into contact, it is necessary for a small number of electrons to pass from GaAs to Ge. This displacement of electrons at the interface bends up the E_c level in the

TABLE 1.1

Values Assumed in Construction of the *n–p* GaAs–Ge Heterojunction
Band Diagram of Fig. 1.3

	GaAs	Ge
Energy gap, E_g	1.45 eV	0.7 eV
Electron affinity, χ	4.07 eV	4.13 eV
Net donor doping, $N_D - N_A$	10^{16} cm^{-3}	—
Net acceptor doping, $N_A - N_D$	—	3×10^{16} cm^{-3}
$E_c - E_F = \delta_{GaAs}$	0.1 eV	—
$E_F - E_v = \delta_{Ge}$	—	0.14 eV
Lattice constant, a	5.654 Å	5.658 Å
Dielectric constant, relative	11.5	16

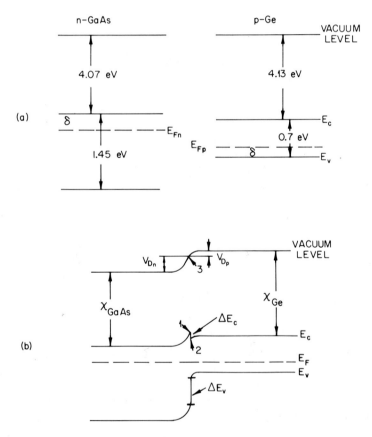

Fig. 1.3. *Construction (not drawn to scale) of the energy band diagram for n–p GaAs–Ge.*

GaAs. The band bending will be assigned the symbols V_{Dn} and V_{Dp}. The amount of the Fermi displacement $E_{F_p} - E_{F_n}$ that must be accommodated, from Fig. 1.3(a), is

$$E_{F_p} - E_{F_n} = (\chi_{Ge} + E_{g(Ge)} - \delta_{Ge}) - (\chi_{GaAs} + \delta_{GaAs}) = V_{Dn} + V_{Dp} \quad (1.1)$$

which in the present example reduces numerically to 0.52 eV. This energy difference corresponds to the total of the band bending, $V_{Dn} + V_{Dp}$. As in a homojunction, the transition regions are assumed to be completely depleted over the distances x_n and x_p, where

$$x_n/x_p = N_A/N_D \quad (1.2)$$

to satisfy charge conservation. The application of Poisson's equation then gives

$$V_{Dn} = N_D x_n{}^2/2\epsilon_{GaAs}$$

and

$$V_{Dp} = N_A x_p{}^2/2\epsilon_{Ge}$$

Whence

$$V_{Dn}/V_{Dp} = N_A \epsilon_{Ge}/N_D \epsilon_{GaAs} \quad (1.3)$$

In our example therefore, the ratio, V_{Dn}/V_{Dp}, is about $4:1$, whence V_{Dn} is 0.42 eV and V_{Dp} is 0.10 eV. The construction of the band diagram can proceed with the result shown in Fig. 1.3(b). The expression for the ΔE_c energy step, from simple geometrical considerations, is

$$\Delta E_c = \delta_{GaAs} + V_{Dn} - (E_{g(Ge)} - \delta_{Ge}) + V_{Dp} \quad (1.4)$$

Substitution from (1.1) gives the much more useful form

$$\Delta E_c = \chi_{Ge} - \chi_{GaAs} \quad (1.5)$$

The validity of this form can also be seen directly from Fig. 1.3(b) since the distance (3)–(1) is χ_{GaAs} and the distance (3)–(2) is χ_{Ge}. Therefore, in our example, ΔE_c is $4.13 - 4.07$ eV, which is 0.06 eV. (It should be noted that the diagrams of Fig. 1.3 are not drawn strictly to scale.)

The energy step in the valence band, again by simple geometrical considerations, can be seen to be

$$\Delta E_v = (E_{g(GaAs)} - E_{g(Ge)}) - (\chi_{Ge} - \chi_{GaAs}) \quad (1.6)$$

From (1.5) and (1.6) it follows that

$$\Delta E_c + \Delta E_v = E_{g(GaAs)} - E_{g(Ge)} \quad (1.7)$$

These conclusions about ΔE_c and ΔE_v and their sum are important in all heterojunction studies and to a first order approximation are valid irrespective of the doping levels. In Fig. 1.3 for the *n–p* GaAs–Ge heterojunction, ΔE_c is 0.06 eV and ΔE_v is 0.69 eV.

These values should be equally valid for Fig. 1.4 where the heterojunction is *p–n* GaAs–Ge. In Fig. 1.4, ΔE_v creates a substantial energy spike between the valence bands of the junction. Such energy spikes, if large, may limit hole injection and allow recombination at the interface to dominate the current flow.

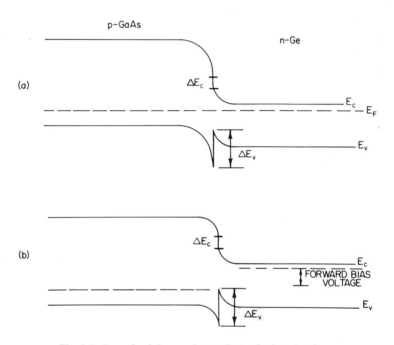

Fig. 1.4. *Energy band diagram for p–n GaAs–Ge heterojunction.*
(a) No external applied voltage bias; (b) with forward bias voltage applied.

The model developed in Fig. 1.3 assumes that there is no charge at the interface between the two semiconductors. Such charge will exist if there are energy states at the interface which may accumulate electrons or holes from one or both semiconductors. We shall see later that interface states are indeed important in heterojunctions between materials that have a difference in lattice constant of more than about 1% or that have large differences in their coefficients of expansion that may cause substantial strain disorder at the interface on cooling down from the growth temperature.

1.3 Semiconductor Heterojunction Pairs

For heterojunction studies to be effective, the semiconductors involved must be under good technological control. The elemental semiconductors, Ge and Si, and the III–V semiconductors are available with both n- and p-type doping, and their properties have been the subject of much study. The III–VI compound semiconductors are not usually available with both n- and p-type doping because of self-compensation effects that occur. For many II–VI compounds the crystal structure is hexagonal instead of cubic. The wide band gaps of many of the II–VI compounds, however, are of interest because of possible optical effects.

Table 1.2 summarizes some of the properties of the most important semiconductors used in heterojunction studies. Properties in this table such as energy gap, dielectric constant, and lattice constant are well established. Other properties such as mobilities depend on the degree of crystal perfection. The electron affinity values for most of the semiconductors listed may rest on a single determination. Even though precautions may have included ultrahigh vacuum conditions and surfaces cleaved *in situ*, there is uncertainty within a few tenths of an electron volt for many of the values quoted. Similar uncertainties exist with respect to the work functions of metals even in recent determinations under carefully controlled conditions.

1.3.1 *Semiconductor Pairs with Good Lattice Matches*

Inspection of Table 1.2 shows that Si and Ge have a lattice constant mismatch of about 4%. This may produce unpaired valence band electrons or dangling bonds of a density of 10^{14} cm^{-2} (Table 4.1) at the interface between the two semiconductors. Two consequences are to be expected: (1) bending of the energy bands at the interface and (2) extensive recombination of excess minority carriers in the interface region (Oldham and Milnes, 1964). With a lattice mismatch of a few percent and therefore interface state densities of 10^{14} cm^{-2}, the heterojunction behavior may be completely dominated by the interface states (see Section 2.5), and transistor action is not expected. Si–Ge junctions that have been fabricated bear this out by showing barriers that are not accounted for by electron affinity considerations. Also the current–voltage characteristics of Si–Ge junctions have temperature dependences that suggest tunneling action through interface states rather than injection over barriers.

If the interface state densities are 10^{13} cm^{-2} or less, the expected effect on the junction becomes much less severe. Current flow by injection can be observed in junctions where the lattice match is a fraction of 1%. This matter is considered more extensively in later chapters. For the present

TABLE 1.2

Properties of Some Semiconductors Used in Heterojunctions

Material	Energy gap 300°K (eV)	Gap transition	Mobility 300°K (cm² V⁻¹ sec⁻¹)		Dielectric constant (relative)	Lattice constant a (Å)	Temperature coefficient of expansion at 300°K ($\times 10^{-6}$ °C⁻¹)	Electron affinity (eV)	Typical dopants	
			Electron	Hole					p-type	n-type
Si	1.11	Indirect	1350	480	12.0	5.431	2.33	4.01	B, Al, Ga	P, As, Sb
Ge	0.66	Indirect	3600	1800	16.0	5.658	5.75	4.13	B, Al, Ga, In	P, As, Sb
AlAs	2.15	Indirect	280	—	10.1	5.661	5.2	—	Zn, Cd	Se, Te
AlSb	1.6	Indirect	900	400	10.3	6.136	3.7	3.65	Zn, Cd	Se, Te
GaP	2.25	Indirect	300	150	8.4 (op)	5.451	5.3	4.3	Zn, Cd	Se, Te
GaAs	1.43	Direct	5–8000	300	11.5	5.654	5.8	4.07	Zn, Cd, Ge, Si	Si, Sn, Ge, Se, Te
GaSb	0.68	Direct	5000	1000	14.8	6.095	6.9	4.06	Zn, Cd, Ge	Se, Te
InP	1.27	Direct	4500	100	12.1	5.869	4.5	4.38	Zn, Cd	Se, Te
InAs	0.36	Direct	30,000	450	12.5	6.058	4.5 (5.3)	4.9	Zn, Cd	Se, Te, Sn
InSb	0.17	Direct	80,000	450	15.9	6.479	4.9	4.59	Zn, Cd	Se, Te, Sn
ZnS (hex)	3.58	Direct	120	—	8.3	3.814	6.2–6.5	3.9	—	Cl, Br, Al
ZnSe	2.67	Direct	530	—	9.1	5.667	7.0	4.09	—	Br, Ga, Al
ZnTe	2.26	Direct	530	130	10.1	6.103	8.2	3.5	Cu, Ag, P	—
CdS (hex)	2.42	Direct	340	—	9.0–10.3	4.137	4.0	4.5	—	Cl, Br, I; Al, Ga, In
CdSe (hex)	1.7	Direct	600	—	9.3–10.6	4.298	4.8	4.95	—	Cl, Br, I
CdTe	1.44	Direct	700	65	9.6	6.477	—	4.28	Li, Sb, P	I
SiC (hex)	2.75–3.1	Indirect	60–120	10–20	10.2	3.082	5.7	—	Al	N
PbTe	0.29	Indirect	2500	1000	17.5 (op)	6.52	—	—	Te, Na, K	Pb, Cl, Br

discussion only semiconductors that match in lattice constant by closer than 1% will be considered as prime heterojunction pairs. Such heterojunction pairs are quite numerous as Table 1.3 shows. To keep the table within reasonable length it does not list heterojunctions involving ternary alloys such as $Al_xGa_{1-x}As$, $GaAs_xP_{1-x}$, or $ZnSe_xTe_{1-x}$. The binary alloy $Ge_{0.9}/Si_{0.1}$ is listed as a heterojunction with Ge since this is perhaps the nearest approach possible to successful use of the Si–Ge pair.

TABLE 1.3

Semiconductor Heterojunction Pairs with Good Lattice Match Conditions

Semi-conductor	Energy gap (eV)	Lattice con-stants (Å)	Energy gap structure	Expansion coefficient at 300°K ($\times 10^{-6}$ °C^{-1})	Hetero-junction preferred doping	Typical dopants	Electron affinity (eV)
$Ge_{0.9}Si_{0.1}$	0.77	(5.63)	Indirect	—	n	P, As, Sb	(4.1)
Ge	0.66	5.658	Indirect	5.7	p	Al, Ga, In	4.13
GaAs	1.43	5.654	Direct	5.8	n	Se, Te	4.07
Ge	0.66	5.658	Indirect	5.7	p	Al, Ga, In	4.13
ZnSe	2.67	5.667	Direct	7.0	n	Al, Ga, In	4.09
Ge	0.66	5.658	Indirect	5.7	p	Al, Ga, In	4.13
ZnSe	2.67	5.667	Direct	7.0	n	Al, Ga, In	4.09
GaAs	1.43	5.654	Direct	5.8	p	Zn, Cd	4.07
AlAs	2.15	5.661	Indirect	5.2	p	Zn	3.5
GaAs	1.43	5.654	Direct	5.8	n	Se, Te	4.07
GaP	2.25	5.451	Indirect	5.3	n	Se, Te	4.3
Si	1.11	5.431	Indirect	2.33	p	Al, Ga, In	4.01
AlSb	1.6	6.136	Indirect	3.7	n/p	Se, Te/Zn, Cd	3.65
GaSb	0.68	6.095	Direct	6.9	p/n	Zn, Cd/Se, Te	4.06
GaSb	0.68	6.095	Direct	6.9	n	Se, Te	4.06
InAs	0.36	6.058	Direct	4.5(5.3)	p	Zn, Cd	4.9
ZnTe	2.26	6.103	Direct	8.2	p	Cu	3.5
GaSb	0.68	6.095	Direct	6.9	n	Se, Te	4.06
ZnTe	2.26	6.103	Direct	8.2	p	Cu	3.5
InAs	0.36	6.058	Direct	4.5(5.3)	n	Se, Te	4.9
ZnTe	2.26	6.103	Direct	8.2	p	Cu	3.5
AlSb	1.6	6.136	Indirect	3.7	n	Se, Te	3.65
CdTe	1.44	6.477	Indirect	—	p/n	Li, Sb, P/I	4.28
PbTe	0.29	6.52	Indirect	—	n/p	Cl, Br/Na, K	—
CdTe	1.44	6.477	Direct	—	p	Li, Sb	4.28
InSb	0.17	6.479	Direct	4.9	n	Se, Te	4.59
ZnTe	2.26	6.103	Direct	—	p	Cu	3.5
CdSe(hex)	1.7	4.3($\sqrt{2}$) (6.05)	Direct	—	n	Cl, Br, I	4.95

If there are preferred doping conditions, dictated by compensation mechanisms or by energy spike considerations, these are shown in Table 1.3. For most of the pairs listed, the energy step, ΔE_c, in the conduction band is less than the step, ΔE_v, in the valence band. It is therefore desirable to have the wider gap semiconductor *n*-type so that injection of electrons into the smaller gap material is feasible as the mode of current flow without being impeded by a large energy spike that may increase interface recombination (compare Figs. 1.2 and 1.4).

Some typical energy band diagrams follow for several of these heterojunction pairs as illustrations of what may be expected from energy-gap and

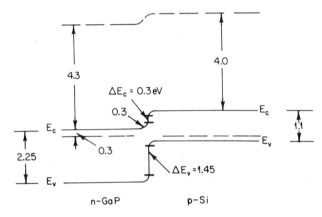

Fig. 1.5. *Energy band diagram for n–p GaP–Si heterojunction.*

electron-affinity differences. For *n–p* GaP–Si the ΔE_c step from the electron-affinity difference is 0.3 eV, and for the doping assumed, the conduction band barrier energy (Fig. 1.5) is a suitably convenient quantity such as 0.7 eV. Conditions are similarly favorable for *n–p* GaAs–Ge, *n–p* ZnSe–Ge, and *n–p* ZnSe–GaAs. These are discussed in Chapters 2 and 3. As a further example, the expected diagrams for AlSb–GaSb are given in Fig. 1.6. From these it seems that the pair might be worth study. Pairs such as *n–p* ZnTe–GaSb suffer from the problem that high resistance layers may form during fabrication which happens if Ga enters ZnTe.

Heterojunctions are possible between hexagonal and cubic crystal forms of different materials. Because of the similarity of hexagonal and cubic crystal forms, good lattice match conditions may be expected if the cubic lattice edge is $\sqrt{2}$ times the hexagonal lattice constant and the *c*-axis is grown perpendicular to a (111) plane. The pair *p–n* ZnTe–CdSe is included in Table 1.3 as an illustration of this and the expected band structure is shown in Fig. 1.7.

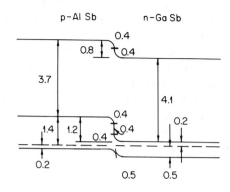

Fig. 1.6. AlSb–GaSb *heterojunction energy diagrams.*

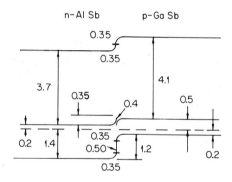

Fig. 1.7. *p–n* ZnTe–CdSe *heterojunction energy diagram.*

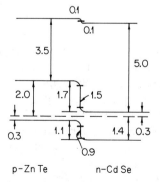

Most of the energy diagrams given are hypothetical since the heterojunction pairs have not been either fabricated or subjected to detailed study. Even for certain pairs which have been carefully fabricated and subjected to detailed I–V and capacitance studies, it is not easy to be sure of the band structure in the junction region. Some methods of fabrication have produced results that are dominated by the process technology rather than by the fundamental properties of the semiconductors themselves.

For favorable heterojunction action, the following appear to be desirable conditions:

(a) The semiconductors should be of the same crystal structure and closely matched in lattice constants—preferably to within $\frac{1}{2}\%$.

(b) The semiconductors should be reasonably matched in the temperature coefficient of expansion so that in cool-down from the growth temperature, high thermal strains do not develop. A slow cool-down period is usually desirable. The grown layer thickness may also be a factor in whether cracking or poor interface structure develops.

(c) The growth system must be carefully chosen to minimize autodoping effects of the heterojunction elements and cross diffusion of the dopants. Diffusion of course is a rapidly increasing function of the growth temperature and this is an additional reason, besides the thermal strain, for preferring low growth temperatures, such as 400–600°C, for heterojunctions. On the other hand, the lower the growth temperature the more risk there is that the surface mobility of the depositing atoms is too low to give good single-crystal interface structure.

A heterojunction pair such as GaAs–Ge is liable to suffer from cross-doping effects since both As and Ga are dopants in Ge, and Ge is a dopant in GaAs. A pair such as $Ge_{0.9}Si_{0.1}$–Ge does not have such problems.

Extreme cleanliness is desirable in the growth process to prevent inclusion of surface contamination at the interface. For example, contamination by vacuum grease should be avoided by the use of dry Teflon clamp seals. Thorough leak testing is desirable before every growth run, and growth boats should be chosen to be nonabsorbing so that out-gassing does not occur. Every precaution should be taken to obtain the seed surface as perfect and as clean as possible and to preserve this during the early stages of growth.

The ohmic contacts to heterojunctions usually require careful study to determine the most effective technology since many of the semiconductors involved in heterojunctions are ones that present special problems in this respect. These and related problems are discussed in more detail in Chapter 9, which reviews heterojunction growth experience.

Many other heterojunction pairs have been fabricated besides those given in Table 1.3. Such studies include: CdS–Si, CdS–SiC, CdS–Ge, CdS–ZnS,

CdS–ZnTe, CdS–Cu$_2$S, CdSe–Ge, CdTe–Cu$_{2-x}$Te, CdTe–HgTe, GaAs–GaP, GaAs–GaP$_x$As$_{1-x}$, GaAs–Ga$_{1-x}$In$_x$As, CdSe–ZnSe, GaAs–InP, GaAs–InSb, GaP–Ge, Al$_x$Ga$_{1-x}$As–GaAs, Ge–PbS, Si–Ge, Si–CdTe, Si–ZnTe, ZnS–Cu$_2$S, ZnO–Cu$_2$O.

References to these studies may be found by use of the index and bibliography at the end of the book.

The results occasionally have been interesting even though one of the semiconductors has been polycrystalline. CdS grown on Si is an example of such a pair and transistor action has been observed with polycrystalline CdS as the emitter (see Section 1.5.4). In general, however, most of the studies with heterojunction pairs that are significantly mismatched in lattice constants have shown tunneling and interface recombination actions that are not particularly useful.

1.4 Performance Features of Heterojunctions

1.4.1 Heterojunction Transistors

A heterojunction transistor is normally one in which the emitter material is of wider energy band gap than the base and collector material. Figure 1.8(a) shows the energy band diagram for a homojunction transistor, and Fig. 1.8(b) shows the energy diagram for a wide-gap emitter transistor under slight bias conditions. In the homojunction transistor the barrier for the movement of electrons, j_n, from emitter to base is the same height as the barrier for the movement of holes, j_p, from the base to the emitter. The injection efficiency of the transistor, defined as the fraction of the total emitter current that is minority injection current, is therefore

$$\gamma = j_n/(j_n + j_p) \tag{1.8}$$

Under normal operating conditions the value of γ is, of course, an upper limit on the value of α the current gain, J_c/J_e, of the transistor in the common–base configuration. In the common-emitter configuration the current gain is given by

$$\beta = J_c/J_b = \alpha/(1 - \alpha) \tag{1.9}$$

and thus an upper limit on β is given by

$$\beta = j_n/j_p \tag{1.10}$$

For high current gain it is therefore highly desirable that the base of the transistor be lightly doped with respect to the doping level in the emitter.

This reduces the number of holes available for reverse injection into the emitter and so the ratio j_n/j_p is suitably large. The relatively light doping in the base creates a number of problems such as lateral bias effects caused by current flow in the base resistance. Lateral bias effects contribute to non-uniform emitter current flow paths, known as emitter crowding effects, that

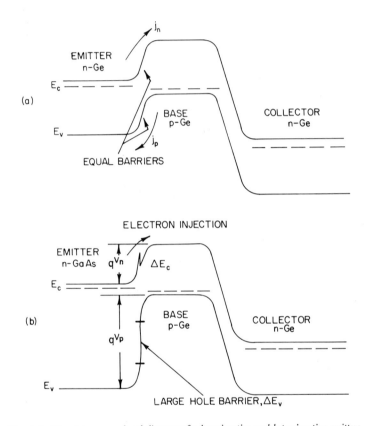

Fig. 1.8. *Transistor energy band diagrams for homojunction and heterojunction emitters.*

must be solved by emitter and base contact interdigitation. Other effects of relatively high base resistance may be "reach-through" of the base depletion region from the collector to the emitter at a low base–collector voltage level. The lateral bias effects may also contribute to "second-breakdown" in a transistor due to current pinch-in effects. High base resistance may also be a factor in the frequency response of a transistor because of the $r_b C_c$ time constant. The transistor designer has to keep all of these factors under control and in balance.

An exciting feature of the heterojunction transistor, however, is that the wide-gap emitter introduces an extra barrier for reverse injection from base to emitter as shown in Fig. 1.8(b). The importance of this feature was first recognized by Shockley (1951) in U.S. Patent 2,569,347. From inspection of Fig. 1.8(b) the extra barrier is of magnitude

$$\Delta E = qV_p - qV_n = \Delta E_c + \Delta E_v \tag{1.11}$$

Therefore from Eq. (1.7),

$$\Delta E = E_{g(GaAs)} - E_{g(Ge)} = \Delta E_g \tag{1.12}$$

and is the difference of the energy gaps of the two semiconductors involved.

The ratio of j_n/j_p may now be determined by the application of the Boltzmann relation. The injected electron density n_b just inside the base region at the base–emitter junction is

$$n_b/n_e = \exp(-qV_n/kT) \tag{1.13}$$

Similarly the injected hole density p_e just inside the emitter region at the base–emitter junction is

$$p_e/p_b = \exp(-qV_p/kT) \tag{1.14}$$

Following the simple Shockley diffusion model, the injected electron current is given by

$$j_n = qD_n\, dn/dx = qD_n n_b/W_b \tag{1.15}$$

where W_b is the base width and D_n is the diffusion constant for the electrons. The reverse injected hole current in the emitter is given by

$$j_p = qD_p\, dp/dx = qD_p p_e/L_p \tag{1.16}$$

where L_p is the diffusion length for holes in the emitter (assuming the emitter width is several times the diffusion length). The ratio of the currents is then from (1.12)–(1.16)

$$\begin{aligned}
j_n/j_p &= L_p D_n n_b/W_b D_p p_e \\
&= (L_p D_n n_e/W_b D_p p_b)\, \exp[(qV_p - qV_n)/kT] \\
&= (L_p D_n N_D/W_b D_p N_A)\, \exp(\Delta E_g/kT)
\end{aligned}$$

where N_D/N_A is the emitter/base doping ratio.

For a homojunction ΔE_g is zero, hence the improvement factor created by the heterojunction is seen to be $\exp(\Delta E_g/kT)$. This factor (Kroemer, 1957a) is normally quite large in size. For example, if ΔE_g is 0.2 eV and kT is 0.026 eV at room temperature, then the factor $\exp(\Delta E_g/kT)$ is about 3000. In a homojunction transistor the emitter is normally doped at least a hundred times larger than the base for good injection efficiency. In a heterojunction transistor this ratio may be inverted allowing base dopings of the order of 10^{19} atoms cm^{-3}. The consequences of this are discussed in Chapter 3 where the performance of heterojunction transistors is considered.

In homojunction transistors with lightly doped base regions, the alpha of the transistor falls off as the injected current is increased. This is partly attributable to the fact that at high injection the electrons cause increased hole density in the base (for space-charge limitation) and therefore increased reverse injection occurs. This conductivity modulation of the base would not be significant in the heavily doped bases of heterojunction transistors. Thus improvement may be hoped for in the high-injection alpha of transistors by the use of heterojunctions.

1.4.2 *Heterojunction Photodiodes and Phototransistors*

When light is incident on a heterojunction from the high energy-gap side, photons of energy between the two energy gaps, E_{g1} and E_{g2}, pass freely through the wide-gap material and are absorbed very near the junction in the low energy-gap side. This is shown schematically in Fig. 1.9(a) for a ZnSe–Ge junction where light of energy 2.6–0.7 eV passes through the ZnSe and creates electron–hole pairs at the interface with the Ge. This is known as the window effect in heterojunctions. For comparison a diffused silicon homojunction photocell is sketched in Fig. 1.9(b). The photons of energy greater than 1.1 eV are absorbed a micron or so within the surface and the lifetime of carriers in the diffused layer tends to be short ($< 10^{-8}$ sec). For good collection efficiency it is therefore necessary to keep the diffused layer thin and this tends to result in lateral resistance problems which influence the collection of induced carriers and reduce the output power available from the photocell.

The open-circuit output voltage that can be achieved from a heterojunction photocell is discussed in Chapter 5. This output voltage is comparable to that of the equivalent small energy-gap homojunction cell. The potential advantage of the heterojunction cell essentially depends on the reduction of surface-recombination losses and of lateral sheet resistance losses that can be hoped for from the window effect.

In heterojunction phototransistors the window effect should also be effective in increasing the efficiency. If a phototransistor, either homojunction or heterojunction, is operated in a floating base connection, carriers generated in the base should cause the sensitivity of the transistor to be increased by a factor that is approximately the current gain β of the transistor since the emitter–base junction assumes a forward bias. (This is the same effect that

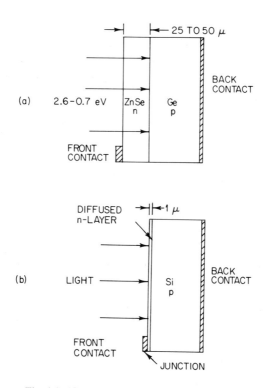

Fig. 1.9. *Heterojunction and homojunction photocells.*
(a) The thick epitaxial ZnSe layer in the ZnSe–Ge heterojunction cell produces window effect and low lateral resistance. (b) Diffused Si homojunction photocell.

causes the common–base I_{c0} leakage current of a transistor to become $I_{c0}/(1 - \alpha)$ in the emitter–collector biased, open-base, mode.) In a hetero-junction transistor the Kroemer factor, $\exp(\Delta E_g/kT)$, should result in a very high injection efficiency and the β of the transistor is potentially large, being limited presumably by base or heterojunction interface recombination. Therefore, the photoquantum efficiency, namely the number of electrons collected as photocurrent per photon of light applied to the junction, should be large for a heterojunction transistor in the window range of energy which allows carrier generation within the base region.

1.5 Other Heterojunction Concepts

In this and later sections, other heterojunction concepts are reviewed. Some of these are already in practical use, for example, $Al_{0.5}Ga_{0.5}As$ layers are added to GaAs injection lasers and lower the threshold currents for laser action to a very important extent.

Many of the concepts and effects described here have been demonstrated. However, some may not operate at the performance levels needed to be

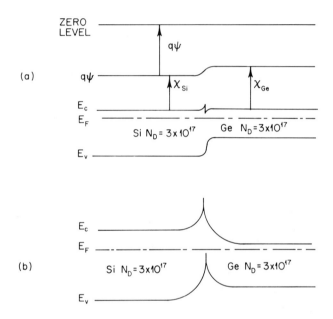

Fig. 1.10. *Energy band diagrams of an n–n Ge–Si heterojunction in equilibrium.*
(a) An electron affinity difference of 0.18 eV is assumed and interface state effects are ignored. (b) Same structure with interface states assumed that accept electrons. (After Oldham and Milnes, 1964.)

competitive with alternative approaches. The technological problems of fabrication seem to be limiting factors, at present, in a number of the concepts. In a few instances the concept is hypothetical and unlikely to be realizable but is included for interest.

A variety of diode concepts will be presented followed by a discussion of transistors of special types, infrared to visible light converters, and laser structures incorporating heterojunctions. Metal–semiconductor device concepts are discussed separately in Chapter 7.

1.5.1 *n–n Heterojunction Diodes*

The structure formed between two different *n*-type semiconductors may be a rectifier if a barrier is developed in the conduction band because of the electron affinity difference. Figure 1.10(a) shows the conduction band barrier expected for a *n–n* Ge–Si heterojunction in the absence of interface states. In the presence of interface states which accept electrons, depletion regions may exist on both sides of the interface and the energy barrier may be considerably increased as shown in Fig. 1.10(b).

Isotype (*n–n* and *p–p*) heterojunctions are discussed in detail in Chapter 4. In reverse voltage rating they tend to be limited by tunneling to a few volts or a few tens of volts, depending on the doping. They have the advantage over *p–n* junctions of being majority carrier devices. Therefore, *n–n* junctions are potentially high-frequency diodes free of minority carrier storage effects. However, this is true also of metal–semiconductor Schottky barrier diodes and such diodes are usually easier to make than *n–n* heterojunction diodes. Good uses for *n–n* semiconductor heterojunction diodes are therefore not known at present. In photosensing applications *n–n* junctions might possibly have some advantage associated with the window effect that could not be matched by metal–semiconductor structures, but this has yet to be of practical importance.

1.5.2 *Heterojunction Tunnel Diodes*

In heavily doped *p–n* heterojunctions, the tunnel action observed is similar to that for homojunction tunnel diodes. The characteristics of a Ge–GaAs tunnel heterojunction are given in Fig. 1.11. The peaks and valleys tend to be somewhat intermediate between those of the related homojunction tunnel diodes. Heterojunction tunnel diodes have not been the subject of detailed study and no particularly interesting or useful feature is expected for them.

The fact that it is possible to fabricate tunnel diodes by certain heterojunction growth processes is taken as an indication that such processes produce very abrupt junctions.

1.5.3 *Heterojunction Diodes Formed by Interaction of a Metal with a Substrate Semiconductor*

Many of the II–VI semiconductors such as CdS, CdSe, and ZnSe may be doped only *n*-type because the possible acceptor dopants are deep-lying and self-compensation occurs by vacancy action. ZnTe is obtainable only in *p*-type form and CdTe may be obtained with both *p*- and *n*-type doping. One obvious approach to achieve injection and, perhaps, interesting optical properties from II–VI semiconductors is to form heterojunctions of them with other semiconductors that can be oppositely doped. The energy band diagram of a *p–n* ZnTe–CdSe heterojunction is given in Fig. 1.7.

A less direct approach involves the deposition of a metal on a semiconductor accompanied by heat treatment that results in interaction of the

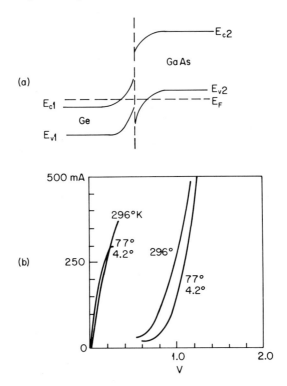

Fig. 1.11. *Characteristics of a* Ge–GaAs *vapor-grown tunnel heterojunction at three temperatures.* (a) Presumed energy band diagram; (b) *I–V* characteristics. [After Marinace (1960b), *IBM J. Res. Develop.* **4**, 280.]

metal with the substrate to produce a semiconductor–semiconductor heterojunction. For example, Te has been evaporated on CdS and the result is interpreted as a CdS/CdTe/Te structure (Fig. 1.12) (Dutton and Muller, 1968).

CdTe solar cells have been prepared (Cusano, 1963) with a copper film that results in heterojunctions between *n*-CdTe and *p*-type Cu_2Te. In related work by Aven and Cusano (1964), injection electroluminescence has been observed in Cu_2S–ZnS and Cu_2Se–ZnSe heterojunctions. The light emission occurs through hole injection from the p-type Cu chalcogenide into n-type ZnS or ZnSe.

Selenium rectifiers, commercially so important before the advent of silicon, have properties that are very much a function of the counterelectrode metals (Zn, Cd) and the electrical-forming process applied to develop the reverse-voltage characteristics. Amorphous selenium is a p-type semiconductor and semiconductor heterojunctions with n–CdSe or other n-type semiconductors probably develop during the forming process. Since inorganic or organic insulating layers are sometimes put on the selenium prior to the application of the counterelectrode metals, the structure is further complicated by these layers.

In work aimed at reducing the capacitance and reverse-bias leakage current of CdS thin film diodes, Muller and Zuleeg (1964) have interposed a thin layer of insulating material (CdTe, SiO_x, or Al_2O_3) between the CdS film and the metallic blocking contact. If the insulator is made thin enough it easily transmits currents of a useful magnitude by Schottky emission. The heterojunction energy band diagram proposed is shown for zero bias and forward bias conditions in Fig. 1.13. Considerable improvements in characteristics were obtained by adding the insulating layer.

1.5.4 *Heterojunction Strain Sensors*

The physical property which is of interest here is the piezoelectric effect that with strain produces a uniform polarization throughout a semiconductor material. This is capable of producing a discontinuity, P, in the field-independent polarization at a heterojunction interface. Such a discontinuity cannot occur in a homojunction since all such field-independent material polarizations are necessarily continuous across the p–n junction in the host crystal. In a heterojunction it may be expected that this discontinuity forces a change in the interface field strengths in the two semiconductors which, in turn, modifies the widths of the space-charge regions in the two materials and alters the diffusion voltages appearing across these regions. These various changes affect the response of the heterojunction to externally applied voltages, and this represents the usable output of the device (Moore 1969).

The materials of primary interest for the polarization discontinuity heterojunction are the III–V and II–VI compounds which crystallize in the cubic zinc blende and hexagonal wurtzite structures. The representatives of these compounds which crystallize in the zinc blende structure exhibit piezoelectric properties, and those which crystallize in the wurtzite exhibit pyroelectric and piezoelectric properties.

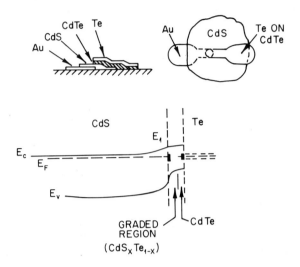

Fig. 1.12. *A heterojunction model of the energy-band diagram for an* Au–CdS–CdTe–Te *diode.* (After Dutton and Muller, 1968.)

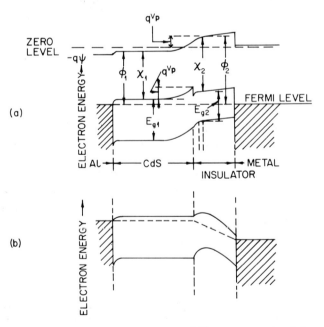

Fig. 1.13. *Assumed energy-band structure for* CdS–*insulator junction diodes.* (a) Zero voltage bias; (b) with external voltage applied. (After Muller and Zuleeg, 1964.)

A second type of material that is of potential interest is the ferroelectric semiconductor. One example of such a material is reduced $BaTiO_3$ which, in addition to its ferroelectric nature, has extremely large piezoelectric constants when in its polarized state.

When a heterojunction is formed between Si or Ge and a zinc blende material so that the interface plane is normal to the cubic body diagonal in both semiconductors, this forms a piezoelectrically active structure. In the simplest case, a stress applied normal to the interface induces a parallel polarization in the zinc blende material. Similarly, for a heterojunction between Si or Ge and a wurtzite compound with the interface normal to the body diagonal of the cubic and the c-axis of the hexagonal material, the resulting combination is both piezoelectrically and pyroelectrically active. For example, a polarization is produced in the wurtzite semiconductor normal to the interface by applying strain either normal or parallel to the interface plane or by uniform heating of the device. Analogous effects can be produced with many other materials including the ferroelectric semiconductors. These are of particular interest not only because of their strong piezoelectric properties but also because the remanent polarization offers a means of attaining a memory in a heterojunction device.

The behavior of such a system has been calculated by Moore (1969) in a study of the barrier height and C–V changes. This study neglects the effect of interface states. It concludes that the sensitivity could be one or two orders of magnitude larger than that available in bulk piezoresistive strain gages. At present, the concept is somewhat hypothetical since few practical studies have been made of such structures.

There are a variety of applications in which the polarization discontinuity heterojunction could have technical importance. As an electrical output device, the most obvious application would be in thermal or mechanical instrumentation where a low-impedance voltage source output characteristic is a significant advantage. The electrical output from a strain-sensitive heterojunction, might be useful in an electromechanical switch where the exponential change in current with applied strain is the output of interest. Utilizing a ferroelectric heterojunction it would be possible to produce an adaptive device or a memory element.

Other possible applications could be based on the use of optical outputs or interactions with optical signals as part of a combination stimulus. If light were used to modulate the carrier density on one side of a heterojunction, the output signal produced by a polarization discontinuity would be influenced by the incident light intensity. In a heterojunction light-emitting diode, a change in P could vary the carrier flux and hence modulate the optical output. These two types of operation would be particularly compatible with a strain-induced piezoelectric polarization acting as the interrogation mechanism.

In addition to these various single-device configurations, the use of element arrays of polarization discontinuity heterojunctions offers possibilities. First of all, the use of strain-sensitive elements would permit sequential excitation of the array via a propagating strain pulse in the substrate. The extremely stable timing possible with such pulses is well known and is an argument in favor of such a technique. A second stimulus could then be applied to the devices as an input signal (e.g., light) and the resulting electrical output used to recover this input information. This same principle could be used in an array of ferroelectric elements to perform the function of a memory array or with an array of light-emitting diodes to produce a sequentially excited optical display.

1.5.5 *Special Heterojunction Transistor Concepts*

Heterojunction transistors are the subject of detailed discussion in Chapter 3. In addition, a few special types will be discussed here briefly. These are: the graded band-gap-base transistor, the space-charge-limited emitter transistor, the "beam-of-light" transistor (photon transfer across the base) and the Auger transistor concept.

1.5.5.1 *Graded Band-Gap-Base Transistor.* To increase the high-frequency cutoff of a transistor, a drift field is usually built into the base by the impurity diffusion which results in heavier doping at the emitter side than at the collector side. At high-injection levels this built-in field may be reduced in its effect because of minority carrier space charge. In a heterojunction transistor the built-in field can usually be made higher than in a homojunction transistor since the base doping may be raised.

Kroemer has proposed quite another way of approaching the problem by the use of a graded band-gap-base as shown in Fig. 1.14. This produces an aiding field that is still effective at high-injection levels. The theory of the structure has been studied by Martin and Stratton (1966). The fabrication of such a transistor, however, presents considerable problems if the base width is very narrow.

1.5.5.2 *Space-Charge-Limited Emitter Transistors.* Single-injection current flow that is space-charge-limited can be obtained in a semiconductor if only one contact is injecting. This provides the basis for the space-charge-limited dielectric triode proposed by Wright (1960). Figure 1.15 shows one version of this device examined by Wright with Brojdo and Riley (Brojdo, 1963; Brojdo *et al.*, 1965). The CdS polycrystalline emitter film is about 10^4 Ω cm in resistivity and is evaporated onto a *p*-type base or gate region diffused into an *n*-type Si collector substrate. The band structure proposed for the triode when biased for normal operation is shown in Fig. 1.15(b).

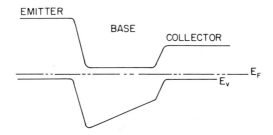

Fig. 1.14. *Energy-level diagram of an unbiased p–n–p graded-band-gap base transistor.*

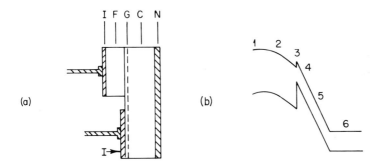

Fig. 1.15. *Space-charge-limited emitter heterojunction transistor.*
(a) Schematic illustration of physical structure of experimental triode: C, single-crystal silicon, 1 Ω cm, *n*-type; N, nickel-plated drain contact; G, surface-diffused gate layer, 0.02 Ω cm, *p*-type; F, vacuum-evaporated, crystallographically orientated, cadmium sulfide film, 10^4 Ω cm, *n*-type; I, vacuum-evaporated indium. (b) Band structure of triode biased for operation: 1, electron space-charge reservoir injected from indium contact; 2, region of space-charge-limited electron current; 3, electron accumulation layer due to conduction band discontinuity; 4, gate region of high mobile hole density and high "built-in" electric field; 5, depletion layer created by applied gate-drain voltage; 6, drain contact formed by silicon substrate. (After Brojdo *et al.*, 1965.)

The electron current in the cadmium sulfide layer is controlled by the potential difference applied across this layer between the source and gate electrodes. Current crosses the small discontinuity at the interface (expected to be rather less than 0.4 eV) and passes through the base layer in the silicon under the influence of the high "built-in" field created by the diffused acceptor density in this layer. Between base (gate) and collector (drain) the current is maintained by the high field existing in the depletion region separating these two electrodes. Mobile holes in the gate layer are constrained to remain in this layer by the large heterojunction discontinuity in the valence band at the interface and by the electric field in the depletion region between gate and drain. Good high-speed operation is expected because the current between virtual source and drain is everywhere carried by drift, and the series input resistance of the gate is low. This possible high-frequency performance is the prime reason for interest in this device.

In the structures fabricated and studied by Brojdo, current amplification factors as high as 10 were observed indicating interfacial transmissivities of up to 90%. The electron mobility in the CdS film was quite low (about 1 cm^2 V^{-1} sec^{-1}) and this limited the transconductances of the experimental devices to about 1 mA V^{-1}. The emitter–base (or source–gate) voltage was rather large compared with that of a normal transistor because the high-resistance emitter exhibited the expected I–V^2 dependence of single-injection space-charge-limited current flow.

CdS–Si transistors have been fabricated by D. J. Page and also by R. Zuleeg (1967) with rather similar results.

Since there is a large lattice mismatch (8%) between CdS and Si, it is noteworthy that interface transmissivities as high as 90% were observed. The reasons for this are probably:

(a) associated with the presence of an SiO$_2$ layer (20–50 Å thick) at the interface between the CdS and the Si. Riley (1966) found that such a layer was an essential part of the structure and reduced the interface recombination.

(b) CdS has smaller interface state tendencies than many other semiconductors. This is discussed in relation to CdS–metal barrier studies in Chapter 6. This may give reduced recombination at the interface. It is probably also helpful that defect structure caused by growth strain is largely located in the CdS where it has the least effect on recombination.

(c) the high fields throughout emitter and base regions of the device tend to improve the transmission to recombination ratios since carriers may be swept rapidly through regions containing recombination centers.

The high-frequency performance of the structure was not established by the Brojdo *et al.* study. However, they point out that at sufficiently high frequencies of operation, the interfacial recombination centers may be unable to retain equilibrium with the input signal and the small signal current amplification factor may increase. In theory, therefore, the performance of the triode may improve as the frequency of operation increases.

The ZnSe–Ge heterojunction transistors that are discussed in Chapter 3 are also somewhat space-charge-limited in their emitter since the ZnSe resistivity tends to be high (10^4–10^6 Ω cm) for certain growth technology. Here the lattice match is good, the ZnSe is single-crystal, and current gains up to about 50 are observed.

1.5.5.3 *Transistors with Photon Transfer across the Base.* Lesk *et al.* (1960) and Rediker *et al.* (1963) have proposed the fabrication of transistors with photon transfer across the base region. Such a structure with an infrared-emitting GaAs *p–n* junction as the emitter, and an *n*–GaAs, *p*–Ge junction as the collector is shown in Fig. 1.16. This beam-of-light transistor (BOLT) might, according to the proposers, have several advantages over conventional transistors. First, very low lifetime material can be used; as a matter of fact, it is advantageous for the radiative recombination lifetime to be very low to

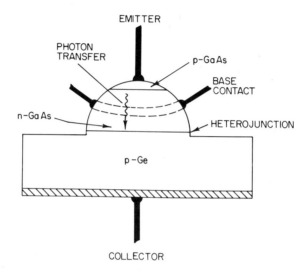

Fig. 1.16. *Schematic diagram of a beam-of-light (photon transfer) transistor.*
The transfer of carriers across the base is by photon emission at the emitter, and photon reabsorption at the collector. The mesa has been shown hemispherical in shape and should be coated with a reflecting coating so as to reflect the generated infrared to the collector junction. (After Rediker *et al.*, 1963.)

ensure that the emitter diffusion capacitance will be small. The transistor field is thus opened up to a number of low-lifetime semiconductors. Second, carrier injection need not be from emitter to base because it does not matter on which side of the *p–n* junction the infrared radiation is generated. Thus the base resistivity may be decreased and the emitter resistivity increased. Because of the geometry of the emitter region, the emitter series resistance will still be negligible and the base resistance can be significantly reduced with no increase in emitter transition-layer capacitance. Third, the transport from emitter to collector is at the speed of light in the material, and essentially all the infrared radiation incident on the collector junction produces carriers within one micron of this junction. As for a heterojunction photodiode, the resistivities of the Ge and GaAs at the collector junction must be adjusted so that on the one hand the photoproduced carriers can traverse the barrier at the heterojunction but on the other hand band-to-band tunneling and/or avalanche multiplication does not take place.

In order to collect a large percentage of the emitted radiation at the collector, the transistor surfaces should be covered with reflecting coatings and the transistor geometry should be such as to reflect the infrared toward the collector. Since laser action is obtainable for GaAs diode infrared sources with unity quantum efficiency, a near unity grounded-base current amplification may be possible without requiring reflection from all the exposed surfaces. It may be possible to produce a coherent mode at the emitter for which all the radiation is aimed at an appropriately placed collector.

A detailed comparison of the high-frequency capabilities of the beam-of-light transistor and the bipolar transistor has been made by Foyt (1963). However, the structures that have been fabricated in attempts to see this transistor action have been disappointing in performance. The problems of achieving the coherent radiation at room temperature needed to give the desired current gain and concentrating this radiation into the heterojunction collector are considerable. There is also the problem that the photon to electron conversion efficiency of the collector depends very much on the perfection of the n–p GaAs–Ge interface.

1.5.5.4 *Auger Transistor*. Under certain hypothetical band-gap conditions for a heterojunction emitter–base junction, the injected carrier could have sufficient energy to create two carriers in the base by the Auger effect.

Consider an abrupt heterojunction with a discontinuity in the conduction band of ΔE_c, which is larger than the narrower of the two band gaps. Assume that the wide-gap side of the junction is n-type while the narrower side is of either conductivity type. If such a heterojunction is biased in the forward direction, electrons will enter the narrow-gap semiconductor with a kinetic energy larger than the gap energy. The electrons will immediately seek to dissipate this excess energy, and one of the processes for doing this is what is commonly called an Auger impact ionization process; that is, the creation of an electron–hole pair, by the excitation of an additional electron from the valence band into the conduction band (Fig. 1.17). For an n–n heterojunction, this would lead to normally absent minority carrier effects due to the holes created in the valence band. For a p–n heterojunction it would lead to an enhancement of the electron injection. This second case is considered in more detail since it appears to be of possible use in transistors. To illustrate this, an n–p–n transistor is shown in Fig. 1.17 with a wide-gap

Auger emitter and a conventional collector. This transistor will inject electrons into the base as in a conventional transistor, but since these electrons are hot electrons in the base, some of them will create additional electrons by an Auger process. Both the primary and the secondary electrons can be collected by the collector, and, under favorable circumstances, the emitter-to-collector current amplification factor α may be larger than one.

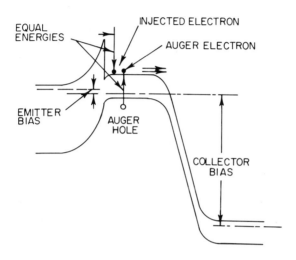

Fig. 1.17. *Energy band diagram for an n–p–n Auger transistor (hypothetical).* (After Kroemer, 1965.)

Transistors with $\alpha > 1$ are, of course, nothing new. They have been achieved by avalanche multiplication in the collector junction—a phenomenon of limited utility—or with an additional junction and an internal feedback action which restricts the operating speed. The Auger effect is essentially instantaneous and it does not require any modifications of the base–collector structure nor any external feedback element. Thus, Auger transistors should be possible for all frequency and power ranges for which conventional transistors can be built. In fact, since in a transistor with $\alpha > 1$, the base resistance provides positive rather than negative feedback, thinner base regions might be employed. Kroemer (1965) comments that for sufficiently large internal or external base resistances, the transistor will have a negative input impedance and it may be used as an oscillator up to microwave frequencies.

The Auger transistor concept suffers from a number of problems including the difficulty of achieving in practice the desired band structure with adequate injection of the hot electrons over the barrier.

1.6 Optical Properties of Heterojunctions in Image Conversion

In addition to the heterojunction photodiodes and phototransistors discussed earlier, there are a number of other optical effects where heterojunctions have potential. The most important are perhaps ideas for the up-conversion of infrared light to the visible range.

One concept is the use of a graded band-gap semiconductor studied by van Ruyven and Williams (1967a,b) and termed an anti-Stokes light converter. The idea is illustrated by the energy diagrams of Fig. 1.18. Infrared light is applied on the low band-gap side of the specimen and the induced minority carriers move towards the high band-gap side by external-field-aided transport and there produce higher energy photons. The few studies that have been made to date with graded crystals such as $Zn_xCd_{1-x}S$ have not shown high conversion efficiency.

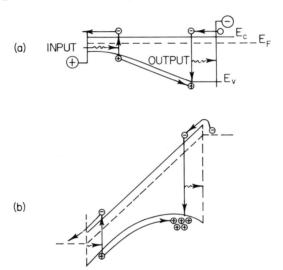

Fig. 1.18. *A graded band-gap semiconductor light converter (infrared to visible).*
(a) Energy band diagram of a constant majority carrier converter showing light conversion by diffusion of minority holes from the low energy gap to the wide energy gap region.
(b) Converter with superlinear valence band grading and applied external field to increase light conversion efficiency by hole accumulation. (After van Ruyven and Williams, 1967a.)

A solid-state infrared-wavelength converter based on a Ge–GaAs hetero-junction has been described by Kruse and co-workers (1967) for conversion of 1.5 μ radiation into 0.9 μ radiation. The structure is shown in Fig. 1.19. Holes created in the Ge by the 1.5 μ radiation are swept into the p–GaAs which is at a floating potential. This causes the p–n GaAs junction to become

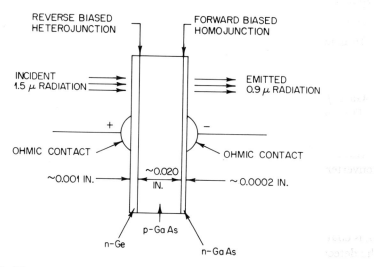

Fig. 1.19. *Three-layer heterojunction structure (n–Ge/p–GaAs/n–GaAs) for conversion of radiation from* 1.5 μ *to* 0.9 μ.
(After Kruse *et al.*, 1967.)

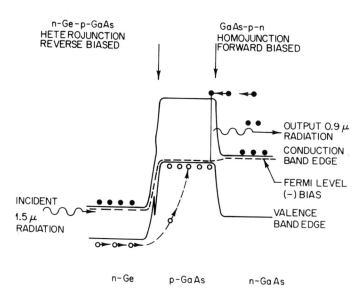

Fig. 1.20. *Energy band diagram for the three-layer radiation converter.*
(After P. W. Kruse.)

forward biased and injection occurs which results in the emission of recombination radiation at about the GaAs band-gap energy, i.e., 0.9 μ. The energy band is shown in Fig. 1.20. The internal wavelength conversion efficiency reported by Kruse was 2.8×10^{-5} and was believed to be limited principally by the low electroluminescent quantum efficiency of the GaAs p–n junction at low injection current densities.

In general, n–p Ge–GaAs heterojunctions are believed to have a valence band spike, ΔE_v, of about 0.7 eV as discussed in Section 1.2, and a conduction band spike of less than 0.1 eV. It is possible therefore that somewhat better results might be obtained with the dopings changed to be p–Ge/n–GaAs/p–GaAs instead of as shown in Fig. 1.19.

The sensitivity of such a system might be raised by adding a further Ge layer at the input side to provide gain increase by the injection effect that then occurs at the p–n Ge junction.

Another approach to the problem of a solid-state infrared to visible light converter is that of Phelan (1967). The structure is a sandwich composed of a capacitor InSb diode detector and a p–n GaAsP diode emitter, as shown in Fig. 1.21.

The radiation is incident on the detector through the semitransparent metal film and oxide layer. During the forward biasing pulses the capacitor, C, is charged and the emitter, E, yields pulses of visible light. Between pulses the detector, D, becomes reverse biased, and the capacitor discharge rate is

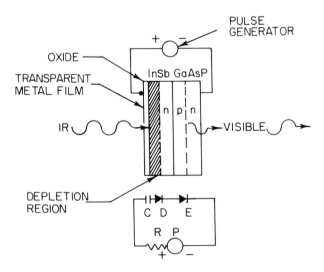

Fig. 1.21. *Structure and equivalent schematic for producing high-current driving pulses from low-level infrared radiation.*
(After Phelan, 1967.)

controlled by the infrared incident on the detector. Essentially, to obtain efficient conversion and allow for a tunable contrast and sensitivity, the low-level current produced by the infrared detector is integrated and stored by the capacitor and then delivered to the light emitter in short high-current pulses. The advantage of converting the low-level detector currents into higher level pulsed currents is that diode light emitters are much more efficient at the higher current levels.

The feasibility of such a pulsed device has been demonstrated using the InSb diode detector and a GaAs–P diode emitter to convert infrared radiation of wavelengths extending to 5.3 μ into visible radiation between 0.6 and 0.7 μ. The overall quantum efficiency of conversion of infrared photons to visible photons was about 10^{-4} (at an infrared power level of 100 μ W cm^{-2}) and was principally limited by the peak efficiency of the emitter.

1.7 Heterojunctions in Injection Lasers: GaAs/Al$_x$Ga$_{1-x}$As Structures

In the last few years heterojunctions have been found to be of great value in connection with semiconductor injection lasers. In 1963, Kroemer proposed that laser action might be improved in direct-gap semiconductors, such as GaAs, if it were possible to supply them with a pair of heterojunction injectors. These should be heavily doped semiconductor layers (one n^+-type and one p^+-type) with a higher energy gap than the radiating semiconductor layer. The essential idea is that the heterojunction barriers provide confinement of the injected carriers in the active GaAs region so that population inversion is more readily achieved and the laser gain is higher. Furthermore, the heterojunction layers, if Al$_x$Ga$_{1-x}$As, are lower in refractive index than the GaAs, and waveguide confinement of the light is also obtained which reduces the laser loss.

Within the last two years the threshold current density for 300°K GaAs injection laser action has been decreased from 25,000 A cm^{-2} to less than 1000 A cm^{-2} and cw operation of GaAs laser diodes has been achieved at and above room temperature. These impressive developments are reviewed in Chapter 5.

Chapter 2 | Semiconductor p–n Heterojunction Models

and Diode Behavior

Various models have been developed for current flow in p–n (n–p) semiconductor heterojunctions. The basic model is that of Anderson (1960a–c, 1962). This considers a heterojunction in which current flow is entirely by injection over the conduction or valence band barriers. This model gives the ideal performance against which we measure the observed behavior of heterojunctions. In practice current flow in a heterojunction is usually a sum of injection, tunneling, and interface state recombination components, and great care in fabrication is needed to minimize the latter two parts.

However, the Anderson model is a fundamental one and deserves close attention for the information it provides on the ideal situation. It also acts as a basis from which to develop a more general discussion that includes current flow by tunneling and interface state recombination.

2.1 Anderson's Model of n–p and p–n Heterojunctions

Consider the energy-band profile of two isolated pieces of semiconductor shown in Fig. 2.1(a). The two semiconductors are assumed to have different band gaps, E_g, different dielectric constants, ϵ, different work functions, ϕ_{WF}, different electron affinities, χ. Work function and electron affinity are defined, respectively, as that energy required to remove an electron from the Fermi level, E_F, and from the bottom of the conduction band, E_c, to a position just outside of the material (the vacuum level). The top of the valence band is represented by E_V. The subscripts 1 and 2 refer to the narrow-gap and wide-gap semiconductors, respectively.

In Fig. 2.1(a), the band-edge profiles E_{c1}, E_{c2}, E_{v1}, and E_{v2}, are shown as "horizontal." This is equivalent to assuming that space-charge neutrality exists in every region. The difference in energy of the conduction-band edges in the two materials is represented by ΔE_c and that in the valence-band edges by ΔE_v. In Fig. 2.1(a) the Fermi level E_{F1} is higher than E_{F2}. This cannot continue to exist if the two materials are brought together to form a junction. For equilibrium in the junction, the Fermi levels must attain the same energy by a transfer of electrons from semiconductor (1) to semiconductor (2). This results in a partial depletion of electrons near the junction in semiconductor (1) and therefore a bending upward of the band edges. There is also a corresponding redistribution of charge in semi-conductor (2) and bending down of the band edges as shown in Fig. 2.1(b). Within any single semiconductor, the electrostatic potential difference

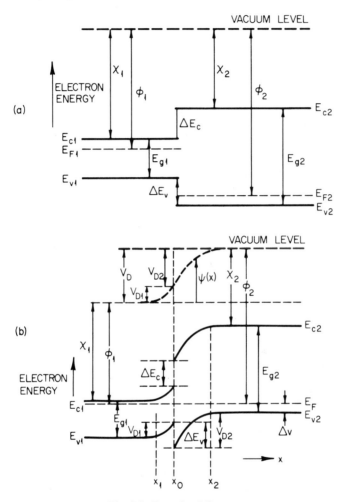

Fig. 2.1. *Energy band diagrams.*
(a) For two isolated semiconductors in which space-charge neutrality is assumed to exist in every region and (b) for n–p heterojunction at equilibrium (no external voltage applied). (After Anderson, 1962.)

between any two points can be represented by the vertical displacement of the band edges between these two points, and the electrostatic field can be represented by the slope of the band edges on a diagram such as Fig. 2.1(b).

Then the difference in the work functions of the two materials is the total built-in voltage, V_D. This is equal to the sum of the partial built-in voltages, $V_{D1} + V_{D2}$ where V_{D1} and V_{D2} are the electrostatic potentials supported at equilibrium by semiconductors 1 and 2, respectively. Since voltage is continuous in the absence of dipole layers and since the vacuum level is parallel to the band edges, the electrostatic potential difference, ψ, between any two points is represented by the vertical displacement of the vacuum level between these two points. Because of the difference in dielectric constants in the two materials, the electrostatic field is discontinuous at the interface.

Since the vacuum level is everywhere parallel to the band edges and is continuous, the discontinuity in conduction-band edges, ΔE_c, and valence-band edges, ΔE_v, is invariant with doping in those cases where the electron affinity and band gap E_g are not functions of doping, i.e., nondegenerate material.

Solutions to Poisson's equation with the usual assumptions of a Schottky barrier give for the transition widths on either side of the interface for a step junction in the presence of an applied voltage V_a

$$(X_0 - X_1) = \left[\frac{2}{q} \frac{N_{A2}\epsilon_1\epsilon_2(V_D - V_a)}{N_{D1}(\epsilon_1 N_{D1} + \epsilon_2 N_{A2})} \right]^{1/2} \tag{2.1a}$$

$$(X_2 - X_0) = \left[\frac{2}{q} \frac{N_{D1}\epsilon_1\epsilon_2(V_D - V_a)}{N_{A2}(\epsilon_1 N_{D1} + \epsilon_2 N_{A2})} \right]^{1/2} \tag{2.1b}$$

and the total width W of the transition region is

$$W = (X_2 - X_0) + (X_0 - X_1) = \left[\frac{2\epsilon_1\epsilon_2(V_D - V_a)(N_{A2} + N_{D1})^2}{q(\epsilon_1 N_{D1} + \epsilon_2 N_{A2})N_{D1}N_{A2}} \right]^{1/2} \tag{2.2}$$

The relative voltages supported in each of the semiconductors are

$$(V_{D1} - V_1)/(V_{D2} - V_2) = N_{A2}\epsilon_2/N_{D1}\epsilon_1 \tag{2.3}$$

where V_1 and V_2 are the portions of the applied voltage V_a supported by materials 1 and 2, respectively. Of course $V_1 + V_2 = V_a$. Then $V_{D1} - V_1$ and $V_{D2} - V_2$ are the total voltages (built-in plus applied) for material 1 and material 2, respectively. We can see that most of the potential difference occurs in the most lightly doped region for nearly equal dielectric constants.

The transition capacitance per unit area is given by a generalization of the result for homojunctions

$$\frac{C}{a} = \left[\frac{qN_{D1}N_{A2}\epsilon_1\epsilon_2}{2(\epsilon_1 N_{D1} + \epsilon_2 N_{A2})} \frac{1}{(V_D - V_a)} \right]^{1/2} \tag{2.4}$$

For the energy band diagram of Fig. 2.1(b) the barrier to hole flow is much less than to electron flow. Therefore only hole flow need be considered. At zero bias voltage the barrier to hole flow from right to left is qV_{D_2} and in the opposite direction is $\Delta E_v - qV_{D_1}$, assuming that the holes do not suffer a collision in the region $x_1 - x_0$. In equilibrium the two oppositely directed fluxes of holes must be equal since the net current flow is zero. The balance equation is therefore, from flux considerations,

$$A_1 \exp[-(\Delta E_v - qV_{D_1})/kT] = A_2 \exp(-qV_{D_2}/kT) \qquad (2.5)$$

where the coefficients A_1 and A_2 depend on the doping levels and carrier effective masses.

Consider now a bias voltage V_a applied to the junction with semiconductor (2) positive, corresponding to a forward bias condition. The portions of the voltage that are dropped on the two sides of the junction are determined by the relative doping levels and are given by

$$V_2 = K_2 V_a, \quad \text{where} \quad K_2 = 1/(1 + N_{D_2}\epsilon_2/N_{D_1}\epsilon_1) \qquad (2.6)$$

and

$$V_1 = K_1 V_a, \quad \text{where} \quad K_1 = 1 - K_2 \qquad (2.7)$$

The expression for K_2 is for low forward bias conditions and neglects the effects of injected carriers on the field. The energy barriers are now $q(V_{D_2} - V_2)$ and $\Delta E_v - q(V_{D_1} - V_1)$ as in Fig. 2.2(a). The net flux of holes from right to left is therefore given by:

$$
\begin{aligned}
\text{Hole flux} = A_1 &\exp[-q(V_{D_2} - V_2)/kT] \\
&- A_2 \exp\{-[\Delta E_v - q(V_{D_1} - V_1)]/kT\} \qquad (2.8)
\end{aligned}
$$

From Eq. (2.5) this simplifies to

$$\text{hole flux} = A_1 \exp[-qV_{D_2}/kT] [\exp(qV_2/kT) - \exp(-qV_1/kT)] \qquad (2.9)$$

Hence the current voltage relationship is of similar form

$$I = A \exp[-qV_{D_2}/kT] [\exp(qV_2/kT) - \exp(-qV_1/kT)] \qquad (2.10)$$

If the current is limited by the rate at which holes can diffuse in the narrow-gap material rather than by space-charge recombination, then

$$A = XaqN_{A_2}(D_p/\tau_p)^{1/2} \qquad (2.11)$$

where the transmission coefficient X represents the fraction of those carriers having sufficient energy to cross the barrier which actually do so; D_p and τ_p are diffusion constant and lifetime, respectively, for holes in the narrow-gap material, and a represents junction area.

The term $\exp(qV_2/kT)$ is dominant in Eq. (2.10) in the forward bias condition. Since $V_2 = K_2V_a$, it follows that the current should vary approximately exponentially with forward voltage according to

$$I = A \exp[-qV_{D2}/kT] \exp(qK_2V_a/kT) \qquad \text{for} \quad K_2V_a \gg kT \qquad (2.12)$$

Later it will be shown that current–voltage behavior with temperature of most of the semiconductor heterojunction diodes that have been carefully

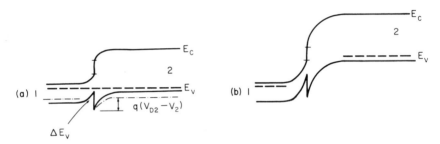

Fig. 2.2. *Heterojunction with applied bias voltages.*
(a) Forward bias (chain-dot lines): semiconductor (1) negative with respect to semiconductor (2). (b) Same heterojunction under reverse bias conditions.

studied does not generally conform with Eq. (2.12). The problem is that tunneling and recombination effects usually cannot be neglected.

A further complication associated with energy barrier spikes is worth pointing out at this stage. Figure 2.3(a) shows the energy diagram for a hypothetical heterojunction without forward bias; Fig. 2.3(b) shows the same junction with moderate forward bias applied (dotted lines). For these bias conditions the conduction band level of the left-hand side has dropped below the spike of energy at the interface. This would not be so at low forward bias conditions.

This change in the nature of the barrier as the bias is increased can be taken account of in the Anderson model and results in a change of the equations governing the *I–V* characteristics (and a slope change, or kink, in the curve). Caution, however, must be exercised in always interpreting such slope changes in observed characteristics of heterojunctions as due to this effect since there are other possible causes of slope changes (Donnelly and Milnes, 1966c).

The transmission factor X in Anderson's model, Eq. (2.11), is an unknown quantity. A study of this problem from a quantum mechanical reflection viewpoint has been made by Price (1962). Perlman and Feucht have considered the implications of a classical kinetic emission model (Perlman, 1964; Perlman and Feucht, 1964).

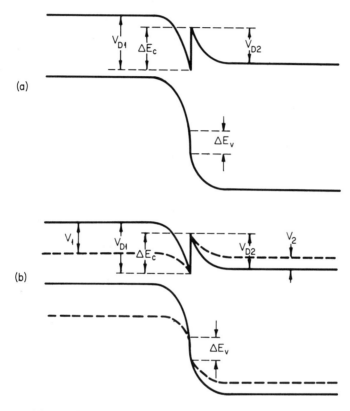

Fig. 2.3. *Energy band diagrams for p–n heterojunction (hypothetical).*
(a) With no forward bias, and (b) with forward bias $(V_1 + V_2)$ the dashed line conditions exist. Notice that the nature of the barrier to the movement of electrons from the right- into the left-hand side has changed considerably.

2.2 Studies of *n–p* Ge–GaAs Heterojunctions*

The energy band model proposed by Anderson can be confirmed by capacitance studies (see Section 2.4). However, when Eq. (2.10) is used to explain

* This notation implies that the Ge is *n*-type and the GaAs is *p*-type.

the observed characteristics of n–p Ge–GaAs heterojunctions, the results are unsatisfactory. Figure 2.4 shows such a comparison of the forward characteristics (Riben and Feucht, 1966b). The lack of agreement is seen to be both qualitative and quantitative. Comparisons of the Anderson model with the reverse characteristics observed are similarly unsatisfactory.

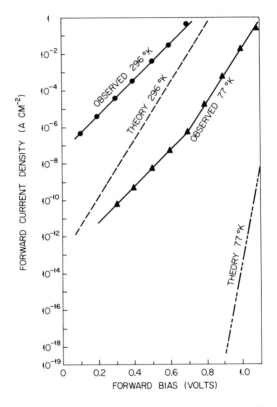

Fig. 2.4. *Comparison of the forward I–V characteristics of a typical n–p Ge–GaAs heterodiode with the theoretical characteristics of Eq. (2.10).*
(After Riben and Feucht, 1966b.)

For the forward characteristics in Fig. 2.4., many differences between the theoretical and experimental curves are evident. The theory predicts that the slope of the curve will change by approximately a factor of four when the temperature is changed from 296 to 77°K, but the experimental curves have essentially the same slope at both temperatures for voltages below 0.7 V and change by only a factor of 1.5 at higher voltages. Also the theory predicts a change in the current of about 16 orders of magnitude from 296 to 77°K whereas the decrease observed is only about 6 orders of magnitude, and the theoretical current is smaller than the experimental current at all voltages. One can conclude from this that although diffusion or emission current may be flowing, there is another current mechanism, which must be predominant, that is responsible for the larger observed current.

The same type of differences is evident in Fig. 2.5 for the reverse currents. The theory predicts an exponential variation of the current followed by

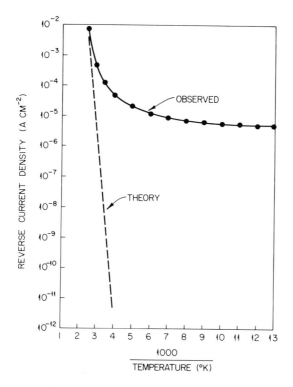

Fig. 2.5. *Comparison of the reverse current as a function of* $1/T$ *at a constant bias of* $-4V$ *of a typical n–p Ge–GaAs heterodiode with the theoretical characteristics of Anderson.*
(After Riben and Feucht, 1966b.)

saturation at higher voltages, but the experiments yield a linear current followed by a power law dependence of the current on voltage. Again, as in the forward direction, the current magnitude only changed by a small fraction of the predicted change as the temperature was reduced from 296 to 77°K. Since the actual currents are again much larger than would theoretically be expected, another current mechanism exists which predominates.

2.2.1 *Forward Bias Tunneling Models*

The basic form of the forward *I–V* characteristics of the *n–p* Ge–GaAs heterojunctions is that the slope of the exponential curve changes relatively slowly as a function of temperature. Also, the temperature dependence of the magnitude of the current is much smaller than is normally expected in semiconductor devices but is very similar to that given by Advani *et al.* (1962) for metal/oxide/metal structures. Since the metal/oxide/metal structure is predominantly a tunneling device and since the only other semiconductor device which exhibits such small temperature dependence is a tunnel diode, one is led to believe that the current mechanism predominant in these particular heterojunctions involves tunneling.

The simplest model that can be considered is that of carriers tunneling through the relatively large spike-shaped barrier in the valence band for *n–p* Ge–GaAs. The spiked barrier arises due to the energy discontinuity, ΔE_v, which was found to be 0.56 eV by capacitance studies. The tunneling current would be carried by holes in the valence bands of the two materials. (Physically, the electron must do the tunneling, but the tunneling of a valence electron in the Ge through the potential barrier into an available state in the GaAs valence band is mathematically identical to the tunneling of a hole from the GaAs into the Ge.) If the doping in the GaAs is uniform, the electric field in the tunneling region will be proportional to $(V_D - V)^{1/2}$.

The forward tunneling current for such a barrier was calculated with the aid of the WKB approximation (Riben and Feucht, 1966a,b). The integrals proved to be too difficult to solve in a closed form so the calculations were made on a computer. The tunneling current was found to be approximately a factor of three larger than the emission current caused by carriers passing over the barrier. The predominant tunnel flux was found to occur very near the peak of the barrier (within a range of about 0.1 eV). Since any current predicted by this model should have essentially the same temperature dependence as the emission current, this tunneling model can be disregarded.

A modification of this model in which the barrier was assumed to have a constant thickness (analogous to the model for metal/oxide/metal structures) was also considered. However, since the capacitance measurements showed that there was a square law dependence for at least 0.3 V forward bias, the current would be expected to vary as $\exp(qV_a/kT)$ for V_a less than 0.3 V. Since this was not observed experimentally, this approach is unsuccessful.

A third model that can be considered for the forward direction is one similar to that proposed by several authors for excess current in tunnel diodes. This model, depicted in Fig. 2.6(a), assumes that the Ge conduction electrons either fall into available band-gap states and then tunnel into the GaAs conduction band (path A) or tunnel into the available band-gap state and then fall into the valence band (path B). Also possible is the staircase path which involves multiple tunneling–recombination steps shown as path C. Assuming that the tunneling process is the rate-limiting process and that

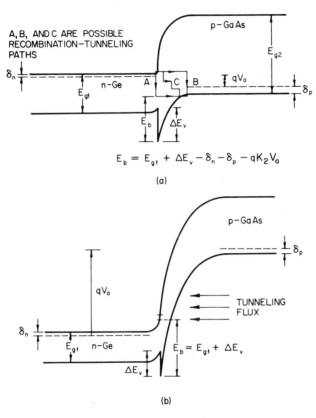

$$E_b = E_{g1} + \Delta E_v - \delta_n - \delta_p - qK_2V_a$$

(a)

(b)

Fig. 2.6. *Proposed tunneling models for the n–p* Ge–GaAs *heterojunction.* (a) Forward bias; (b) reverse bias. (After Riben and Feucht, 1966b.)

path A is the appropriate carrier path, Riben and Feucht (1966b) have shown that the current is

$$J_{ex} = BN_t \exp[-4(2m^*)^{1/2}E_b^{3/2}/3q\hbar\mathscr{E}] \qquad (2.13)$$

where B is a constant, N_t is the density of traps in the forbidden region, m^* is the electron effective mass in the forbidden region, E_b is the energy barrier that the electron must tunnel through, and $\mathscr{E}$ is the electric field in the depletion region on the GaAs side of the junction.

For nondegenerate materials, the energy barrier is given by [see Fig. 2.6(a)]:

$$E_b = E_{g1} + \Delta E_v - \delta_n - \delta_p - qK_2V_a$$

and

$$qV_D = E_{g1} + \Delta E_v - \delta_n - \delta_p$$

The field $\mathscr{E}$, can be written as $H(V_D - V_a)^{1/2}$ where $H = (2qN_{A2}/\epsilon_2)^{1/2}$. It follows that, for voltages greater than a few kT/q, the forward current is

$$J_f = BN_t \exp[-4(2m^*)^{1/2}q^{1/2}(V_D - K_2V_a)/3\hbar H] \qquad (2.14)$$

This equation agrees well with the experimental data of Fig. 2.4. The theory predicts an exponential function of voltage and a slope which is only slightly dependent on temperature. The change in magnitude of the current as a function of temperature is primarily a result of the temperature dependence of V_D. Although Riben's analysis was based on path A, it is not expected that the functional dependence of the current on voltage and temperature would be different for paths B or C.

2.2.2 Reverse Bias Tunneling Models

The tunneling of carriers through the energy spike in the valence band was also considered as a means of explaining the observed reverse current, but this explanation must be rejected since it results in I–V and I–T characteristics that do not conform with those usually seen.

The typical reverse current in n–p Ge–GaAs heterojunction appears to be proportional to the voltage raised to some power (greater than three). This dependence is not unique to heterojunctions. Several workers have seen similar effects in homodiodes made by various methods, and the authors have also seen this behavior in epitaxially grown Ge homodiodes and GaAs homodiodes. The Zener tunneling model, used by McAfee et al. (1951) for Ge homodiodes, yields a characteristic that appears to be functionally correct. Based on this model [shown in Fig. 2.6(b)], the reverse current density of a heterojunction is given by:

$$J_r = q\nu N_v x_p P \qquad (2.15)$$

where ν is the number of electron oscillations per second in the valence band of GaAs and is given by $\nu = qa\mathscr{E}/h$, where a is the lattice constant of GaAs, $\mathscr{E}$ is the electric field, N_v is the density of valence electrons in GaAs, x_p is the depletion layer width in the GaAs, and P is the tunneling probability given by Franz (1956):

$$P = \exp[-4(2m^*)^{1/2}E_b^{3/2}/3q\hbar\mathscr{E}] \tag{2.16}$$

where the tunneling barrier E_b for an electron tunneling from the GaAs valence band to the Ge conduction band is $E_b = E_{g1} + \Delta E_v$. (It is assumed that most of the tunneling takes place near the Ge conduction band due to the exponential dependence of the tunneling probability on the barrier height.)

Since in a diamond lattice the number of electrons per unit cell is 8, N_v will be $8/a^3$. Making the above substitution and assuming that the electric field is constant across the depletion region, Eq. (2.16) becomes

$$J_r = (8q^2/ha^2)(-V_a) \exp[-4(2m^*)^{1/2}(E_{g1} + \Delta E_v)^{3/2}/3q\hbar H(V_D - V_a)^{1/2}] \tag{2.17}$$

When Eq. (2.17) is plotted on a log–log scale, it can appear to be a power law function. It should be mentioned that hidden in the derivation of Eq. (2.17) is the condition that no reverse Zener current will flow until the electrons in the GaAs valence band are aligned with the available states in the Ge conduction band. Thus, no tunneling current should be seen (on the basis of this model) until a critical voltage equal to $(\delta_n + \delta_p)/q$ is applied. The linear region at low reverse bias may be a result of surface leakage or due to a complex tunneling mechanism.

The interesting feature of this reverse tunneling model is that it not only explains the high power law variation of the current with voltage but also provides the correct dependence of the current on temperature by virtue of the temperature dependence of E_{g1} and ΔE_v.

2.3 Heterojunction Recombination and Tunneling Mechanisms

This chapter opened with the development of the Anderson model for energy barriers in a heterojunction with current flow by injection over the barriers. Experimental results were then presented for n–Ge/p–GaAs structures which showed tunneling to be an important factor in the current flow in these junctions. It seems highly desirable therefore that we consider now the possible role of tunneling and other mechanisms, such as interface-recombination, in heterojunctions.

Six major transport processes possible in n–p heterojunctions are shown in Fig. 2.7. To be specific, the narrow- and wide-gap materials have been chosen n-type and p-type, respectively. Junctions in which the narrow- and

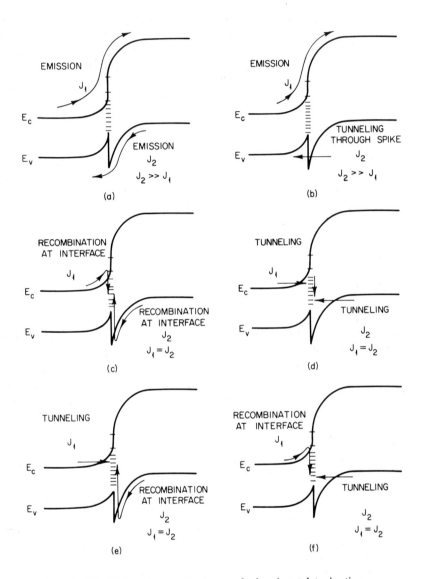

Fig. 2.7. *Six possible current-transport mechanisms in n–p heterojunctions.*

(a) Anderson's model; (b) Rediker's model; (c) Dolega's model; (d) model applicable for *n–p* Ge–GaAs junctions; (e) model applicable for *n–p* Ge–Si and *p–n* Ge–Si junctions; (f) an alternative model to (e). This model, however, is not compatible with variations which result from impurity concentration changes. (After Donnelly and Milnes, 1966c.)

wide-gap materials are *p*-type and *n*-type, respectively, will have completely analogous possibilities (in the sense that this complementary situation has a conduction-band spike and notch with hole current flowing analogously to the electron currents of Fig. 2.7 and vice versa). Actually, a combination of the currents shown in Fig. 2.7 may flow, but normally only one is dominant.

Figure 2.7(a) has emission or diffusion currents flowing over the barriers. This may be accompanied by some depletion region space-charge recombination. This is essentially the model used by Anderson and by Perlman, whose theories have already been discussed. The model shown in Fig. 2.7(b) has been treated by Rediker *et al.* (1964). Although this model incorporates tunneling, it is not in agreement with the experimental results usually obtained (Riben and Feucht, 1966a, b; Donnelly and Milnes, 1966c).

Dolega (1963) has discussed the recombination model depicted in Fig. 2.7(c) assuming a lifetime approaching zero at the interface. He found the forward electrical characteristics to be proportional to $\exp(qV/\eta kT)$, where η ranges from 1 to 2 depending on the ratio of the impurity concentrations.

In Fig. 2.7(d), tunneling currents are dominant across the entire junction. As already discussed, this is believed to be the case for *n–p* Ge–GaAs heterojunctions because of the large barriers inherently present at these junctions. This is basically the model developed by Riben.

In Fig. 2.7(e) a thermal-emission or diffusion current recombines at the interface and flows in the wide-gap material while a tunneling current flows through the narrow-gap material barrier. These two currents flow in series and are related by the interface parameters. The total current may therefore be limited by either current and may exhibit either thermal- or tunneling-current characteristics. Since the recombination current is usually the more rapidly increasing function of voltage, a transition from a recombination-limited to a tunneling-limited current with increasing forward bias is possible. Although the current-transport model depicted in Fig. 2.7(f) is also capable of generating the same general type of characteristic, experimental evidence indicates that the model shown in Fig. 2.7(e) is the appropriate one for *n–p* Ge–Si and *p–n* Ge–Si junctions studied by Donnelly (1965). Figures 2.8 and 2.9 show the forward characteristics obtained for *n–p* Ge–Si and *p–n* Ge–Si solution-grown heterojunctions. At normal current densities the diodes are seen to be characterized by an equation of the form $J = J_0 \exp(AV)$, where A is relatively independent of temperature. On the basis of the model [Fig. 2.7(e)] for recombination–tunneling via interface states assuming an effective interface density $N_{IS} \approx 10^{13}$ cm^2, the characteristics of the *n–p* Ge–Si junctions suggest values between 10^{-15}–10^{-14} cm^2 for the interface-state capture cross sections.

Donnelly concludes that in Ge–Si heterojunctions the effective recombination velocity at the interface is high owing to the inherently large density of interface states, and there is little hope that any injecting current will ever be measured in these devices. *n–p* Si–Ge heterojunctions have been studied

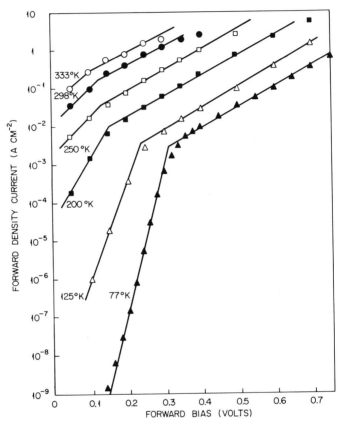

Fig. 2.8. *Forward current–voltage characteristics of an n–p Ge–Si (1 Ω cm) solution-grown diode (28-5-IL).*

Key: (○) 333°K, $\eta = 1.890$ (region 1), $A = 12.2$ (region 2); (●) 298°K, $\eta = 1.860$ (region 1), $A = 12.7$ (region 2); (□) 250°K, $\eta = 1.870$ (region 1), $A = 12.85$ (region 2); (■) 200°K, $\eta = 1.540$ (region 1), $A = 12.54$ (region 2); (△) 125°K, $\eta = 1.602$ (region 1), $A = 12.60$ (region 2); (▲) 77°K, $\eta - 1.64$ (region 1), $A = 12.2$ (region 2). $J_1 = J_{01} \exp(qV/\eta kT)$, $J = J_{02} \exp(AV)$. (After Donnelly and Milnes, 1966c.)

also by Hampshire and Wright (1964). Diode reverse recovery tests did not show the presence of injected minority carriers in their devices.

The large barrier existing in *n–p* Ge–GaAs junctions seems to have the same effect. In *p–n* Ge–GaAs junctions, the interface density is not as great as in Ge–Si junctions, and the injecting barrier is smaller than in *n–p* Ge–GaAs junctions. Since the electron recombination rate at the interface depends on the rate at which holes can be supplied to the interface by Ge, it is advantageous to grow the n–GaAs on the p–Ge instead of Ge on GaAs. In this way, the Ge should be crystallographically more perfect, reducing the probability

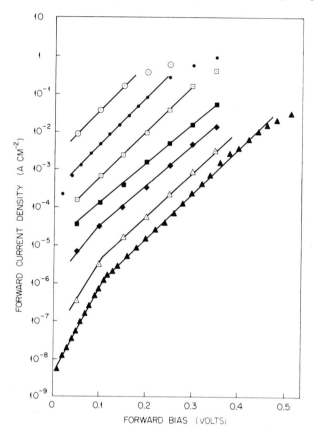

Fig. 2.9. *Forward current–voltage characteristics of a p–n* Ge–Si *solution-grown diode* (26-9-IL).
Key: ($\odot$) 333°K, $A = 26.7$ (region 2); ($\bullet$) 298°K, $A = 27.8$ (region 2); ($\boxdot$) 250°K, $A = 26.8, 29.4$ (region 2); ($\blacksquare$) 200°K, $A = 24.9$ (region 2); ($\blacklozenge$) 154°K, $n = 2.6$ (region 1), $A = 24.9$ (region 2); ($\triangle$) 100°K, $n = 2.6$ (region 1), $A = 24.5$ (region 2); ($\blacktriangle$) 77°K, $n = 2.7$ (region 1), $A = 27.8$ (region 2). $J_1 = J_{01} \exp(qV/\eta kT)$, $J_2 = J_{02} \exp(AV)$. (After Donnelly and Milnes, 1966c.)

of holes tunneling into the interface states. Furthermore, most of the interface states will probably now be in the GaAs, and any thermal flux of holes will have the additional barrier ΔE_v to surmount in order to reach the interface states. The subsequent reduction of electron recombination at the interface can result in electron injection as the principal mode of current flow.

As discussed in Chapter 3, transistor behavior has been observed for n–GaAs/p–Ge/n–Ge, n–ZnSe/p–Ge/n–Ge and n–ZnSe/p–GaAs/n–GaAs structures. Figure 2.10 shows the forward current–voltage characteristics of the

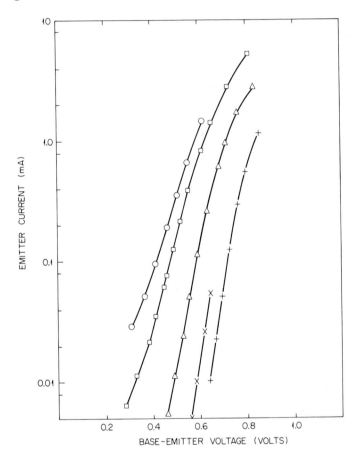

Fig. 2.10. *Forward current–voltage characteristics of an n–p GaAs–Ge emitter–base junction at several temperatures.*
     Key: (○) 323°K, $\eta = 2.6$; (□) 297°K, $\eta = 2.1$; (△) 233°K, $\eta = 2.1$; (×) 166°K, $\eta = 2.2$; (+) 117°K, $\eta = 3.0$. (After Jadus and Feucht, 1969.)

n–p GaAs–Ge emitter–base junction of such a transistor. For the temperature range 323–117°K the parameter, η, is seen to be fairly constant although somewhat influenced by series resistance voltage drops in the junction. The main conclusion that can be drawn from Fig. 2.10 and similar curves for other transistors is that these junctions behave quite differently than the tunnel current dominated heterojunctions of Figs. 2.4, 2.8, and 2.9. The dominant transport mechanism is most probably minority-carrier injection óver a barrier. The limited current and temperature range of the measurement does not exclude the possibility, however, that some part of the current is a tunnel current under certain bias or temperature conditions.

The parameter η in Fig. 2.10 was determined by fitting the curves to the expression $I = I_0 \exp(qV/\eta kT)$. The collector–base junction was reverse biased (near 0.5 V) for these measurements, so that I_b is less than I_e, to reduce the $I_b r_b$ voltage drop that is included in the base–emitter voltage.

2.4 Capacitance Studies of *n–p* and *p–n* Heterojunctions

By measuring the capacitance of a heterojunction as a function of voltage, the value of the diffusion voltage V_D [see Fig. 2.1(b)] can be found. If the junction is abrupt and the doping of one side of the junction is known, this value of V_D can be used to obtain values of ΔE_v and ΔE_c.

Applying Poisson's equation with the assumptions that the junction is abrupt and that the doping is uniform in both materials, the capacitance as a function of voltage is given by Eq. (2.4), repeated here for convenience,

$$\frac{C}{a} = \left[\frac{q N_{D1} N_{A2} \epsilon_1 \epsilon_2}{2(\epsilon_1 N_{D1} + \epsilon_2 N_{A2})} \frac{1}{(V_D - V_a)} \right]^{1/2} \tag{2.4}$$

This equation predicts that C^{-2} will be a linear function of the applied voltage, V_a; Fig. 2.11 shows this type of behavior for a typical *n–p* Ge–GaAs heterojunction at room temperature and at 77°K. This device was grown by a GeI_2 disproportionation reaction using a cleaved seed (Riben and Feucht, 1966b). The process is described in Chapter 9.

Taking the derivative of C^{-2} as obtained from Eq. (2.4) with respect to V_a, the slope of the line in Fig. 2.11 can be obtained as

$$dC^{-2}/dV_a = 2(\epsilon_1 N_{D1} + \epsilon_2 N_{A2})/(a^2 q N_{D1} N_{A1} \epsilon_1 \epsilon_2) \tag{2.18}$$

Using the values of the constants: $\epsilon_1 = 15.7\epsilon_0$, $\epsilon_2 = 11.1\epsilon_0$, $\epsilon_0 = 8.85 \times 10^{-14}$ F cm^{-1}, $q = 1.6 \times 10^{-19}$ C, and the measured values: $a = 1.26 \times 10^{-3}$ cm^2, $N_{D1} = 1.5 \times 10^{18}$ donors cm^{-3}, and $dC^{-2}/dV_a = 4.96 \times 10^{-19}$ F^{-2} V^{-1}, the acceptor density in the GaAs can be calculated. This gives a value of $N_{A2} = 1.77 \times 10^{17}$ acceptors cm^{-3}, which compares favorably with the value of 9.6×10^{16} acceptors cm^{-3} given by the supplier. The difference could easily be attributed to a slight variation in doping along the ingot. The extrapolation of the C^{-2} versus V_a curve to $C^{-2} = 0$, gives the value of V_D. As shown in Fig. 2.11, the value of V_D at room temperature is 1.06 V. Once V_D is known, the difference in electron affinity in the valence band is given by

$$\Delta E_v = V_{D1} + V_{D2} + \delta_{n1} + \delta_{p2} - E_{g1} = V_D + \delta_{n1} + \delta_{p2} - E_{g1} \quad (2.19)$$

Substituting in the known values, Eq. (2.19) gives the value $\Delta E_v = 0.56$ eV, which in turn implies a value of 0.19 eV for ΔE_c. Recent photoelectric

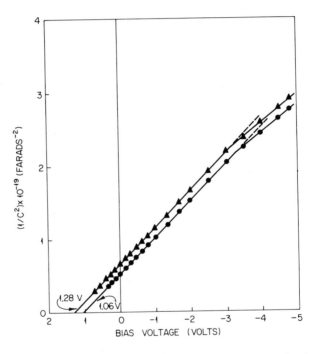

Fig. 2.11. *Capacitance as a function of voltage for a typical n–p Ge–GaAs heterojunction at* (●) 296 *and* (▲) 77°K.
(After Riben and Feucht, 1966b.)

threshold measurements by Gobeli and Allen (1963) indicate a difference of electron affinity $\chi_{Ge} - \chi_{GaAs} = 4.13 - 4.07 = 0.06$ eV. Thus the value of ΔE_c obtained from heterojunction measurements is in moderate agreement with that expected from electron affinity studies.

The ΔE_c value was found to be independent of temperature for these capacitance studies. Using the extrapolated value of 1.28 V for V_D at 77°K and taking into account the temperature dependence of δ_{n1}, δ_{p2}, and E_{g1}, the value of ΔE_v at liquid nitrogen temperature is found to be 0.60 eV. The change in ΔE_v can be explained by the fact that the band gaps of the two materials do not change by the same amount. The value of $\Delta E_v = 0.60$ eV leads to a value of 0.17 eV for ΔE_c.

These values of ΔE_v and ΔE_c are in good agreement with those obtained by Perlman (1963, 1964). Anderson's (1962) values for ΔE_c and ΔE_v between Ge and GaAs at room temperature are 0.15 and 0.55 eV. [For degenerately doped Ge, Anderson reports a substantially increased $\Delta E_c(0.56$ eV$)$, but further studies are perhaps needed before this result is generally accepted.]

The capacitance measurement of Fig. 2.11 does not give any information about the germanium properties since almost all of the depletion region is in

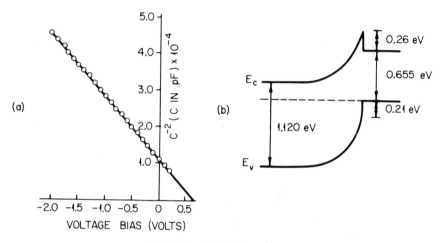

Fig. 2.12. *n–p$^+$ Si–Ge heterojunction.*

(a) Incremental capacitance characteristic. Diode area 0.85 mm^2; $N_{D_1} = 10^{15}$ cm^{-3}; $V_{D_1} = 0.65$ V. (b) Equilibrium band diagram of the *n–p$^+$* Si–Ge heterojunction. $N_{D_1} = 10^{15}$ cm^{-3}; $N_{A_2} = 5 \times 10^{15}$ cm^{-3}. (After Hampshire and Wright, 1964.)

the GaAs. However, since the measurements do indicate an abrupt junction with the correct uniform doping in the gallium arsenide and since energy discontinuities are necessary to account for the observed values of the diffusion voltage, it appears that the Anderson model is a good approximation for the equilibrium band diagram of this *n–p* Ge–GaAs heterodiode.

From these studies and from studies with *n–n* Ge–Si heterojunctions described in Chapter 4, it may be said that the available evidence supports the concept that ΔE_c is the electron affinity difference. However, there is some evidence that the barriers measured may be different if the junctions are not carefully prepared to have as perfect an interface as possible.

Figure 2.12(a) shows the C^{-2} versus V plot for a *n–p$^+$* Si–Ge heterojunction, and Fig. 2.12(b) is the energy diagram inferred from this by Hampshire and

TABLE 2.1

Summary of 1 MHz Capacitance Data on Ge–Si Heterojunctions

Diode[a]	Type	Impurity conc. (cm^{-3}) Ge	Si	298°K V_{D_i}[b] (V)	ΔE_c (eV)	ΔE_v (eV)	N_{IS} (cm^{-2})	77°K V_{D_i}[b] (V)	ΔE_c (eV)	ΔE_v (eV)	N_{IS} (cm^{-2})
41-1L	n–p	2.0×10^{18}	1.6×10^{16}	0.640	0.240	0.210	1.6×10^{12}	0.910	0.190	0.231	7.0×10^{11}
52-1L	n–p	5.0×10^{17}	1.6×10^{16}	0.600	0.235	0.215	7.4×10^{11}	0.930	0.170	0.251	Need dipole
50-1L	n–p	1.6×10^{17}	1.6×10^{16}	0.606	0.222	0.228	4.1×10^{11}	—	—	—	—
43-1L	n–p	2.5×10^{17}	5.0×10^{17}	—	—	—	—	1.17	−0.059	0.480	Need dipole
49-1L	n–p	1.3×10^{17}	5.0×10^{17}	—	—	—	—	1.23	−0.177	0.538	Need dipole
51-1L	n–p	2.0×10^{17}	4.5×10^{14}	0.520	0.230	0.220	3.8×10^{11}	0.810	0.253	0.168	5.1×10^{11}
28-5-1L	n–p	8.0×10^{16}	1.6×10^{16}	—	—	—	—	0.700	0.391	0.030	5.6×10^{11}
25-1-2L	n–p	1.0×10^{17}	1.6×10^{16}	—	—	—	—	0.790	0.293	0.128	4.9×10^{11}
28-6-1L	n–p	8.0×10^{16}	5.0×10^{17}	—	—	—	—	0.790	0.317	0.104	1.0×10^{12}
26-6-1L	p–n	1.0×10^{18}	4.0×10^{18}	—	—	—	—	0.660	−0.049	0.470	4.0×10^{12}
26-3-2L	p–n	1.0×10^{18}	1.0×10^{17}	0.498	0.018	0.432	1.6×10^{12}	0.655	0.010	0.412	1.9×10^{12}
26-9-1L	p–n	1.0×10^{16}	5.0×10^{15}	0.478	0.090	0.340	1.2×10^{12}	0.692	0.070	0.352	1.7×10^{12}

[a] First symbol in conductivity type represents Ge type.   [b] V_{D_i} obtained by extrapolating $1/C^2$ to zero.

Wright (1964). Here ΔE_c is somewhat larger than the electron affinity difference.

The effect of interface states on the capacitance of p–n heterojunctions has been studied by Donnelly and Milnes (1967). The analysis is too lengthy to present here, but the apparent diffusion voltage, V_{Di}, obtained by extrapolating the slope of C^{-2} at low reverse bias and forward bias to zero, is given by

$$V_{Di} = V_D - \phi_m - [Q_{IS}^2/2q(\epsilon_1 N_1 + \epsilon_2 N_2)] \tag{2.20}$$

where Q_{IS} is the charge per unit area of interface states and ϕ_m represents a dipole effect. Electric dipole effects in heterojunctions, for example, due to interface or dislocation states being dipolar and distributed in space on either side of the junction, are a postulate of van Ruyven *et al.* (1965). However, the dipole effect, ϕ_m, is a concept that has not been greatly explored with respect to heterojunctions.

Table 2.1 summarizes the results of capacitance measurements on Si–Ge heterojunctions. The Ge was epitaxially grown on the Si (often on substrates cleaved *in situ*) by either the GeI_2 vapor transport process or a solution growth process. The Table gives some idea of the spread of values typically to be expected for a range of diodes. The room temperature values of ΔE_c and ΔE_v are seen to be fairly compatible with the 0.26 eV and 0.21 eV obtained by Hampshire and Wright (1964).

The density of interface states inferred from these measurements is of the order of 10^{12} effective states cm^{-2}. The density of dangling bonds at the Ge–Si interface expected from the 4% lattice mismatch is 6×10^{13} cm^{-2}.

Capacitance measurements on Ge–GaAs heterojunctions show very little effect that can be attributed to interface states. For these materials the lattice mismatch is only 0.08%.

The capacitance of a heterojunction diode is often found to decrease as the measurement frequency is increased, say from 1 kHz to 1 MHz. Similar effects have been observed in homodiodes of many materials including silicon. The effect has been studied by Sah and Reddi (1964) and Schibli and Milnes (1968) and others. The effect in homojunctions is caused by the inability of deep-impurity centers in the depletion region to follow the frequency as it is increased. For heterojunctions the capacitance variation with frequency may also reflect the relaxation time-constants of interface states.

2.5 Recombination via Interface States

The problem of the role of interface states in semiconductor heterojunctions is a difficult one. Simple experiments are not available that will unambiguously measure the number of interface states, their distribution within the mutual energy gap of the semiconductors if a sheet of states, their spacial distribution on either side of the interface if this is significant, and their capture cross sections as recombination centers. In heterojunctions between semiconductors with large lattice mismatches ($>1\%$) the evidence seems to be that interface states dominate the current flow mechanisms. Very little, if any, minority carrier injection is observed in these circumstances. Usually this means that the heterojunctions are of limited practical interest since they can not be made into transistors or efficient photodiodes.

It is interesting to assume a density of interface states and consider what the effect is likely to be on recombination in a heterojunction. To be specific, consider an n-type ZnSe wide-gap semiconductor which is emitting electrons over a barrier into the conduction band of a p-type base Ge semiconductor. Assume that n_B is the density of such electrons at the interface region. Further assume that high field conditions exist at the interface (due to impurity doping profiles) so that these electrons can be considered to be moving into the base region with a drift velocity, v_d, approaching the saturated drift velocity, v_{ds}, for the material. The minority carrier current flow into the base is therefore

$$J_n = q n_B v_d \quad \text{A cm}^{-2} \tag{2.21}$$

If v_d is about 5×10^6 cm sec^{-1} and the current density is about 10 A cm^{-2}, then n_B is about 10^{13} electrons cm^{-3}.

For the surface-recombination velocity describing the recombination of electrons at the interface, the expression proposed by Hovel and Milnes (1968) is

$$S = \int (K_n^{-1} + K_p^{-1} (n_B/p_B))^{-1} N_{IS} \, dE \tag{2.22}$$

where K_n, K_p = $v_{th}\sigma_n$, $v_{th}\sigma_p$, respectively (σ_n, σ_p, are the capture cross sections and v_{th} is the thermal velocity for electrons and holes, cm sec^{-1}), n_B and p_B are the electron and hole concentrations in the ZnSe conduction band and Ge valence band at the interface, and N_{IS} is the interface state density as a function of energy. Large variations in S may occur for a given N_{IS} from the possible variations in K_n and K_p, which depend strongly on the particular charge state involved; large interface densities do not necessarily lead to high values of S if the capture rates are low.

If $K_p p_B \gg K_n n_B$, which is probable for n–p^+ ZnSe–Ge diodes, (2.22) reduces to

$$S = \int K_n N_{IS} \, dE \tag{2.23}$$

The recombination current density is then given by

$$J_S = q n_B S \tag{2.24}$$

Hence the ratio of injected current to recombination current given by (2.21) and (2.24) is

$$J_n/J_S = v_{\mathrm{d}}/S \qquad (2.25)$$

The effective capture cross section for interface states is probably in the range of 10^{-14}–10^{-15} cm^2; Donnelly and Milnes' studies (1966c) for Ge–Si junctions gave such values. These values also conform with the capture cross sections typically found for deep impurities in Ge or Si (Schibli, 1967; Schibli and Milnes, 1967). From (2.23), using 10^{12} cm^{-2} for the effective interface state density and 10^7 cm sec^{-1} for the thermal velocity, S is then between 10^5 and 10^4 cm sec^{-1}. Hence from (2.25), J_n/J_S is between 50 and 500. These values (admittedly wide in range) would represent an upper limit for the gain of a transistor with this heterojunction as a wide-gap emitter (neglecting all other sources of carrier-recombination loss).

For a heterojunction with a larger interface state density (say, $>10^{13}$ cm^{-2}, as for Si–Ge junctions) there seems little prospect for an acceptable injection efficiency. On the other hand, there are many heterojunction pairs with close lattice match conditions (see Table 1.3) for which good injection characteristics should be achievable.

Chapter 3 | Heterojunction Transistors

Since wide-gap emitter transistors potentially have advantages over normal transistors, many attempts have been made to fabricate such structures to prove the principle of operation and to determine the extent to which the advantages can be achieved in practice. One of the first successful attempts was that by Brojdo *et al.* (1965). They fabricated space-charge limited structures with current gains of 10 from amorphous CdS deposited upon Si. More recently, Jadus (1967) and Jadus and Feucht (1969) obtained current gains of about 15 in GaAs–Ge transistors, and Hovel and Milnes (1967) measured gains of 20–35 for ZnSe–Ge structures. These GaAs–Ge and ZnSe–Ge devices differ from the CdS/Si structures in that the emitters are epitaxial, single-crystal material, and therefore in better control than amorphous layers. Similar transistor performance has also been obtained recently from ZnSe/GaAs structures (Sleger *et al.*, 1970).

Other heterojunctions that should give interesting transistor performance include GaP/Si and $Si_{0.1}Ge_{0.9}$/Ge structures. Suitable growth methods for these are presently under development.

3.1 The Advantages of Idealized Heterojunction Transistors

The unique advantages of heterojunction devices arise from the dissimilar barriers which exist for electrons and holes at the interface between the two materials. Figure 3.1 shows the energy band diagram of an *n–p–n* GaAs–Ge–Ge transistor compared with that of a Ge homojunction transistor. The large barrier in the valence band effectively prevents any holes from reaching the GaAs. This effect gives the circuit designer a new degree of freedom. The base doping level may be high and the emitter level low and an injection efficiency of close to unity can still be maintained. The consequences of this are a higher gain–bandwidth product, f_{max} (because of reduced base resistance), and possible improvements in radiation resistance, optical quantum efficiency, and second breakdown performance.

3.1.1 *Common-Emitter Current Gain* (h_{FE})

In the absence of collector multiplication, the h_{FE} of a transistor is determined by the product of the base transport factor, B, and the emitter injection efficiency, γ. Limitations on the transport factor arise at all injection levels from bulk and surface recombination; these limitations can be reduced by making the base transit time as short as possible. Reduced gain at high injection levels also occurs from the reduced injection efficiency caused by base-conductivity modulation. Such modulation means that more base majority carriers are available for reverse injection into the emitter.

In the design of high-speed homojunction transistors, devices are fabricated

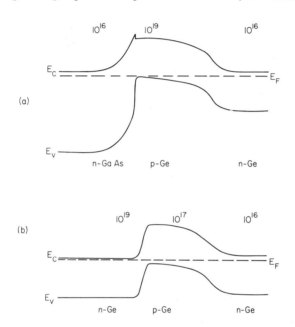

Fig. 3.1. *Energy band diagram comparison.*
(a) An *n–p–n* GaAs–Ge–Ge heterojunction transistor, and (b) an *n–p–n* Ge homojunction transistor. The emitter is on the left. Typical doping levels are indicated. (After Jadus and Feucht, 1969.)

with diffused bases 0.1–0.4 μ wide and base doping levels of 10^{18} cm^{-3} or greater at the emitter side of the base region. The drift fields thus produced aid electron transport across the base resulting in transit times of 10^{-11} sec and reducing both the bulk and surface recombination effects. The high base doping density also raises the current density at which base conductivity modulation occurs. The recombination–generation current is reduced, though not eliminated, by the narrow depletion widths. On the other hand h_{FE} is low (10–30) at all injection levels since the emitter doping level is only a factor of 10 to 50 higher than the base density.

In considering heterojunction transistors, comparable base widths may be assumed, but considerably higher drift fields may be expected from the higher doping levels in the base, 10^{19}–10^{20} cm^{-3}, resulting in possibly smaller transit times and improved base transport factors. The improvement to be hoped for will depend on the degree of impurity scattering expected at high field strengths. The alpha falloff at high injection levels (10^3–10^4 A cm^{-2}) should be eliminated since virtually no reverse injection of holes into the emitter occurs, whether the base conductivity is modulated or not. In an ideal heterojunction transistor, the current gain, i_c/i_b should be high (several hundred) at all injection levels since the base transport factor is high and the emitter injection efficiency is very close to one regardless of the relative base and emitter doping densities. In actual heterojunction transistors, as discussed later, practical factors such as emitter–base interface recombination and heat dissipation may limit current gains and permissible injection levels.

3.1.2 Frequency Response, f_{max}

Since a detailed analysis of heterojunction transistor frequency response appears later in this chapter, only some general comments are presented here.

The gain-bandwidth figure of merit of a transistor is given, approximately, by

$$f_{max} = \tfrac{1}{4}\pi(r_b{}'C_c\tau_{ec})^{1/2} \tag{3.1}$$

where $r_b{}'$ is the base resistance, C_c is the collector capacitance, and τ_{ec} equals $\tau_e + \tau_b + \tau_{cs1} + \tau_c$, in which τ_e is the emitter diode charging time, τ_b is the base transit time, τ_{cs1} is the collector depletion layer transit time (saturated-drift-velocity limited), and τ_c is the collector diode charging time.

The emitter diode charging time, τ_e, is $r_e C_e$ where these are the emitter resistance and capacitance. The very high doping levels in the emitter and base of a homojunction device produce a high C_e, and under forward bias it becomes increased even further. A reduction in this product may be obtained for a heterojunction device since the emitter is doped lightly while r_e is kept low by decreasing the total emitter thickness. With the highly advanced technology of Si transistors, r_e is reduced to the degree where τ_e is only several picoseconds. However, a heterojunction transistor may pick up some advantage here.

The transit time through the base, τ_b, in high-speed homojunction transistors is of the order of several picoseconds. The reduction to be expected in the heterojunction device because of the increased base field will depend on the drift-velocity–field relationship appropriate to the doping levels used.

The collector depletion layer transit time, is given by

$$\tau_{cs1} = W_c/2v_{ds} \tag{3.2}$$

where W_c is the depletion width and v_{ds} the saturation limited velocity. This transit time is shortened by designing for a narrow depletion layer width although the heavier dopings required are at the cost of lower breakdown voltage. A heterojunction structure does not have any advantage here.

The collector diode charging time, τ_c, is given by $r_c C_i$ where r_c is the collector bulk resistance and C_i the inner collector capacitance, largely determined by geometry. No appreciable change is expected in this term by use of a heterojunction transistor. Some improvement is expected from the $r_b'C_c$ term in Eq.(3.1) since the base resistance is dependent on the doping density and may thus be improved by a factor of 4 to 5 for the heterojunction device.

In summary then, the overall f_{max} may be increased for heterojunction transistors relative to present high-frequency Si or Ge transistors. This is developed in more detail in Section 3.4 where the improvement is shown to be possibly a factor of 2.

3.1.3 *Switching Characteristics*

Good switching characteristics are also expected with heterojunction transistors, stemming mainly from three factors:

(a) As mentioned previously, the emitter doping level is greatly reduced relative to homojunction transistors resulting in a low C_e and a low $r_e C_e$ time constant. (A low r_e is obtained by the use of emitter thicknesses only slightly larger than the depletion width.)

(b) Lower base lifetime may be tolerated due to the improved base transit time thus reducing the minority carrier storage time in the base without reducing the gain. Lower lifetimes may be expected in heterojunction base regions due to dislocations and other recombination centers produced by the strains of lattice and thermal mismatch.

(c) Since in an n–p–n heterojunction transistor holes are prevented from entering the emitter valence band by the hole barrier, ΔE_v, no minority carrier storage time is expected in the emitter. On the other hand, some trapping effects may be present in wide-gap emitter materials that would not be present in Si or Ge transistors.

3.1.4 *Device Operating Limitations*

Operating limitations imposed by emitter crowding, second breakdown, reach–through, etc., create difficulties for the device designer. Some advantages may be expected for heterojunction transistors in these respects:

(*a*) *Emitter crowding.* The higher doping level in the base reduces the lateral base resistance, and thus the emitter crowding, proportionally. Device fabrication technology is therefore simplified since larger and fewer emitter stripes are tolerable.

(*b*) *Second breakdown.* One of the modes of second breakdown is caused by "pinch-in," the opposite of emitter crowding, since current is forced to concentrate into smaller and smaller areas toward the emitter center by positive feedback effects. The reduced lateral base resistance minimizes this "pinch-in" in the same way that it reduces emitter crowding.

Second breakdown may be reduced in homojunction transistors by making the individual devices smaller and isolating them from each other, or by adding "ballast" resistance in series with the emitter to provide negative feedback. In heterojunction transistors resistance may be provided automatically by the lower emitter doping thus eliminating extra steps in fabrication.

(*c*) *Reach–through.* The extremely narrow base widths used in high frequency devices may limit the collector–base voltage rating by reach-through of the depletion region in the base. The higher doping levels in the bases of heterojunction transistors should ease this problem.

(*d*) *Emitter breakdown.* The extremely high doping levels in the emitter and base of high frequency homojunction devices result in very low breakdown voltages (1–2 V) which hamper the circuit designer. For heterojunction transistors, the lower emitter doping allows emitter–base breakdown voltage ratings of tens of volts.

(*e*) *Radiation effects.* The most serious effect of radiation is to lower the lifetime in the base thus reducing the base transport factor and the gain. The heterojunction transistor, with its reduced base transit time and higher initial gain, should be able to sustain much higher radiation doses before this effect becomes important.

3.2 The "Defect" Components of Current Expected in Heterojunction Transistors

In actual heterojunction transistors it must be recognized that other components of current, aside from the desired injection current in the base, may flow at the emitter–base junction. The magnitude of these defect currents will be a function of the character of the interface between the two materials. The discussion of defect components that follows has been presented as specific to an n–p–n ZnSe–Ge–Ge transistor. However, it represents a general treatment that should be applicable, with relatively little change, to a range of heterojunction transistors. The treatment considers in some detail the expected dependence of the defect-current components on the current density and on the temperature. When the observed results for ZnSe–Ge transistors are

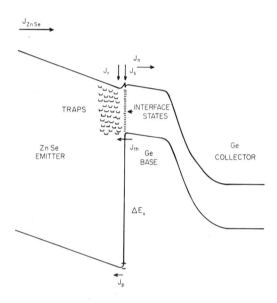

Fig. 3.2. *Energy band diagram of an* $n(large)$–$p(small)$ *heterojunction* (ZnSe–Ge).
The various particle currents are defined in the text. Quantum mechanical reflection at the interface and image effects have been neglected. The "spike barriers" ΔE_c and ΔE_v are 0.04 and 1.9 eV, respectively. (After Hovel and Milnes, 1968.)

presented in Section 3.3, it is then possible to see general agreement between the model and the actual behavior.

Figure 3.2 shows the energy band diagram of a heterojunction transistor emitter–base diode, ZnSe–Ge, with the expected components of current resulting from forward bias. These components are: J_n, J_p = injected electron and hole currents, J_{th} = hole current from the Ge valence band to band-gap states to match the electron current J_r, J_r = electron capture from the ZnSe depletion region into traps, followed by recombination with J_{th} possibly involving tunneling, J_S = recombination of injected electrons at states at the metallurgical interface between the ZnSe and Ge (excluding the J_r component), and J_{ZnSe} = total diode current as limited by the bulk ZnSe.

The injection efficiency, γ, is the fraction of the total emitter current injected into the base and is given by

$$\gamma = J_n/(J_n + J_p + J_r + J_S) \tag{3.3}$$

The hole component of tunneling from the Ge valence band, J_{th}, is necessary to supply unoccupied states in the ZnSe depletion region, or to empty interface states, to allow for the existence of the recombination current, J_r. Without this hole component, electrons captured by traps would either have to cross the depletion region by many tunneling steps or be reemitted to the conduction band. Therefore, J_{th} and J_r form a type of generation–recombination current in the ZnSe depletion region. This may be a significant process competing with electron injection and interface recombination J_S as one of the principal mechanisms of current transport through the junction region.

The electron current, J_n, injected into the Ge may perhaps conform to the Shockley expression for a simple diode model

$$J_n = \frac{qD_{n(Ge)}}{L_{n(Ge)}} \frac{n_{i(Ge)}^2}{N_{A(Ge)}} \left[\exp\left(\frac{qV_n}{\eta k T}\right) - 1 \right] \tag{3.4}$$

assuming that the current is limited by the diffusion of the injected electrons (diffusion length L_n) away from the junction into the Ge. However, if the ZnSe is lightly doped or recombination at the interface is very rapid or drift fields exist in the Ge which move electrons away from the interface very quickly, the injected current may be limited by the rate at which electrons can be supplied from the ZnSe. The current is then given by electron flux and energy considerations and may have a form such as

$$J_n = q\mu_n(2qN_D/\epsilon)^{1/2}(V_D - V_{an})^{1/2}\{N_D \exp[-q(V_{Dn} - V_{an})/kT] - n_B\} \tag{3.5}$$

where μ_n, and ϵ are the mobility of electrons, doping level, and permittivity on the ZnSe side, and V_{Dn} and V_{an} are the portions of the diffusion and applied voltages appearing on the ZnSe side. The currents given by (3.4) and (3.5) differ in magnitude but have the same qualitative temperature dependence.

The electron current injected into the Ge may also be written as

$$J_{n(x=0)} = q\mu_n n_B E_0 f(K) \tag{3.6}$$

where μ_n is the electron mobility, n_B the electron density at the interface, E_0 the constant electric field resulting from an exponential impurity distribution, $K = (qE_0W/kT)$ is the field factor, W the base width, and $f(K)$ a factor accounting for the relative magnitudes of the base width, diffusion coefficient, and drift velocity (Hovel, 1968). Although constant mobility and field assumptions are invalid, use of more exact descriptions leads to impracticable equations.

The component of hole current, J_p, injected from the Ge into the ZnSe valence band is assumed to be zero because of the very large barrier for this process ($\Delta E_v = 2.0$ eV).

Of course, if the ZnSe is of high or moderate resistivity, the entire diode current may be limited by space charge processes in the bulk emitter. When the dielectric relaxation time is short, no mobile space charge can be built up in the solid, but if the transit time of the electrons is shorter than the relaxation time, appreciable space charge can accumulate. The ratio of these times is

$$\tau_{\text{transit}}/\tau_{\text{dielectric}} = d^2/\mu_n V \rho \epsilon \tag{3.7}$$

and space charge effects will be observed if the resistivity, ρ, is greater than $d^2/(\mu V \epsilon)$, where d is the sample thickness, V the voltage across the sample, μ the mobility, and ϵ the dielectric constant. For a $1~\mu$ thick emitter with a mobility of $100~\text{cm}^2~\text{V}^{-1}~\text{sec}^{-1}$, space charge effects are expected if the resistivity is greater than $139~\Omega$ cm ($N_D < 4.5 \times 10^{14}~\text{cm}^{-3}$) for 1 V applied across the sample. For a resistivity ten times larger, space charge effects for a $1~\mu$ thick emitter should begin for 0.1 V applied. In terms of current density, this represents $1.4~\text{A cm}^{-2}$ in both cases. Usual current densities for diode or transistor action are in the range of hundreds of amperes per square centimeters. Therefore resistivities of less than a few ohm-centimeters are desirable for the ZnSe (corresponding to $N_D > 5 \times 10^{16}~\text{cm}^{-3}$) in order to eliminate space charge effects. Although such doping levels are readily achievable in ZnSe bulk-doped at high temperatures, the best that can presently be obtained in thin films of ZnSe epitaxially grown on a Ge heterojunction substrate appears to be between one and two orders of magnitude lower in effective doping level. Therefore in such structures (Section 3.3), space charge effects are normally seen.

Interface states at the metallurgical junction result in an interface recombination velocity (Hovel and Milnes, 1968) given by

$$S = \int \frac{K_n K_p p_B N_I s}{K_n n_B + K_p p_B}\, dE \tag{3.8}$$

The current due to the interface recombination of electrons is therefore given by

$$J_S = q n_B \int \frac{1}{K_n^{-1} + K_p^{-1}(n_B/p_B)} N_{IS} \, dE \tag{3.9}$$

where N_{IS} is the density of interface states at an energy E below the conduction band.

How the defect current, J_r, is sustained is somewhat hypothetical. One assumption is that it consists of the capture of electrons within the ZnSe depletion region and subsequent tunneling through the remaining distance into interface recombination states. This component therefore might be given by

$$J_r = q \int_{E_t} \int_{y=0}^{l} \frac{K_n n N_t N_{IS} P}{K_n(n + n_1) + N_{ISt} P} \, dy \, dE_t \tag{3.10}$$

where n is the electron density at the position y within the ZnSe, N_t is the density of capture centers at energy E_t below the conduction band, N_{ISt} is the density of interface states at the same level E_t, $K_n = v_{th}\sigma_n$ where σ_n is the trap capture cross section, n_1 is the electron density when the Fermi level lies at E_t, and P is a tunneling probability factor, cm² sec^{-1}.

The injection efficiency is then given by

$$\gamma = (1 + J_S/J_n + J_r/J_n)^{-1} \tag{3.11}$$

and the common–emitter current gain, for base transport factors B close to one, is

$$h_{FE} = B/[(J_S/J_n) + (J_r/J_n)] \tag{3.12}$$

3.2.1 *Injection Level*

The variation of the gain with injection level is determined by the injection efficiency since the base transport factor is assumed independent of the current. If the interface-recombination current, J_S, is much greater than the capture-tunneling term, J_r, then from (3.6), (3.9), and (3.12) the gain is

$$h_{FE} = B\mu_n E_0 f(K)/ \int (K_n^{-1} + K_p^{-1}(n_B/p_B))^{-1} N_{IS} \, dE \tag{3.13}$$

If $K_p p_B \gg K_n n_B$, (3.13) becomes

$$h_{FE} = B\mu_n E_0 f(K)/ \int K_n N_{IS} \, dE \tag{3.14}$$

$$= B\mu_n E_0 f(K)/S_n \tag{3.15}$$

where S_n is the interface-recombination velocity when limited by the electron capture rate. In this case the gain is constant, independent of the current level, and the value of this constant depends on B and the ratio of the effective drift velocity in the base near the interface to the recombination velocity. If the capture rates are reversed, such that $K_n n_B \gg K_p p_B$, then

$$h_{FE} = BJ_n/q \int p_B K_p N_{IS} \, dE \qquad (3.16)$$

and if $h_{FE} > 4$ or 5, so that the emitter current $\approx J_n$, the gain becomes proportional to the emitter current.

If the capture-tunneling component is the dominant defect mechanism, the variation will be determined by (3.10). A large fraction of J_r occurs at some distance within the ZnSe depletion region since a balance is established between increased tunneling rates and decreased electron concentration as the interface is approached. If the "maximum" occurs far from the interface, the tunneling probability will be low, and (3.10) becomes

$$J_r = \exp[-\alpha(V_{Dn} - V_{an})] \int N_t N_{IS} \, dE \qquad (3.17)$$

The gain is then

$$h_{FE} = \frac{B(\mu_n E_0 n_0 f(K))^{\alpha kT/q}}{\int N_t N_{IS} \, dE} J_n^{1-\alpha kT/q} \qquad (3.18)$$

In these equations, α is the tunneling constant, n_0 is the electron density in the bulk ZnSe, V_{Dn} is the portion of the diffusion voltage on the emitter side of the junction, and V_{an} is the portion of the applied voltage appearing on the emitter side. The gain now varies as a fractional power of the emitter current, and the value of this power depends on the tunneling parameters and on temperature. Similar behavior is observed in homojunction transistors at low current densities due to recombination–generation currents in the space charge region.

If the maximum of J_r occurs near the interface, the current is limited by the electron concentration

$$J_r = q n_B \int K_n N_t \, dE \qquad (3.19)$$
$$= q n_0 \exp(-q(V_{Dn} - V_{an})/kT) \int K_n N_t \, dE \qquad (3.20)$$

and the gain is once again independent of injection level

$$h_{FE} = B\mu_n E_0 f(K)/\int K_n N_t \, dE \qquad (3.21)$$

Both defect components are therefore capable of leading to dependence of gain on injection level, and both may lead to power law dependence. Powers between the fraction given by (3.18) and unity from (3.16) might be expected from the various possible combinations of the two currents, J_r, J_S.

3.2.2 Drift Field and Base Width

The two defect components are virtually independent of base width and field and only slightly dependent on the base doping level (specifically p_B) at high injection levels where $K_n n_B \geqslant K_p p_B$. The injected component, J_n, however, is dependent on these parameters. The injection efficiency is nearly proportional to $\mu_n E_0$. The base width is contained in $f(K)$ in Eq. (3.18). In the limit of very large fields, however, $f(K)$ is equal to unity and γ is independent of the base width.

3.2.3 Effect of Emitter Doping Level on the Current Gain

Both J_n (3.6) and J_S (3.9) are proportional to n_B and hence related to the emitter doping level. The ratio, J_n/J_S, however, is independent of the emitter level as long as the emitter doping level is much less than the base level. The capture-tunneling component, J_r, has an additional dependence on doping through the injection and tunneling probabilities. When the emitter doping level, N_D, is lowered, the depletion region is widened and the increased distances over which the tunneling has to take place greatly lowers the tunneling probability and moves the position of the maximum closer to the interface. At the same time the injection barrier, V_{Dn}, is decreased. The ratio of J_n to J_r, and therefore the injection efficiency and the gain, might be expected to increase with increasing emitter resistivity as long as J_r is an important defect component in the device. This is opposite to the behavior of homojunction transistors where γ and h_{FE} improve with decrease of the emitter resistivity.

The second effect of increasing the ZnSe resistivity is to increase the electric field in the bulk ZnSe for a given current raising the electron drift velocity through the emitter and altering the band profile. The result of the increased band bending is to effectively lower the barrier for injection. In addition, at high drift fields the electrons may become "heated" raising their distribution in energy and increasing their probability of injection.

3.2.4 Expected Temperature Dependences

The injected component of current, J_n, depends on temperature through E_0, μ_n, K, and n_B. The electric field in the diffused base is proportional to kT/q

[see Eq. (3.27)]. The mobility near the interface is determined by ionized impurity scattering and is decreased with decreasing temperatures as $T^{3/2}$. The product $\mu_n E_0$ may therefore decrease by up to an order of magnitude from 350 to 100°K. The parameter K, and the factor $f(K)$ in (3.6) are relatively independent of temperature for the devices presented here. The largest dependence arises from n_B which decreases exponentially with decreasing temperature.

The dependence of J_S on temperature arises through S and n_B. The recombination velocity is dominated by the capture rates, K_n and K_p. Although the thermal velocity decreases as $T^{1/2}$, the cross sections, σ_n and σ_p, increase substantially at lower temperatures; variations of T^{-0} to T^{-4} have been observed in In and Au doped silicon (Schibli and Milnes, 1967/1968; Bullis, 1966). The interface-recombination velocity may therefore increase by more than an order of magnitude from 350 to 100°K.

As a result of these variations, the ratio of J_S to J_n,

$$J_S/J_n = S/\mu_n E_0 f(K) \tag{3.22}$$

which is independent of n_B, increases somewhat with decreasing temperature, lowering the injection efficiency and the gain h_{FE} in those devices where J_S is an important defect component.

For the capture-tunneling component, J_r, the temperature dependence depends on whether the main part of this current occurs from near the edge of the depletion region rather than from near the interface. As discussed in Section 3.2.1, if J_r is from the depletion edge and is tunneling limited, it should decrease only weakly with lower temperatures. The injected component of current, J_n, however (being an over-the-barrier component) should decrease strongly with lower temperatures. For constant emitter current conditions, therefore, J_r/J_n may be expected to increase by several orders of magnitude from 350°K to 100°K, and the transistor gain h_{FE}, from Eq. (3.12), may be expected to decrease substantially. On the other hand if the defect component J_r occurs mainly relatively near the interface, it would have approximately the same dependence as J_S, determined by the capture rate of the traps, K_n, and the electron density at the interface n_B. The ratio J_r/J_n and the gain h_{FE} might then decrease less than an order of magnitude at the lower temperatures.

3.2.5 Base Transport Factor

The transport of electrons across the base is reduced by bulk and surface recombination. Since the base region is determined by the Ge p–n junction, the transport is qualitatively the same as in a homojunction transistor; however, quantitative differences may result for the heterojunction device from the higher base doping levels and fields and possibly from the low lifetimes caused by the lattice and thermal mismatch between the two materials.

Using constant field, mobility, and lifetime approximations the base transport factor for diffused base structures has been given (Iwerson *et al.* 1962; Kroemer, 1956b; Das and Boothroyd, 1961; Lee, 1956; Rollett, 1959; Mayburg and Smith, 1962) as

$$B = J_n(W)/J_n(0)$$
$$= \exp(K/2)/(\{K \sinh \phi/2\phi\} + \cosh \phi) \tag{3.23}$$

where

$$\phi = \{(K/2)^2 + (W^2/D_n\tau_n) + (j\omega W^2/D_n)\}^{1/2} \tag{3.24}$$

and K is the field factor, W the base width, D_n and τ_n the minority carrier diffusion coefficient and lifetime, and ω the angular frequency. An alternative expression for B which does not require these approximations is

$$B = \exp\{-\int_0^W [t(x)/\tau(x)] \, dx\} \tag{3.25}$$

where $t(x)$ represents the time elapsed for the injected electron to reach the position x, and $\tau(x)$ is the lifetime as a function of position. The approximation of constant lifetime leads to the simple form

$$B = \exp(-\tau_{tr}/\tau_B) \simeq \tag{3.26}$$

where τ_{tr} is the total transit time across the base and τ_B is the lifetime.

The diffusion technique used to fabricate the Ge *p–n* junction typically results in a complementary error function impurity distribution; the drift field in the base is then given by (Nanavati, 1963)

$$E = \frac{2kT}{\sqrt{\pi}\,q} \frac{1}{L} \frac{\exp(-x^2/L^2)}{\mathrm{erfc}(x/L)} \tag{3.27}$$

where $L = 2(Dt)^{1/2}$ is the effective diffusion length for the time t. Typically the field across the base increases by about a factor of 5 from the emitter to the collector.

Since relatively high fields and impurity densities exist in the base, the mobility and diffusion constant are complicated functions of distance. Since impurity scattering decreases with increased field strength, at high fields the drift velocity is considerably higher than might be expected from the low-field mobility corresponding to the impurity density in the base. The base transit time is therefore very short, in the range of 10^{-11} to 10^{-10} sec for the ZnSe–Ge devices discussed in Section 3.3.

The lifetime in the base is determined by the initial material imperfections and the dislocations, defects, etc., introduced by the heterojunction epitaxy. At the emitter edge the lifetime is equal to the electron lifetime in heavily p-type material,

$$\tau_{n0} = (v_{\text{th}}\sigma_n^+ N_r)^{-1} \tag{3.28}$$

where N_r is the density of recombination centers and σ_n^+ the electron capture cross section. At the collector edge of the base region, where the Fermi level crosses the intrinsic level,

$$\tau_B = \tau_{n0} + \tau_{p0} \tag{3.29}$$

where τ_{p0} is the hole lifetime in heavily n-type material. The lifetime varies between these extremes according to the error function distribution of the acceptor density; this variation will be large or small depending on the relative values of τ_{n0} and τ_{p0}. In addition the density of recombination centers may vary across the base with a consequent effect on τ_{n0}.

3.2.6 *Possible Factors Influencing Base Transport*

The base transport factor of a homojunction transistor is independent of current (Iwerson *et al.*, 1962) up to very high injection levels. When conductivity modulation occurs, the lifetime at the emitter edge increases to the value $\tau_{n0} = \tau_{p0}$. The electric field, mobility, and drift velocity are decreased, however, which increases the transit time. The transport factor may improve or worsen depending on the relative magnitudes of these changes.

The transit time is approximately proportional to the base width and inversely proportional to the velocity. The balance between drift field and impurity concentration in determining the drift velocity in heterojunction base regions is discussed elsewhere (Hovel, 1968). For devices with low fields (<800 V cm^{-1}) the doping level assumes greater importance; the transit time increases with N_A since the mobility decreases more than the field increases. For devices with high fields the opposite is true; the transit time decreases and the transport factor improves at higher base doping levels.

Considering now the probable temperature dependence, the drift field (3.27) is proportional to kT/q and therefore decreases with decreasing temperature. The low-field mobility with impurity scattering varies as $T^{3/2}$. The transit time through the first 20 or 30% of the base, where the low field mobility is important, might therefore increase substantially from 350 to 100°K. In addition the decrease in field extends the importance of impurity scattering farther into the base, which further increases the transit time. In the remainder of the base, the drift velocity is essentially saturated, and therefore does not vary appreciably with temperature. (It should be mentioned that at fields this high, i.e., where the drift velocity saturates, the Einstein relation is no longer valid. This produces some error in using (3.27) to estimate the field at the collector side of the base region.)

The lifetime depends on temperature mainly through the capture cross sections of the recombination centers. These are expected to behave similarly to the cross sections involved in the interface recombination and capture-tunneling mechanisms (see Section 3.2.4); the lifetime therefore decreases with decreasing temperature. As a result of the increasing transit time and decreasing lifetime, the transport factor should decrease at lower temperatures.

3.3 Characteristics of ZnSe–Ge Heterojunction Transistors

The characteristics of some actual heterojunction transistors will now be described, and the gains examined as functions of the doping parameters, the injection current density, and the temperature for comparison with the predictions above.

The heterojunction transistors whose performance and temperature characteristics will be described were fabricated by Hovel and Milnes (1967). Single-crystal ZnSe emitter layers (about 4 μ thick and doped 10^{11}–10^{14} donors cm^{-3}) were grown by an HCl transport process onto p–n Ge–Ge diffused base-collector substrates. The base surface concentrations C_S were 4×10^{18} to 6×10^{19} cm^{-3}; the base widths were typically 0.2 to 1.4 μ; and the collector (substrate) doping levels were 10^{15} to 10^{17} cm^{-3}.

3.3.1 *Injection Level Effects*

The characteristics of a typical n–p–n ZnSe–Ge–Ge transistor are shown in Fig. 3.3. The numbers in the caption refer to the base doping level (surface concentration) and field at the emitter edge, the base width, the estimated emitter doping level, and a base parameter τ to be described. In Fig 3.3(a) the device is shown at low injection levels; the gain is constant, independent of the emitter current. This is expected whenever interface recombination is dominant or the capture-tunneling dominates and is limited by electron

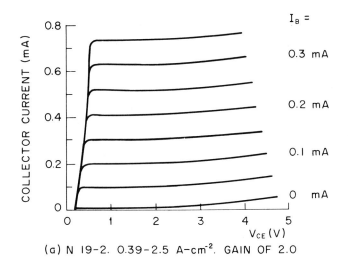

(a) N 19-2. 0.39-2.5 A-cm^{-2}. GAIN OF 2.0

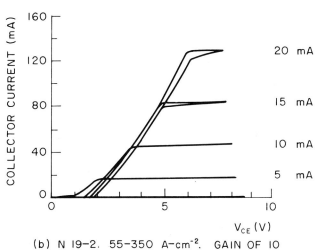

(b) N 19-2. 55-350 A-cm^{-2}. GAIN OF 10

Fig. 3.3. *Variation of current gain with injection level at* 27°C.
$\beta = h_{FE}$ is constant up to 10 A cm^{-2}, then increases rapidly with the emitter current.
$C_S = 4 \times 10^{18}$ cm^{-3}, $E(0) = 550$ V cm^{-1}, $W = 1.0\,\mu$, $N_D = 5 \times 10^{13}$ cm^{-3}, $\tau = 2.8 \times 10^{-10}$
sec. (After Hovel and Milnes, 1967.)

capture near the interface. Constant gain at low injection levels was observed
for most of the transistors tested.

In Fig. 3.3(b), the device is shown at higher injection levels, where the
gain has begun to increase with the emitter current. Although this could be

caused by a decrease of the interface-recombination velocity, such a decrease
is not expected until current levels several orders of magnitude greater have
been reached. The increase of gain is probably due to a change in the

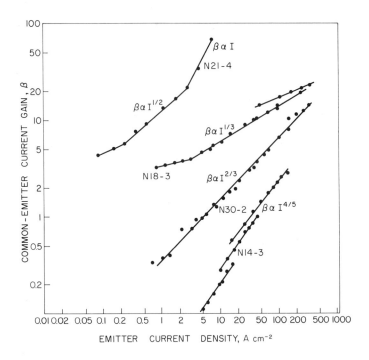

Fig. 3.4. *Variation of current gain with injection level at 27°C, showing the fractional power dependence on the emitter current.*

N21-4: $C_S = 2 \times 10^{19}$ cm^{-3}, $E(0) = 1120$ V cm^{-1}, $W = 0.6$ μ, $N_D = 1 \times 10^{11}$ cm^{-3},
$\tau = 2.1 \times 10^{-10}$ sec. N18-3: $C_S = 4 \times 10^{18}$, $E(0) = 935$ V cm^{-1}, $W = 0.6$ μ, $N_D = 5 \times$
10^{13} cm^{-3}, $\tau = 1 \times 10^{-10}$ sec. N30-2: $C_S = 2 \times 10^{19}$, $E(0) = 760$ V cm^{-1}, $W = 1.0$ μ,
$N_D = 1 \times 10^{13}$ cm^{-3}, $\tau = 5.3 \times 10^{-10}$ sec. N14-3: $C_S = 2 \times 10^{19}$, $E(0) = 470$ V cm^{-1},
$W = 1.4$ μ, $N_D = 5 \times 10^{13}$ cm^{-3}, $\tau = 1.2 \times 10^{-9}$ sec. (After Hovel and Milnes, 1967.)

behavior of J_r from capture limited to tunnel limited. This is further sub-
stantiated in Fig. 3.4, which in log–log form shows the gain versus the
emitter current density. The fractional power dependence is expected from
(3.18), where the value of the fractional power is determined by the temperature and the tunneling parameters. The kinks in the straight lines for
several of the devices might indicate a shift in the defect mechanism, an
increasing influence of the high field in the bulk of the emitter, or some
effect of the traps in the ZnSe.

3.3.2 Drift Field and Base Width Effects

The characteristics of two devices with identical ZnSe resistivities are shown in Fig. 3.5. The factor τ is used to characterize the base parameters and is equal to the base width divided by the mobility and the field

$$\tau = W/\mu_n E \qquad (3.30)$$

where the values of μ_n and E chosen are those at the emitter end of the base region since they appear in the expression for J_n (3.6) and since they represent the rate of removal of electrons into the Ge relative to their rate of recombination at the interface.

In Fig. 3.5(a), the value of τ is high, corresponding to a large base width and low field; the gain is therefore low. In Fig. 3.5(b) τ is reduced by an order of magnitude and the gain has increased by a factor of 8. Similar behavior was observed for devices with much higher ZnSe resistivities and consequently higher gains.

The drift field and base width affect both the injection efficiency through the ratio J_n/J_S and the transport factor through the transit time. Since the devices of Fig. 3.5 show low power variation of gain with injection level (see Fig. 3.4), J_r should be the dominant defect current, rather than J_S, for these devices. Therefore the variation of β with τ is due to an improvement in the base transport factor, and the effective lifetime is therefore of the same order as the transit time. Using the values of τ from Fig. 3.5 and the fact that the transit times are roughly an order of magnitude less, the base lifetimes are estimated to lie in the range 10^{-10}–10^{-9} sec. Such low values are not unreasonable because of the defects, dislocations, strain, etc., resulting from the thermal and lattice mismatches between the two materials.

3.3.3 Effect of Emitter Resistivity

The variation of gain with emitter resistivity can be seen in Figs. 3.3–3.5 and more directly in Fig. 3.6 for two devices with identical base–collector junctions. In Fig. 3.6(a), the resistivity is about 10^3 Ω cm and the gain slightly greater than 1. In Fig. 3.6(b), the resistivity is several orders of magnitude higher and β has increased to 34. From Section 3.2.3, the higher gain with increased emitter resistivities is due to a decrease of J_r and an increase in the field in the bulk emitter, both of which improve the injection efficiency.

This result can be seen in a different way by using the self-compensation property of ZnSe. Heating a low resistivity layer to moderate temperatures

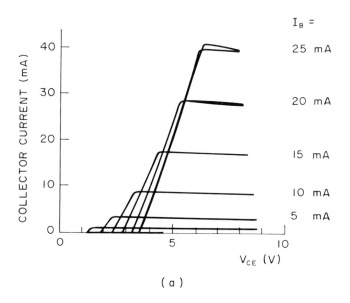

(a)

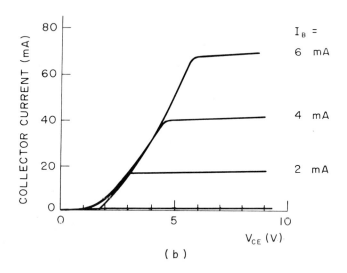

(b)

Fig. 3.5. *Variation of current gain with base width and field at* 27°C, *showing an increase in β of* 8.5 *for a decrease in* τ *of* 10.

ZnSe resistivity = 10^3 Ω cm. (a) N14-3: $\tau = 1.2 \times 10^{-9}$ sec, gain of 2 at $I_E = 65$ mA ($\equiv 180$ A cm^{-2}). (b) N18-3: $\tau = 1 \times 10^{-10}$ sec, gain of 17 at $I_E = 76$ mA ($\equiv 180$ A cm^{-2}). (After Hovel and Milnes, 1967.)

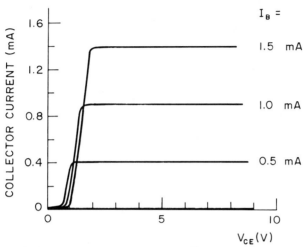

(a) N 33–2. N(ZnSe) = 5 x 10^{13}. GAIN OF 1.0 AT
I_E = 1.9 mA = 4.8 A cm^{-2}

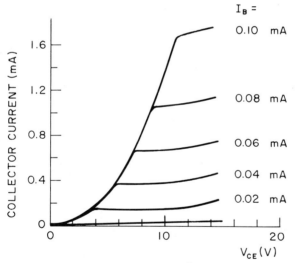

(b) N 21–4. N(ZnSe) = 1 x 10^{11}. GAIN OF 34 AT
I_E = 1.8 mA = 4.5 A cm^{-2}

Fig. 3.6. *Variation of gain with emitter resistivity at* 27°C, *β increases by* 34 *while* N_D *decreases by*
5 × 10^2.
Identical Ge base–collector substrate: $C_S = 2 \times 10^{19}$ cm^{-3}, $E(0) = 1120$ V cm^{-1},
$W = 0.6$ μ, $\tau = 2.1 \times 10^{-10}$ sec. (After Hovel and Milnes, 1967.)

for a few minutes causes the formation of Zn vacancies which act as acceptors and drives the material up in resistance. When the ZnSe emitter of a transistor was converted from relatively low resistivity (5×10^3 Ω cm) to high resistivity ($\sim 10^5$ Ω cm) by heating at 400°C for 30 min, the current gain increased from 1.2 to 35.

3.3.4 Temperature

The variation of the gain of a ZnSe–Ge transistor with temperature at relatively high injection levels was typically that β decreased by 20 to 30% from 350 to 100°K. This implies that γ and B are either temperature insensitive at these current levels or that they vary in opposite but compensating directions. From Section 3.2.4 the injection efficiency might be reduced at lower temperatures by the increasing interface-recombination velocity and decreasing drift field and mobility in the base. The ratio of J_r to J_n may also increase, depending on the limiting mechanism of J_r. The transport factor is also expected to decrease; however, the nature of the base lifetime is uncertain, and the lifetime may actually be increased at lower temperatures.

At lower injection levels, a wider behavior of gain with temperature was observed. For several devices, β decreased by an order of magnitude from 25 to −180°C, while for others it was independent of temperature or even increased slightly at lower temperatures. Similar behavior has been reported for Ge–GaAs heterojunction transistors by Jadus and Feucht (1969). Variations in trap densities, interface densities, ZnSe mobilities, etc., may be expected from device to device, and varying temperature behavior may be expected from the effects of these parameters on the relative magnitudes of J_S, J_r, and J_n. At the higher injection levels the barrier is nearly deleted, the traps are nearly filled, and the field in the ZnSe is high; hence variations in these quantities are probably of less importance in determining the gain.

3.3.5 Other Heterojunction Transistors

The n–p–n GaAs–Ge–Ge transistors reported by Jadus and Feucht (1969) have common–emitter characteristics similar to those of ZnSe–Ge previously discussed. The base and collector dopings are very similar, but the GaAs emitter was doped to 5×10^{15} cm^{-3} resulting in current gains of near 15. Because of the higher emitter doping the emitter voltage drop is somewhat less than for the ZnSe–Ge devices. The emitter–base I–V characteristics of the GaAs–Ge junction show less of an effect due to defect currents at the junction. Recent work (Ladd, 1969) has resulted in GaAs–Ge devices with more heavily doped GaAs emitters (5×10^{17} cm^{-3}) but these have shown current gains of only 1 to 2. This reduction in gain with an increase in doping of the emitter is similar to a trend seen in ZnSe–Ge junctions. The effect could be attributed to an increase of tunneling defect currents in the narrower depletion regions of more heavily doped emitter–base junctions.

Heterojunction transistors of *n–p–n* ZnSe–GaAs–GaAs have also recently been fabricated in our laboratory with current gains up to 70 (Sleger *et al.*, 1970). The grown ZnSe emitter is doped in the range 10^3–10^4 Ω cm, and the devices are space-charge limited in the emitter region. Base region widths of 0.2 to 1 μ were obtained by Zn diffusion. Initial studies of these transistors show the performance to be very similar to that of *n–p–n* ZnSe–Ge–Ge transistors.

Other heterojunction transistors that are expected to be of interest include *n–p–n* GaP–Si–Si, *n–p–n* Si$_{0.1}$Ge$_{0.9}$–Ge–Ge, and *p–n–p* Al$_x$Ga$_{1-x}$As–GaAs–GaAs.

3.4 Performance Potential of Heterojunction Transistors

When revived by Kroemer in 1957 (Kroemer, 1957a), the heterojunction transistor concept promised large advantages over the transistor structures then in use. Since that time homojunction transistors have improved considerably so that commercial high-frequency units can be made to work in the low GHz range. However, many of the factors involved in the improvement of homojunction transistors, such as narrow base widths and stripe geometries, are equally applicable to heterojunction transistor fabrication (although requiring development effort). The purpose of this section is to show that heterojunction transistors have the potential of surpassing the present high-frequency performance of Si or Ge homojunction transistors by a useful factor. For the GaAs–Ge model considered here the factor in frequency is 2 and the factor in power gain is 4. Heterojunctions gain their advantage from the fact that the base doping may be very high, and the emitter doping low. Thus, the base resistance and emitter-base capacitance are both lower, which contributes to improved frequency response.

Since it is only relatively recently that heterojunction transistors have been shown to be feasible, knowledge of their performance is limited, at present, to GaAs–Ge, ZnSe–Ge, and ZnSe–GaAs types. At present these are fabricated by processes that introduce limitations of doping and geometry. In this discussion fundamental limitations are considered, but other problems of the present heterojunction structures (such as trapping effects in the emitter) that are probably fabrication limitations at this time and not intrinsic to future heterojunction transistors are neglected.

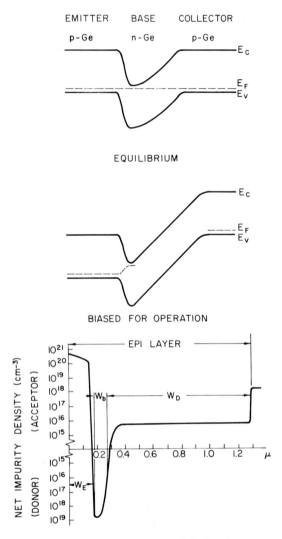

Fig. 3.7. *Band diagrams and impurity profile of a high-speed* Ge *homojunction transistor (assumed).*
(After Ladd and Feucht, 1970b.)

3.4.1 *Description of Transistor Types*

The emphasis of this analysis is on frequency response rather than power output, the economics of fabrication, or special features such as optical properties. As a model for comparison, a Ge homojunction transistor is considered with a band diagram and a doping profile as shown in Fig. 3.7.

The structure might be fabricated by growing a lightly doped epitaxial layer on a heavily doped *p*-substrate. The *n*-base region is then diffused and the emitter alloyed into the diffused layer to produce the thin base region under the emitter. This structure has a high field in the base region aiding minority

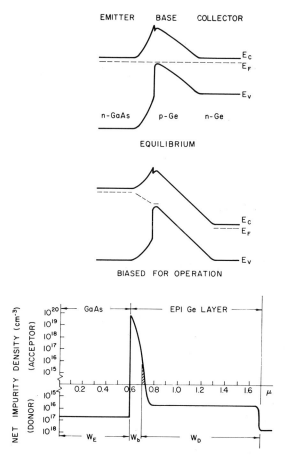

Fig. 3.8. *Band diagrams and impurity profile for a heterojunction transistor with normal-resistance emitter (assumed).*
(After Ladd and Feucht, 1970b.)

carrier transport from the emitter and has a collector depletion region which is bounded by two regions of heavy doping. Under bias the collector region is "swept out," depleted to the epi-substrate, and the high field necessary for short transit times can be obtained without an extended collector depletion region.

A p–n–p homojunction structure has been chosen for comparison rather than a n–p–n structure for two reasons. First, the p–n–p transistor is expected to be slightly faster since for the same base doping, the base region may be thinner for the same base resistance. Second, the solubilities of typical acceptors in

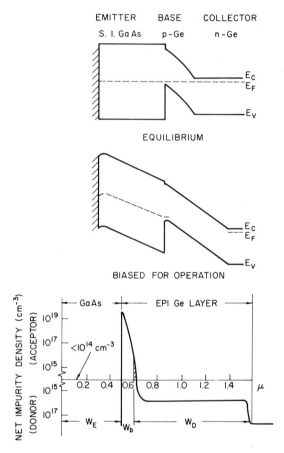

Fig. 3.9. *Band diagrams and impurity profile of a space-charge-limited emitter heterojunction transistor* (*assumed*).
(After Ladd and Feucht, 1970b.)

Ge are about twice that of the most soluble donor, so assuming the emitter is doped to the maximum solubility limit, the injection efficiency would be lower in the n–p–n transistor.

The GaAs–Ge heterojunction devices to be considered have the band diagrams and doping profiles shown in Figs. 3.8 and 3.9. The heterojunction device of Fig. 3.8 has a base doping level at the emitter, N_{be}, of 5×10^{19}

cm^{-3} while the emitter doping is only 5×10^{16} cm^{-3}. Such structures have been fabricated with current gains, h_{FE}, much larger than unity. For these calculations we assume that $h_{FE} = 30$ for the heterojunction devices.

The device depicted in Fig. 3.9 is the space-charge-limited (SCL) triode proposed by Wright (1962) with a GaAs emitter, a p–Ge base, and an n–n^+ collector.

The frequency response of the dielectric triode with a high resistivity collector is the subject of a detailed treatment by Brojdo (1963). The calculated curves, however, are for the case where the collector transit time is less than that of the emitter which is not so for the practical example considered here. The n–n^+ collector approach leads to a more useful comparison of the three model devices. GaAs–Ge was chosen as the pair to be considered as GaAs has a very high mobility and can be made either low resistivity or semi-insulating. The high mobility of the GaAs emitter leads to a lower emitter time constant than for ZnSe–Ge or CdS–Si transistors.

3.4.2 *Figure of Merit*

In evaluating the transistor models, the maximum frequency of oscillation, f_{max}, has been used as one criterion. This figure of merit is given by

$$f_{max} = (4\pi)^{-1}[r_b C_c \tau_{ec}(1 + \nu)]^{-1/2} \tag{3.31}$$

where r_b is the effective base resistance, C_c is the effective collector capacitance, τ_{ec} is the total emitter-to-collector delay time, and ν is the excess phase factor associated with the graded base doping (Beadle *et al.*, 1969). Also

$$f_t = [2\pi\tau_{ec}(1 + \nu)]^{-1} \tag{3.32}$$

where f_t is the frequency at which the common–emitter current gain becomes unity. The delay time, τ_{ec}, can be written as

$$\tau_{ec} = \tau_b + \tau_e + \tau_{sl} + \tau_c \tag{3.33}$$

where τ_b is the base transit time, τ_e is the $r_e C_e$ time-constant for charging the emitter capacitance, τ_{sl} is $\frac{1}{2}$ the collector transit time, and τ_c is the rC time-constant for charging the collector capacitance. The excess phase factor, ν, is associated with the frequency dependence of the phase shift of the common emitter current gain. A good approximation for this has been given by teWinkel (1959) as $\nu = 0.22 + 0.098 \ln(N_{be}/N_{bc})$, where N_{be} and N_{bc} are the base doping levels at the emitter and collector junctions respectively.

The figure of merit, f_{max}, is the frequency at which the unilateral power gain of the transistor becomes unity. The unilateral power gain is the maximum available power gain of the transistor when the reverse gain is zero and is independent of the terminal configuration. In practical cases where the reverse gain is neutralized by an external network, the unilateral power gain of the neutralized amplifier may be greater or less than the maximum available gain of the original unneutralized device depending on the overall modification of the two-port network parameters by the neutralizing network. The transistor is unconditionally stable when unilateralization has been employed.

3.4.2.1 *Collector Time Constant.* By comparison of the band diagrams and doping profiles of Fig. 3.7–3.9 we see that the collector depletion region is the same for all devices. The time constant τ_c is given by

$$\tau_c = r_c C_i \tag{3.34}$$

where r_c is the ohmic resistance of the episubstrate, and C_i is the inner collector transition region capacitance. In calculating r_c and C_i, the area is assumed to be that associated with the emitter stripes.

The collector transit time τ_{sl} is given by

$$\tau_{sl} = W_D/2v_{sl} \tag{3.35}$$

where W_D is the width of collector depletion region, and v_{sl} is saturated drift velocity in the collector. For Ge, v_{sl} is 6×10^6 cm sec^{-1}. The factor of 2 is a result of analyzing the drift limited flow of carriers across the depletion region (Pritchard, 1967, p. 321).

3.4.2.2 *Base Transit Time.* The base transit time must be calculated separately for the heterojunction and homojunction transistors. For a uniform base doping

$$\tau_b = W_b^2/2D \tag{3.36}$$

where W_b is the base width and D is the diffusion constant of minority carriers in the base.

The formula for τ_b in the presence of an arbitrary impurity gradient cannot be given explicitly. An exact analysis would take account of the electric field as well as the variation of D with impurity concentration. The usual way of calculating τ_b is to approximate the typical complementary error function or gaussian distribution by an exponential. The base doping profile is then

$$N_b = N_{be} \exp(-mx/W_b) \tag{3.37}$$

and

$$m = \ln(N_{be}/N_{bc}) \tag{3.38}$$

where N_{be} is the base impurity concentration at the emitter, and N_{bc} is the base impurity concentration at the collector. If D is taken as constant, this results in

$$\tau_b = (W_b{}^2/D)(m - 1)/m^2 \tag{3.39}$$

The value of D used in the calculations below is taken as the bulk value corresponding to N_{be}, the highest concentration in the base. The effect of this approximation is that the calculated τ_b will be greater than the exact value, leading to a conservatively low estimate of f_t.

Since D decreases as the doping of the base increases, it follows that τ_b will increase if r_b is lowered by raising the doping of the base. In general there will be an optimum value of N_{be}, for a particular base width and N_{bc}, leading to the shortest base transit time. For homojunction transistors the value of N_{be} must be low enough to permit a useable injection efficiency. The parameters for small τ_b shift in the direction of larger values of N_{be} for decreasing base widths.

Inspection of the f_{max} expression (3.31), shows that r_b has a large effect. In fact, r_b is much more important than τ_b in determining f_{max} because in general τ_b is only about 10–15% of τ_{ec}. However, note that the excess phase factor ν also increases with the base doping gradient. These considerations lead to the notion for a heterojunction structure of an optimum value of N_{be}, for a particular value of W_b and N_{bc}, leading to an optimum f_{max}. The value of τ_b thus obtained will be larger than the shortest τ_b which could be obtained for a given W_b and N_{bc}. This idea is not as useful for homojunction structures because of the injection efficiency restriction on N_{be}. Ignoring the effect of doping on τ_b but including the variation of ν it can be shown that the optimum base doping lies above 10^{20} cm^{-3} for the heterojunction transistors of the geometry considered here. However, our comparison is based on a value of 5×10^{19} for N_{be} since this is more typical of transistors we have fabricated.

3.4.2.3 *Emitter Time Constant.* The emitter time delay τ_e is given by

$$\tau_e = r_e C_e \tag{3.40}$$

where the differential emitter resistance r_e is kT/qI_e, C_e is the emitter transition capacitance, and the series resistance of the emitter bulk is neglected.

The emitter capacitance calculation for the homojunction is complicated by the fact that the depletion region spreads into a retrograde doping profile. The calculation may also be affected by a significant grading of the emitter to base doping at the junction. When this occurs, the value of C_e is partly determined by a linearly graded behavior near the junction and by a retrograde behavior some distance into the base. For the homojunction model used here the emitter to base transition is taken as abrupt and the junction is treated as retrograde. The values of C_e are found from the curves of Nathanson and Jordan (1962) for an exponential distribution of base impurity concentration.

The calculation of C_e for the doped heterojunction device makes use of the formula for an abrupt junction with one side heavily doped. The depletion region extends into the emitter bulk and the capacitance is determined by the emitter area A_e, the emitter junction bias, and the emitter impurity concentration. The formula is then

$$C_e = A_e(\epsilon_q N_e/2 V_{\text{eff}})^{1/2} \tag{3.41}$$

The capacitance of the heterojunction emitter will be much lower than for the homojunction device due mostly to the difference in doping levels. A slight further advantage accrues because the dielectric constant is lower for GaAs than Ge. For the models presented the value of C_e of the heterojunction is less than one-sixth of that for the homojunction.

In order to calculate r_e and the emitter bias voltage for a heterojunction emitter the forward bias voltage–current characteristic must be known. For the purposes of this study a simple exponential characteristic of the form

$$J = J_0 \exp(-qV/\eta kT) \tag{3.42}$$

has been assumed with η equal to 1. The GaAs–Ge heterojunction transistors reported by Jadus and Feucht (1969) exhibit exponential characteristics with η between 2 and 3. With improved growth technology it is expected that GaAs–Ge junctions should eventually exhibit about the same η characteristics as for homojunctions. In our calculations therefore the values of η and V_{eff} are assumed to be the same as for a Ge emitter.

In order to calculate a value of τ_e for the SCL heterojunction emitter, the analysis of Shao and Wright (1961) is used. They give the admittance of a SCL diode as

$$G = g + j\tfrac{3}{4}\omega C \tag{3.43}$$

where

$$g = 2I_e/V_e \quad \text{and} \quad C = \epsilon A_e/W_E$$

The expression is valid for

$$f < 1/2\pi t_r, \quad \text{where} \quad t_r = \tfrac{4}{3}W_E{}^2/\mu V_e$$

The value of τ_e is given by

$$\tau_e = 3C/4g = 3A_e V/8W_E I \tag{3.44}$$

The equation for the trap-free space-charge limited current is

$$I_e = \tfrac{9}{8}\epsilon\mu A_e V_e^2/W_E^3$$

and substitution of this in (3.46) gives

$$\tau_e = \tfrac{1}{3}W_E^2/\mu V_e = t_r/4 \qquad (3.45)$$

The accuracy of the analysis can be tested by comparing this value of τ_e with the rC product obtained by Brojdo's (1963) exact calculations. His results give

$$rC = t_r/\pi \approx t_r/3 \qquad \text{for} \quad \omega t_r = \pi$$
$$rC = 3t_r/2\pi \approx t_r/2 \qquad \text{for} \quad \omega t_r = 2\pi$$

The value $t_r/4$ is considered to be applicable because practical devices are normally operated below $f_t < (2\pi t_r)^{-1}$. If this is not the case a value of r_e given by $t_r/3$ or $t_r/2$ may be substituted as is appropriate. The consequences of this will be discussed below in connection with the numerical calculations.

3.4.2.4 *Calculation of* $r_b C_c$. Calculation of the $r_b C_c$ product is relatively straight forward as long as the resistance of the contact to the base does not have to be taken into account. The r_b formula for uniform current density over the emitter stripe area is

$$r_b = R_{sbe}d_e/12L \qquad (3.46)$$

where R_{sbe} is the base sheet resistance under the emitter in ohms per square, d_e is the emitter stripe width, and L is the emitter stripe length. The collector capacitance is simply

$$C_c = C_i = \epsilon A_e/W_D \qquad (3.47)$$

when the wafer parasitics are neglected.

An analysis of the $r_b C_c$ product where the base contact resistance cannot be ignored in a stripe geometry, has been made by Arnold and Pritchett (1965) and is illustrated in Fig. 3.10. This $r_c C_c$ product is given by

$$r_b C_c = r_{con}C_{ct} + \tfrac{1}{2}r_s C_s + r_b'C_i \qquad (3.48)$$

where the total collector capacitance C_{ct} is $C_a + C_i + C_s$, and r_s, C_s are the resistance and capacitance in the stripe spacing region. The resistance r_b' is $r_{be} + r_s$, and C_i is the collector capacitance under the emitter. The factor of

2 in the $r_s C_s$ product arises from the distributed nature of the resistance and capacitance.

The contact resistance r_{con} is calculated from

$$1/r_{con} = L_B \ (g_i/R_{sb})^{1/2} \tanh d_b (R_{sb} g_i)^{1/2} \tag{3.49}$$

where R_{sb} is the sheet resistance of the base diffused layer in the space between stripes, d_b is the width of the base contact, g_i is the interfacial contact conductance in Ω^{-1} cm^{-2} of the base contact area, and L_B is the effective base

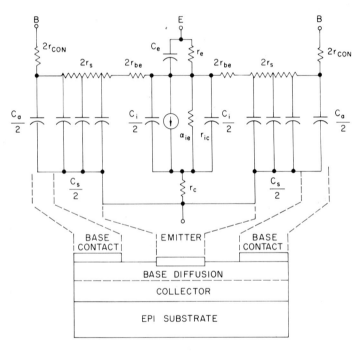

Fig. 3.10. *Cross-section view and equivalent circuit of stripe geometry transistor.*
(After Ladd and Feucht, 1970b.)

stripe length. For the geometry considered the effective base length is between 3 and 4 times the length of one stripe (L). On these calculations an effective base length of $3L$ has been assumed.

By inspection of the geometry of the device it is obvious that the value of r_{con} can be important because of the large value of C_e. The other important part of $r_b C_e$ will be the product $r_{be} C_i$ because R_{sbe} will be quite large relative to R_{sb}. For our purposes R_{sb} and R_{sbe} can be calculated once the diffusion profile is established.

A value for g_i can be found only from experiment. For the computations below a value of g_i of 5 Ω^{-1} mil^{-2} (7.75 $\times$ 10^5 Ω^{-1} cm^{-2}) has been used. This value is appropriate for a low resistance contact to n-type Ge using a combination layer of Cu, Ti, and Al (Beadle *et al.*, 1969). Although the same value was used for p-type Ge, it might be expected that g_i would be lower for p–Ge contacts such as aluminum than for n–Ge contacts.

3.4.3 *Model Calculations Neglecting Base Contact Resistance*

Base contact resistance is presently an important part of the $r_b C_c$ product in high frequency transistors. The contribution is, however, about the same as for the $r_b' C_i$ product. One of the results of heavy doping of the base region of a

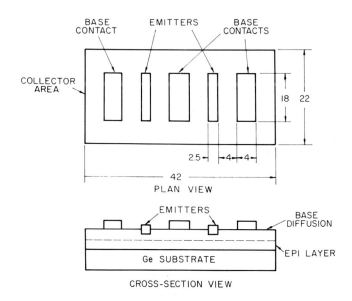

Fig. 3.11. *Geometry of microwave transistor structure. All dimensions in microns.* (After Ladd and Feucht, 1970b.)

heterojunction device is that r_{be} is lowered (by a factor of 6 in our example). The trend of transistor technology is towards the fabrication of transistor structures with very small base contact resistance. Therefore calculations have been made of the frequency response of the three transistor models of Figs. 3.7–3.9 using the emitter area of Fig. 3.11 but neglecting the base contact resistance and the parasitic r's and c's of the stripe spacing region.

Structural details of the transistor models are given in Table 3.1.

TABLE 3.1

Details of the Three Transistor Models

W_b (base width)	$0.1\ \mu$
W_D (collector depletion width)	$1.0\ \mu$
Emitter area	$90\ \mu^2$
Epitaxial substrate: resistivity thickness	$0.01\ \Omega$ cm
Collector doping	5×10^{15} cm^{-3}
Collector bias	5 V
SCL emitter width	$0.5\ \mu$
SCL emitter mobility	5000 cm^2 V^{-1} sec^{-1}

The doping and bias levels along with the calculated values leading to f_{max} are given in Table 3.2, which shows that the f_{max} of either type of heterojunction transistor is twice that of the homojunction transistor. This advantage is obtained in the doped heterojunction device at the same emitter current density. In the SCL device, interestingly enough, the same advantage is obtained at only 20% of the J_e of the homojunction device.

TABLE 3.2

Circuit Constants and Time Delays of the Three Idealized Transistor Models

	Ge homojunction	Doped emitter heterojunction	SCL emitter heterojunction
N_{be} (cm^{-3})	2×10^{18}	5×10^{19}	5×10^{19}
N_{bc} (cm^{-3})	5×10^{15}	5×10^{15}	5×10^{15}
I_e (mA)	5	5	0.930
J_e (A cm^{-2})	5560	5560	1030
V_e (V)	0.35	0.35	0.15
Emitter doping (cm^{-3})	$\sim 10^{20}$	5×10^{16}	$< 10^{14}$
m	5.99	9.2	9.2
D (cm^2 sec^{-1})	11	5.4	5.4
R_{sbe} (Ω/square)	1250	330	330
r_b (Ω)	7.3	1.9	1.9
C_c (pF)	0.013	0.013	0.013
r_c (Ω)	111	111	111
C_e (pF)	0.668	0.098	—
r_e (Ω)	5.2	5.2	—
ν	0.81	1.12	1.12
τ_b (psec)	1.26	1.74	1.74
τ_c (psec)	1.41	1.41	1.41
τ_{sl} (psec)	8.34	8.34	8.34
τ_e (psec)	3.47	0.51	1.11
τ_{ec} (psec)	14.48	12.00	12.60
f_t (GHz)	6.1	6.26	6.04
f_{max} (GHz)	50.6	100.4	98.6

The magnitude of the unilateral power gain is given by

$$U = [f_{max}/f]^2 \tag{3.50}$$

Hence the power gain of the heterojunction transistor will be about four times that of the equivalent homojunction structure at high frequencies.

A comparison of the components of τ_{ec} and the values of r_b for the homojunction and heterojunction transistors shows that the reduction of r_b is by far the most important factor contributing to the advantage of the heterojunction. For the model considered, the advantage in r_b is a factor of 3.8 while the advantage in τ_{ec} due to decrease in emitter capacitance by a factor of almost 7 yields only a 20% decrease in τ_{ec}. The effect of the phase factor, ν, is to make the values of f_t almost the same. Obviously the limiting component in f_t is the collector transit time τ_{sl} which depends on collector depletion width and the saturated drift velocity.

The assumptions made earlier concerning the emitter time delay and doping level of the SCL emitter can now be examined. From the discussion in Section 3.4.2.3, recall that at very high frequencies the value of τ_e became larger. The transit angle corresponding to 98.6 GHz is

$$\omega t_r = 2\pi(9.86 \times 10^{10})(4.44 \times 10^{-12}) = 2.75 < \pi$$

Thus the more accurate value of τ_e for operation near f_{max} is $t_r/3 = 1.48$ psec. However, this will not change the value of f_{max} significantly.

For space charge flow conditions to prevail the emitter doping density need only be less than about a tenth of the stored charge density in the emitter. This gives a value for the doping density of

$$N_e \leqslant \tfrac{1}{10} Q_s/q = \tfrac{1}{10} J_e t_r/q d_e = 2.9 \times 10^{14} \quad cm^{-3}$$

3.4.4 *Model Calculations Including Contact Resistance*

Although the performance of heterojunction transistors is clearly superior when compared to homojunction devices in the idealized geometry described in the preceding section, it is also clear that the advantage will be nullified to some extent by the presence of base contact resistance. This problem is explored by calculating values of f_{max} for the three transistor models when used in the geometry of Fig. 3.11 with contact resistance included. Table 3.3 gives the values assumed and the calculated results. All other parameters are the same as those given in Table 3.2.

TABLE 3.3

Circuit Constants and Time Delays of the Transistor Models with Contact Resistance Included

	Ge homojunction	Doped heterojunction	SCL heterojunction
R_{sbe} (Ω/square)	1250	330	330
R_{sb} (Ω/square)	115	115	115
g_i (Ω^{-1} cm^{-2})	7.75×10^5	7.75×10^5	7.75×10^5
r_{con} (Ω)	2.3	2.3	2.3
C_{ct} (pF)	0.13	0.13	0.13
r_s (Ω)	6.4	6.4	6.4
C_s (pF)	0.029	0.029	0.029
r_b' (Ω)	13.6	8.3	8.3
C_i (pF)	0.013	0.013	0.013
$r_{con}C_{ct}$ (psec)	0.299	0.299	0.299
$\frac{1}{2}r_sC_s$ (psec)	0.093	0.093	0.093
$r_b'C_i$ (psec)	0.177	0.108	0.108
r_bC_c (psec)	0.569	0.5	0.5
τ_{ec} (psec)	14.48	12.00	12.6
f_t (GHz)	6.1	6.26	6.04
f_{max} (GHz)	20.7	22.3	21.8
For r_{con} reduced by a factor of 10:			
f_{max} (GHz)	29.4	34.5	33.8

It is seen that the presence of base contact resistance eliminates most of the advantage of the heterojunction transistors. The advantage remaining in f_{max} is 8% and in unilateral power gain 17%. If the contact resistance was one-tenth of the value originally used, the results are as shown in the last line of Table 3.3. The advantage of the heterojunction devices rises to about 17% in f_{max} and 38% in power gain.

This calculation of f_{max} including the effect of the base contact resistance shows that geometry is at present the controlling factor in the frequency response of microwave transistors. The value of f_t itself has practically reached the limit of its development with the advent of epitaxial collector techniques. Collector transit time for a particular material is limited by avalanche breakdown considerations. From this it is clear that the advantages of low base resistances in heterojunction devices will only be exploited if suitable geometries can be developed.

Beadle *et al.* (1969) have studied the effects of contact resistance and stripe width and spacing on the frequency response of a germanium device very similar to the one considered here. They show by measurements of devices that the simple formula for f_{max} is accurate to within about 15% if the parasitic wafer resistances and capacitances are included in the equivalent circuit. These have been included in the present calculations. As an example of special processing to reduce the sheet resistance of the base diffusion between the emitter and base stripes, they have made use of an additional high concentration diffusion during the emitter alloying cycle. The effect of including this step in the present homojunction device where all the parasitics are included would probably be to make the performance of the three types of transistors about equal.

Because of the difference in fabrication methods between the homojunction and heterojunction transistors, the thickness of the diffused region between the emitter and base stripes is larger for the homojunction transistor. This occurs because the aluminum alloying cycle which forms the homojunction emitter penetrates about half way into the base diffused layer. The heterojunction emitter, on the other hand, has been assumed to be epitaxially deposited onto the surface of the base diffusion. The alloyed emitter homojunction geometry could be obtained in a heterojunction device by growing the emitter in a pocket etched into the surface of the base.

3.4.5 *Conclusions*

The figures of merit of a germanium homojunction transistor, a doped emitter heterojunction transistor, and a SCL emitter heterojunction transistor have been calculated and compared for an equivalent geometry which is close to the state of the art. If the parasitic wafer resistances and capacitances are ignored it is shown that the heterojunction transistors would have about twice the frequency response and four times the power gain of the homojunction transistor. If the wafer parasitics are included in the calculation the performances of the three types of transistors are roughly equivalent. This means that an advance in the state of the art of transistor geometry control which permits very narrow line widths coupled with superior base contact structures will be needed if heterojunction transistors are to have a sizable advantage in frequency response over homojunction transistors.

This discussion has related merely to frequency response. The properties of heterojunction transistors as optical sensors have not been examined extensively. However, this may well be an area where significant advantages are available over homojunction devices even with present contact technology.

Chapter 4 | Isotype (n–n, p–p) Heterojunctions

4.1 Introduction

In an n–n heterojunction the conduction-band energy barrier is determined by the electron affinity difference for the two semiconductors (since $\Delta E_c = \chi_2 - \chi_1$) and by the doping levels of the semiconductors since these position the conduction band edges with respect to the common Fermi level at zero applied voltage. A typical energy band diagram is shown in Fig. 4.1(a) where a barrier, V_{D2}, exists in the conduction band for electron flow from 2 to 1. Similarly in a p–p heterojunction a barrier to the movement of holes can exist in the valence band, as shown in Fig. 4.1(b). Both of these structures, therefore, can act as rectifiers. No minority carrier injection is involved and the devices should be fast in switching response, being rC limited. Studies of such structures have concentrated on n–n junctions since the ohmic contact problem for the materials studied is less severe than for p–p junctions. For this reason the discussion that follows concentrates on n–n junctions although much of the comment should be readily transferable to the p–p situation.

The energy step at n–n junctions of Ge–Si from electron affinity considerations should be quite small, $\Delta E_c = 0.12$ eV ($4.13 - 4.01$ eV from Table 1.2). For Ge–GaAs the step, ΔE_c, should be 0.06 eV from Table 1.2, or 0.11 eV as inferred from measurements on n–p Ge–GaAs heterojunctions (Section 2.4). In fact electron affinity values are not known with the confidence that Table 1.2 might suggest, and the inaccuracy may be substantial when taking the difference of two values. Furthermore, the presence of interface states can have a dominant effect on the barrier in n–n heterojunctions.

In semiconductor heterojunctions fabricated from two semiconductors with different lattice constants, dislocations will occur at the interface in order to compensate for the lattice mismatch. Assuming both semiconductors have a diamond lattice and the epitaxy is on a (111) plane, the density of dangling bonds (unpaired covalent electrons per square centimeter) (Oldham, 1963; Donnelly, 1965; Holt, 1966b) is given by

$$N_{DB} = (4/\sqrt{3})(a_1^{-2} - a_2^{-2}) \qquad (4.1)$$

where a_1 and a_2 are the cube edge lattice constants of the two semiconductors. For (110) plane epitaxy the density is $\sqrt{3/2}$ times greater and for the (100)

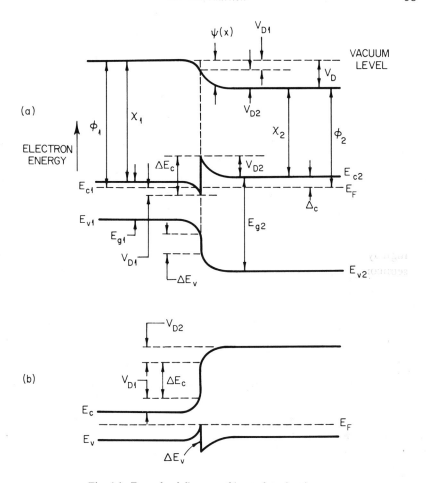

Fig. 4.1. *Energy band diagrams of isotype heterojunctions.*
(a) *n–n* heterojunction, barrier to current flow is in the conduction band; (b) *p–p* heterojunction, barrier to current flow is in the valence band. (After Anderson, 1962.)

plane $\sqrt{3}$ times greater. Table 4.1 presents the calculated values for a number
of important heterojunction pairs.

The single dangling electrons are presumably unpaired. It is reasonable to
expect that an electron paired with one of these dangling electrons would have
less energy than a single free electron in the conduction band. It would,
however, probably have more energy than an electron in a complete valence
band. The energy of an electron accepted by the unpaired electron to form a
dangling pair is therefore expected to lie somewhere in the energy gap. The
unpaired atom can also act as a donor. The single dangling electron is less

TABLE 4.1

Densities of Dangling Bonds in Heterojunctions

Heterojunction and lattice parameters	Plane	Dangling bonds (cm^{-2})
Ge–GaAs, AlAs–GaAs	(111)	1.2×10^{12}
5.658 Å, 5.6535 Å	(110)	1.4×10^{12}
	(100)	2.0×10^{12}
ZnSe–Ge	(111)	2.3×10^{12}
5.667 Å, 5.658 Å	(110)	2.8×10^{12}
	(100)	4.0×10^{12}
ZnSe–GaAs	(111)	3.4×10^{12}
5.667 Å, 5.653 Å	(110)	4.1×10^{12}
	(100)	5.9×10^{12}
GaP–Si	(111)	5.6×10^{12}
5.4506 Å, 5.431 Å	(110)	6.9×10^{12}
	(100)	9.7×10^{12}
Ge–Si	(111)	6.2×10^{13}
5.658 Å, 5.431 Å	(110)	7.5×10^{13}
	(100)	1.1×10^{14}

tightly bound than a valence electron but has less energy than a free electron in the conduction band. On the basis of this discussion, it is believed that high lying acceptors and low lying donors will occur at the interface of a semiconductor heterojunction.

Bardeen (1947) in dealing with metal–surface state–semiconductor contacts has shown that the role of the metal in determining junction properties becomes secondary if the surface state density is about 10^{13} cm^{-2} or greater. If the same considerations hold here, it is expected that interface states will be a major factor in determining the energy band diagram of Ge–Si heterojunctions while having only a secondary influence in Ge–GaAs heterojunctions. However, though interface states may not significantly affect the band structure of Ge–GaAs heterojunctions, they are still capable of dominating the current transport mechanism in n–p Ge–GaAs junctions as discussed in Chapter 2.

Usually the interface states are considered as localized states in a single infinitesimal plane, and therefore only the net charge on these states will have any effect on the energy band diagram. It is not improbable, however, that the defects at the interface will create two sets of localized states, one in each material. An electric dipole may then be formed at the interface similar to those that exist at metal–semiconductor contacts in which the chemically prepared surface of the semiconductor possesses a large surface state density (see Table 6.4). The electric dipole creates a quasi discontinuity in the electrostatic potential, i.e., an electrostatic potential difference across a few lattice constants. Any barriers associated with it will have equivalent thicknesses and may be neglected because of high tunneling probabilities. An electric dipole therefore has the same effect as a difference in electron affinity. Depending on the strength of the dipole, the energy band diagram will become less dependent on work function differences and may be rather more dependent on fabrication techniques and doping conditions.

Other possible effects at $n-n$ interfaces such as quantum mechanical reflections, impurity segregation, and hot-carrier effects are neglected.

Table 4.1 predicts the minimum number of unpaired atoms at an abrupt heterojunction interface. If a structure is not abrupt but graded, the required number of dangling bonds is not reduced. Thus, the number of dislocations is not changed significantly. This may be readily visualized as a bubble raft with a gradual increase in bubble size.

There is also some possibility of a chemically abrupt heterojunction having a gradient in lattice constant, i.e., a smearing out of the dislocations. Calculations (Oldham, 1963) for systems with large lattice mismatches indicate that after a very few layers of growth, it becomes energetically favorable to produce dislocations at the interface rather than grow the layer in uniform sheet strain with the substrate lattice constant. In particular for growth on the (111) plane, it would be expected that the dislocations in Ge–Si junctions would start to form in the overgrowth or at the interface after about 15 Å of growth. In Ge–GaAs this number would be increased to about 2000 Å. It will be noticed that for Ge–Si ($\sim 4\%$ lattice mismatch) the dislocations would from well within any junction depletion or accumulation region.

The dislocation morphology in graded heterojunctions of $GaAs_{1-x}P_x$ has been studied by Abrahams *et al.* (1969) by transmission electron microscopy. Formation of a heterojunction, whether abrupt or graded,

"gives rise to not only a set of misfit dislocations lying in the plane of growth, but also a set of inclined dislocations, n_1, which propagate throughout the growing crystals. This occurs because the misfit dislocations are segmented, and since they cannot end within the crystal, must bend upwards into the plane of growth. The value of n_1 remains constant throughout the crystal, since no additional misfit dislocations are formed because the inclined dislocations can bend in and out of subsequently formed misfit planes.

"It is found that these inclined dislocations will propagate through an ungraded region grown onto a graded heterojunction and this is of importance in device fabrication. Thus, to decrease the value of n_1, the compositional grading at the heterojunction must be decreased, since it is not sufficient to merely grow additional material at constant composition. In the $GaAs_{1-x}P_x$ system, the value of n_1 decreases from 4×10^7 cm^{-2} to

10^6 cm^{-2} with a decrease in compositional gradient from 5% P/μm to 0.2% P/μm. It is also seen that the dislocation densities can be quite high, orders of magnitude higher than the densities in the substrates.

"Finally, the data indicate that the misfit dislocations are a combination of pure-edge and mixed. This affects the interaction that occurs when two misfit dislocations cross, and cause the dislocations to appear to bend back and forth in the plane. Three possible cases are (1) no interaction occurs, (2) interaction eliminates the intersection, (3) a triple node is formed. All three cases are found to occur."*

4.2 *n–n* Energy Band Diagrams with Interface States Present

Consider an *n–n* Ge–Si junction with uniform doping of 3×10^{17} cm^{-3}. If the electron affinity of germanium is greater than the electron affinity of silicon by about 0.12–0.18 eV,† then the equilibrium energy band diagram would be as shown in Fig. 4.2 in the absence of interfacial defects.

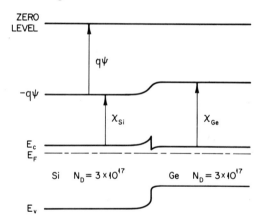

Fig. 4.2. *The energy band diagram of an* n–n *Ge–Si heterojunction in equilibrium if interface state effects are ignored.*
(After Oldham and Milnes, 1964.)

The band diagram is determined by the requirement that the net charge be equal to zero. In the absence of charge at the interface the usual Schottky boundary condition requires that the conduction band be discontinuous by the amount of the difference in electron affinity of the two materials in order

---

* Quoted from Abrahams *et al.* (1969), Dislocation morphology in graded heterojunctions, *J. Mater. Sci.* **4**, 223.

† The electron affinity difference, χ, of 0.18 eV agrees with the photoelectric emission measurements of Allen and Gobeli (1962) and Haneman (1959).

that the electrostatic potential, ψ, be continuous. The positive charge in the depletion region (Si) just balances the negative charge in the accumulation region (Ge).

If, however, interface states are present, the energy bands at the interface are free to move up or down with the necessary charge being supplied by electrons (or their absence) in the interface states. The discontinuity in the conduction band is still equal to the difference in electron affinities; however, the height of the conduction band edge above the Fermi level at the interface is determined primarily by the interface states. As an example, Fig. 4.3(a) shows the energy band diagram of the same Ge–Si junction as Fig. 4.2 but with a distribution of interface states similar to that for a free Si surface. It is observed that both sides of the junction are depleted, a situation made

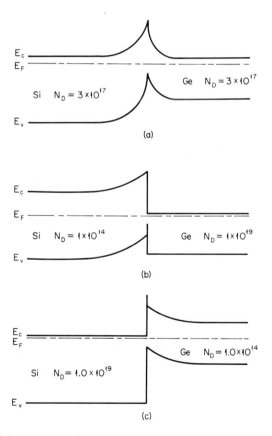

Fig. 4.3. *Equilibrium energy band diagrams at n–n Ge–Si heterojunctions including interface states.* (a) Constant doping of 3×10^{17} cm^{-3}; (b) Si doping of 10^{14}, Ge of 10^{19}; (c) Si doping of 10^{19}, Ge of 10^{14}. (After Oldham and Milnes, 1964.)

possible by the acceptor nature of the interface states. The energy band diagram of Fig. 4.3(b) is for a Ge–Si junction with the same values of χ and interface state distribution as Fig. 4.3(a) but with doping levels of 10^{19} cm^{-3} in the Ge and 10^{14} cm^{-3} in the Si. Figure 4.3(c) is similar except that the doping levels are 10^{14} cm^{-3} in the Ge and 10^{19} cm^{-3} in the Si.

The specific choice made for the interface state distribution is not particularly important; any dense set of interface states near or below the center of the Si band gap would produce the double depletion found here. An *n–n* or *p–p* device with double depletion may display rectification in either direction or saturation in both directions much as two metal–semiconductor diodes joined metal to metal. In fact, if all current flow were from semiconductor to interface state to semiconductor, the structure would be indistinguishable from two metal–semiconductor contacts joined externally in series. However, in an *n–n* structure we must recognize that current may flow from one semiconductor to the other without coming to equilibrium with the interface states.

The main features of *n–n* heterojunction device behavior may be illustrated by a kinetic treatment. Long carrier mean free paths and small deviations from equilibrium are assumed, and tunneling and image effects are neglected. The energy band diagram of an *n–n* heterojunction with the interface region magnified is shown in Fig. 4.4. It is assumed that the interface states lie in a thin layer sandwiched between the two depletion regions. In Fig. 4.4(a), F_{1S} and F_{2S} are emission fluxes from the interface states into the bulk regions. The bulk emission fluxes toward the barriers, F_1 and F_2, are given by

$$F_1 = \tfrac{1}{4}\bar{c}_1 n_1 \tag{4.2}$$

$$F_2 = \tfrac{1}{4}\bar{c}_2 n_2 \tag{4.3}$$

in which n_1 and n_2 are the free electron concentrations and $\bar{c}_1$ and $\bar{c}_2$ the mean thermal velocities of the electrons in regions (1) and (2). The voltage division across the structure and some other parameters are defined in Fig. 4.4(b). The voltages V_1 and V_2 give the reductions in barrier potential in regions (1) and (2). F_1' and F_2' are those portions of F_1 and F_2 sufficiently energetic to surmount the barriers in region (1) or region (2). Therefore, assuming Boltzmann statistics,

$$F_1' = F_1 \exp[-q(V_{D1} - V_1)/kT] \tag{4.4}$$

$$F_2' = F_2 \exp[-q(V_{D2} - V_2)/kT] \tag{4.5}$$

The total "junction" voltage drop is $V_a = V_1 - V_2$; R_1 and R_2 are the interface reflection coefficients for the fluxes F_1' and F_2'; α_1 and α_2 are the interface state transmission coefficients for the fluxes, $(1 - R_1)F_1'$ and $(1 - R_2)F_2'$. The interface state emission fluxes F_{1S} and F_{2S} are assumed constant,

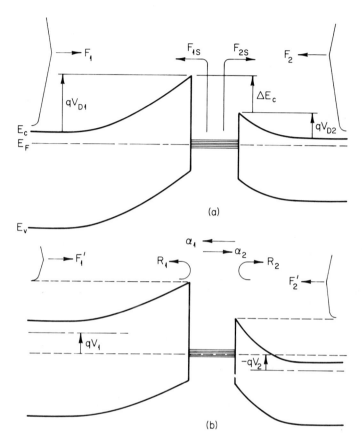

Fig. 4.4. *Energy-band diagrams and electron fluxes in an n–n heterojunction with interface states.* (a) A definition of various fluxes; (b) voltage division across the structure. (After Oldham and Milnes, 1964.)

independent of bias. For a very dense set of interface states F_{1S} and F_{2S} would be very nearly constant and a slight bias dependence of these quantities does not alter the basic nature of the V–I characteristics.

With this approach it may be shown (Oldham and Milnes, 1964) that

$$[\exp(qV_2/kT) - 1]/[\exp(qV_1/kT) - 1] = \alpha_2(1 - \alpha_1)/\alpha_1(1 - \alpha_2) \quad (4.6)$$

From (4.6), if

$$V_a \gg (kT/q) \ln[\{\alpha_1(1 - \alpha_2)/\alpha_2(1 - \alpha_1)\} + 1]$$

then the current saturates at the value

$$F \cong F_1 \exp(-qV_{D1}/kT)(1 - R_1)[\{\alpha_1(1 - \alpha_2)/\alpha_2(1 - \alpha_1)\} + \alpha_1] \quad (4.7)$$

For the opposite polarity of bias, if

$$-V_a \gg (kT/q) \ln[\{\alpha_2(1 - \alpha_1)/\alpha_1(1 - \alpha_2)\} + 1]$$

then similarly, the current saturates at the value

$$F \cong -F_2 \exp(-qV_{D2}/kT)(1 - R_2)[\{\alpha_2(1 - \alpha_1)/\alpha_1(1 - \alpha_2)\} + \alpha_2] \quad (4.8)$$

Equation (4.7) indicates the saturation current for positive V_a and Eq. (4.8) for negative V_a.

Evaluation of the equations requires knowledge of α_1, α_2, R_1, and R_2. Although these quantities are not readily determinable, it may be concluded that if the interface state transmission coefficients α_1 and α_2 are small, the device characteristics would be expected to be the same as for two metal–semiconductor diodes joined externally in series.

For a structure with an energy band diagram similar to Fig. 4.3(a), a double saturation characteristic is expected where the term "saturation" is used loosely to describe a pronounced sublinear dependence of current on voltage. However, a structure with energy bands as in Fig. 4.3(b) or 4.3(c) would very likely not show double saturation. The heavily doped side would not support a reverse voltage drop since tunneling would occur. Only the Si–interface diode would rectify for the device of Fig. 4.3(b) and only the Ge–interface diode for Fig. 4.3(c).

Oldham has grown n–n Ge–Si junctions by iodine transport of Ge onto Si surfaces cleaved *in situ*. Typical characteristics are given in Fig. 4.5 for three sets of doping levels which correspond roughly to the energy diagrams of Fig. 4.3. As expected the diode of Fig. 4.5(a) shows current saturation in both directions, and the voltage direction for easy current flow changes from diode 4.5(b) to diode 4.5(c).

The barrier heights were estimated in two ways: (1) by a comparison of observed current density with theory (assuming α is small), and (2) by thermal activation energies measured in reverse bias. It can be seen from the diode Eqs. (4.7) and (4.8) that if the device has a large applied bias (saturated in either direction), then the dominant temperature dependence of the current at constant bias arises from the exponential dependence of the interfacial carrier concentrations on $-qV_D/kT$. Thus, a plot of $\ln I$ versus q/kT yields a straight line with a slope equal to $-V_D$.

In the n–n units the barrier heights in the Si fall in the range 0.46–0.56 eV and in the Ge, 0.36–0.48eV. Though there is considerable scatter in the data, the two methods of estimating the barrier heights agree within the scatter. The difference in electron affinity can be estimated directly from the double saturating units. The average difference was 0.15 eV from current

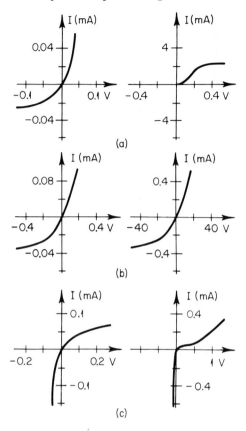

Fig. 4.5. *V–I characteristics of n–n* Ge–Si *heterojunctions.*
(a) 4×10^{16} cm^{-3} Ge on 10^{17} cm^{-3} Si; (b) 10^{18} cm^{-3} Ge on 1.2×10^{14} cm^{-3} Si; (c) 3.5×10^{16} cm^{-3} Ge on 5×10^{19} cm^{-3} Si. Positive voltage corresponds to positive potential on the Ge side of the junctions. (After Oldham and Milnes, 1964.)

density measurements and 0.16 eV from thermal activation energy measurements.

Oldham (1963) also grew *p–p* Ge–Si heterojunctions and observed double saturation as for the *n–n* units.

4.3 Confirmation of the Energy Band Model from Capacitance Measurements

The barriers caused by interface states accepting electrons and therefore creating depletion regions in both the Ge and Si can be seen from *n–n* capacitance measurements (Donnelly and Milnes, 1965).

The room temperature capacitance measured at a frequency of 1 MHz of a double saturation n–n Ge–Si diode is shown in Fig. 4.6. The diode was fabricated by epitaxially growing the Ge on a cleaved silicon substrate by germanium diiodide disproportionation. The impurity concentrations are approximately 4×10^{16} cm^{-3} and 10^{17} cm^{-3} in Ge and Si, respectively.

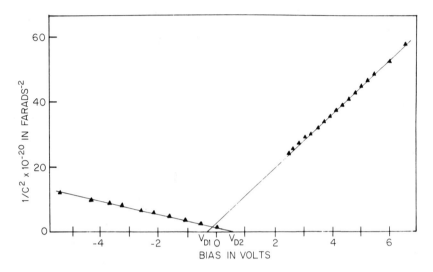

Fig. 4.6. *Capacitance of a double saturation n–n* Ge–Si *heterodiode. Positive bias refers to* + *on the germanium* (1 MHz).
(After Donnelly and Milnes, 1965.)

Following the work of Taylor *et al.* (1952) for grain boundaries, the net negative charge on the interface states of a double saturation diode will increase with an applied bias in either direction so that the forward bias side of the junction will remain practically unchanged. Under these conditions the capacitance per unit area of a n–n Ge–Si double saturation diode is given by

$$C = (q/2\epsilon_1 N_1)^{1/2}(V_{\mathrm{D1}} + V_{\mathrm{a}})^{-1/2}, \quad \text{for} \quad V_{\mathrm{a}} > 0 \qquad (4.9)$$

$$C = (q/2\epsilon_2 N_2)^{1/2}(V_{\mathrm{D2}} - V_{\mathrm{a}})^{-1/2}, \quad \text{for} \quad V_{\mathrm{a}} < 0 \qquad (4.10)$$

where V_{a}, the applied bias is considered positive in the forward direction; V_{D1} and V_{D2} represent the built-in voltage in the germanium and silicon, respectively; N_1 and N_2 are the corresponding impurity concentrations; ϵ_1 and ϵ_2 are the dielectric constants; and q is the unit electronic charge.

Using these equations and the experimental results in Fig. 4.6, $V_{\mathrm{D1}} \cong 0.31$ and $V_{\mathrm{D2}} \cong 0.43$ V. With these values the total equilibrium negative charge density on the interface states may be determined from

$$Q_{IS} = - \{[2q\epsilon_1 N_1 V_{D1}]^{1/2} + [2q\epsilon_2 N_2 V_{D2}]^{1/2}\}$$
$$\cong -1.97 \times 10^{-7} \quad C \text{ cm}^{-2}.$$

The total number of interface states is then

$$N_{IS} \geq -Q_{IS}/q = 1.23 \times 10^{12} \quad \text{cm}^{-2}$$

This number is compatible with the total possible number of interface states expected in Ge–Si heterojunctions from simple lattice mismatch considerations which is 6.2×10^{13} cm^{-2}.

Because of the differences in the forbidden energy gap between the two materials, discontinuities in the band edges are expected in abrupt heterojunctions. Using the previous values of V_{D1} and V_{D2}, the conduction band edge discontinuity is given by

$$\Delta E_c = q(V_{D2} + \delta_{c2} - V_{D1} - \delta_{c1}) \cong 0.12 \quad \text{eV}$$

where δ_c is the potential difference between the Fermi level and the conduction band edge in the bulk of the respective semiconductor. This value of ΔE_c compares with the value of 0.15 eV found by Oldham.

4.4 Studies of *n–n* Ge–Si Heterojunctions as Double Schottky Barriers

Van Opdorp and Kanerva (1967), in an extension of the Oldham model, have considered the implications of representing an *n–n* heterojunction as two Schottky metal–semiconductor diodes connected in series-opposition. The starting equations of the double Schottky model are

$$I_1 = I_{S1}[\exp(qV_1/kT) - 1] \tag{4.11}$$

and

$$I_2 = -I_{S2}[\exp(-qV_2/kT) - 1] \tag{4.12}$$

where the saturation currents I_{S1} and I_{S2} are given, if emission theory is used, by an equation of the form (Spenke, 1958, p. 82)

$$I_S = 4\pi mqk^2 T^2 h^{-3} S \exp(-E/kT) \tag{4.13}$$

where E is the barrier height.

Neglecting series resistance, the overall *I–V* relation for the total junction is found from equating the currents I_1 and I_2 through the two diodes to the junction current I and by putting the total voltage drop V_a across the junction equal to $V_1 + V_2$. This gives

$$I = \frac{2I_{S_1}I_{S_2}\sinh(qV_a/2kT)}{I_{S_1}\exp(qV_a/2kT) + I_{S_2}\exp(-qV_a/2kT)} \tag{4.14}$$

From this equation double saturation is seen to occur in the limits $V_a \to \infty$ and $V_a \to -\infty$.

An experimental determination of the respective barrier heights E_1 and E_2 is in principle possible from the activation energies of the saturation currents I_{S_1} and I_{S_2}. Because of early breakdown, however, the saturation currents are difficult to determine directly or via an extrapolation from the saturating regions of the *I–V* characteristics. A simple way to derive these values, which overcomes this difficulty, follows from the van Opdorp and Kanerva treatment.

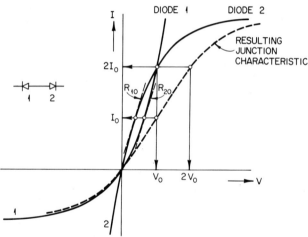

Fig. 4.7. *Schematic diagram of the respective contributions of the two Schottky diodes to the junction voltage.*
On the junction characteristic the point of inflection is (V_0, I_0), and the point $(2V_0, 2I_0)$ corresponds to the point of intersection of the individual diode characteristics. For the actual junctions the ratio I_{S_2}/I_{S_1} is much larger than in this diagram. (After van Opdorp and Kanerva, 1967.)

The *I–V* characteristic has a point of inflection for the polarity where the diode with the higher barrier is biased in its forward direction (see Fig. 4.7). The junction voltage, V_0, current, I_0, and differential resistance, R_0, at the point of inflection are given by

$$V_0 = (kT/q)\ln(I_{S_2}/I_{S_1}) \tag{4.15}$$

$$I_0 = \tfrac{1}{2}(I_{S_2} - I_{S_1}) \tag{4.16}$$

$$R_0 = 4kT/q(I_{S_1} + I_{S_2}) \tag{4.17}$$

The expression for the differential resistance at the origin R_{00} is also useful

$$R_{00} = (kT/q)(I_{S_1}^{-1} + I_{S_2}^{-1}) \tag{4.18}$$

If one of the saturation currents is much larger than the other one (e.g., $I_{S_2} \gg I_{S_1}$) Eqs. (4.16)–(4.18) can be simplified

$$I_0 = 1/2I_{S_2} \tag{4.16a}$$

$$R_0 = 4kT/qI_{S_2} \tag{4.17a}$$

$$R_{00} = kT/qI_{S_1} \tag{4.18a}$$

The numerical values of V_0, R_0, I_0, and R_{00} can be derived directly from the experimental I–V characteristics even for very low breakdown voltages; V_0 and R_0 may be determined more accurately from the minimum in the low frequency differential resistance $(R_{\omega L})$ versus voltage curves. According to the equations the barrier heights E_1 and E_2 follow the slopes of $\ln(R_{00}T)$ versus $1/T$ and $\ln(I_0 T^{-2})$ or $\ln(R_0 T)$ versus $1/T$ plots, respectively. The difference between E_1 and E_2 gives the discontinuity in the conduction band ΔE_c in Fig. 4.4(a).

With the aid of (4.13), (4.15) can be transformed to

$$V_0 = (\Delta E_c/q) + (kT/q)\ln(m_2/m_1) \tag{4.19}$$

Thus, a straight line should be obtained by plotting V_0 versus T. The slope of this line gives m_2/m_1, while the extrapolation to $T = 0$ yields another method for the determination of ΔE_c. In the van Opdorp and Kanerva studies the temperature interval in which the I–V characteristics was measured was too short and the spread of the measured points in the V_0 versus T plots was too large to obtain reliable values for m_2/m_1 and ΔE_c using Eq. (4.19). However, the value 0.5 for m_2/m_1, calculated by Crowell for the materials and orientation used, may be accepted as giving the slope of the straight line. With the slope known, a line may be drawn through the experimental points and extrapolated to $T = 0$ to obtain ΔE_c. The results of the E_{Si}, E_{Ge}, and ΔE_c determinations are presented in Table 4.2.

The van Opdorp diodes were alloyed junctions and the Oldham diode was grown by disproportionation of GeI_2. It is apparent that the barrier heights E_{Si} and E_{Ge} in the alloyed junctions are not, in practice, simple quantities determined by the doping levels but may exhibit variations of ± 0.1 eV from specimen to specimen. However, the values of the discontinuity in the

TABLE 4.2

Barrier Heights Determined for n–n Ge–Si Junctions

| Doping (cm^{-3}) (starting materials) | | Barrier heights (eV) | | | From $E_{Si} - E_{Ge}$ | ΔE_c (eV) | | Reference |
| Si(P) | Ge(Sb) | E_{Si} | | E_{Ge} | | | | |
		From capacitance measurements	From I–V characteristics	From I–V characteristics		From capacitance measurements	From V_0 versus T [Eq. (4.19)]	
2.5×10^{17}	4.2×10^{14}	—	0.46	0.31	0.15	—	—	van Opdorp and Kanerva (1967)
2.5×10^{17}	4.2×10^{14}	—	0.49	0.34	0.15	—	—	van Opdorp and Kanerva (1967)
2.7×10^{15}	8.5×10^{17}	0.58	0.57	0.42	0.15	0.15	0.16	van Opdorp and Kanerva (1967)
2.7×10^{15}	8.5×10^{17}	0.66	0.67	0.52	0.15	0.15	0.15	van Opdorp and Kanerva (1967)
1.0×10^{17}	4.0×10^{16}	—	0.59	0.43	0.16	—	—	Oldham and Milnes (1964)

conduction band ΔE_c derived from the three kinds of measurements are in good agreement; ΔE_c has the constant value 0.15 ± 0.01 eV for all junctions independent of the n-type doping concentrations of the starting materials and independent of the preparation procedure of the samples. This excludes the presence of a variable dipole at the junction interface. In the absence of a dipole layer ΔE_c equals the difference in electron affinity between Ge and Si. This leads to a directly determined value of the affinity difference of 0.15 ± 0.01 eV which is in acceptable agreement with the value expected $(0.12–0.18$ eV) from measurements of χ_{Ge} and χ_{Si}.

A further interesting feature of the double Schottky barrier model is that it explains the sharp minimums observed (Fig. 4.8) in capacitance–voltage curves of $n–n$ diodes with the Ge slightly forward biased. The treatment is somewhat lengthy and reference should therefore be made to the van Opdorp and Kanerva paper (1967).

For alloyed $p–p$ Ge–Si heterojunctions Hampshire (1970) has made a careful study of the $1/C^2$ and the conductance versus voltage peaks and concludes that ΔE_v is 0.1 eV. On the otherhand, the E_c value of 0.15 eV from Table 4.2, which is compatible with electron affinity difference considerations, would correspond to a ΔE_v value of 0.25 eV.

Hampshire has also modeled the behavior of an $n–n$ isotype junction, with interface states of a single time constant, and has found that only one of the two $1/C^2–V$ extrapolations (of the kind shown in Fig. 4.6) gives the correct partial built-in voltage. This is for the material with the smallest electron affinity in the $n–n$ case. The minimum in the low frequency capacitance is at a voltage that corresponds to ΔE_c as in van Opdorp's work.

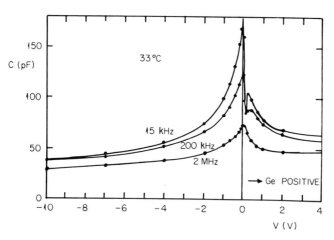

Fig. 4.8. *n–n* Ge–Si *junction capacitance versus bias voltage for various frequencies.* (After van Opdorp and Kanerva, 1967.)

4.5 Germanium–Gallium Arsenide n–n Heterojunctions

The lattice match for Ge–GaAs is much closer than that for Ge–Si (0.08% mismatch compared with 4% mismatch) and the probable density of inter-face states ($\sim 10^{12}$ cm^{-2}) therefore is about a factor of 50 less. The effect of the interface states on the energy band diagram for n–n Ge–GaAs junctions may therefore be expected to be rather small. A diagram of the form shown in Fig. 4.1(a) therefore may be expected where ΔE_c from the difference of electron affinities should be only 0.06 eV. In fact the barriers observed in practice seem to vary from a negligible value (no rectification) up to sizable values such·as 0.4 eV. The barrier heights that are obtained reflect the quality of the growth at the interface.

4.5.1 Theoretical Treatments

Analyses of the current flow in n–n heterojunctions have been made by Anderson (1962), Fang and Howard (1964), Chang (1965a,b), Gribnikov and Melnikov (1966a,b), and others. The treatments generally represent Schottky-type approaches with neglect of interface state effects. Fang and Howard obtain the result

$$J = (8\pi)^{1/2} \frac{(n_{+\infty})^{5/2} \mu q^{5/2} K_2^{1/2}}{n_{-\infty} kTK_1} (V_a + V_D)^{3/2} \exp\frac{-qV_D}{kT}\left[\exp\left(\frac{qV_a}{kT}\right) - 1\right]$$

(4.20)

where $n_{+\infty}$ is the electron density in the GaAs bulk and $n_{-\infty}$ is the electron density in the Ge bulk. From this expression Fang and Howard expect a somewhat stronger voltage dependence of reverse current for an n–n hetero-junction than for a metal–semiconductor contact.

Chang, in an emission model approach of oppositely-directed electron fluxes, obtains

$$J = J_0(1 - (V_a/V_D)[\exp(qV_a/kT) - 1]$$

(4.21)

where

$$J_0 = q^2 N_{D2} V_D (2\pi m_2 kT)^{-1/2} \exp(-qV_D/kT)$$

Equation (4.21) is different from that which governs the conduction in a metal–semiconductor junction. The value of J_0 is different and so is its temperature dependence. The reverse current never saturates but increases linearly with the voltage at large V_a. In the forward direction, if the depend-ence of J on qV_a/kT is approximated by an exponential function, its slope is no longer unity but somewhat smaller.

4.5.2 *Experimental Studies of n–n Ge–GaAs Junctions*

A rectifying *n–n* Ge–GaAs heterojunction conducts in a forward bias condition when the GaAs side is made negative with respect to the Ge. This lowers the barrier qV_{D2} shown in Fig. 4.1(a), and the electron emission from the GaAs into the Ge is greatly increased. The forward voltage drop may be from 0.2 to 0.4 V for typical diodes at moderate current densities.

If the applied voltage is reversed, the barrier in the GaAs is increased, and for normal doping conditions the bulk of the applied voltage appears in the GaAs depletion region. The temperature dependence of the forward characteristics of a *n–n* junction is shown in Fig. 4.9. Most diodes exhibit a soft

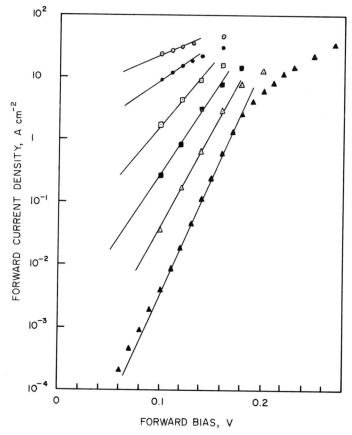

Fig. 4.9. *Forward current–voltage characteristics of an n–n Ge–GaAs (116-2-1L) heterojunction.* The term η in the equation $J = J_0 \exp(qV/\eta kT)$ is seen to be relatively independent of temperature. Key: (▲) 77°K, $\eta = 1.70$; (△) 100°K, $\eta = 1.60$; (■) 125°K, $\eta = 1.57$; (▫) 154°K, $\eta = 1.67$; (●) 200°K, $\eta = 1.88$; (○) 250°K, $\eta = 1.96$. (After Donnelly, 1965.)

breakdown in this reverse bias condition with substantial reverse current flow beginning to occur at anywhere from a few volts to a few tens of volts of reverse bias, depending on the GaAs doping level. Fang and Howard report breakdown voltages of 0.5, 4, and 20 V for 3×10^{17}, 6×10^{16}, and 10^{15} cm^{-3} doping levels. The mechanisms contributing to the soft breakdown may include (1) electron tunneling from the conduction band of Ge to the conduction band of GaAs through the GaAs depletion layer; (2) interband tunneling between Ge and GaAs (somewhat speculative); (3) surface leakage effects; and (4) ionization multiplication. Tunneling between the conduction bands of Ge and GaAs is mainly responsible for the low-voltage excess current. The carrier multiplication may also take place in GaAs at higher voltages. The tunneling probability between conduction bands of Ge and GaAs has been treated by Fang and Howard (1964), and by Chang (1965a,b).

From studies of the temperature dependence of their characteristics, and from capacitance measurements, Fang and Howard infer the values in Table 4.3 for the barrier heights of their n–n Ge–GaAs diodes.

Although the table shows a dependence of the barrier heights on crystal orientation and a difference between the (111) Ga and (111) As face, these results may be primarily a function of the growth process and should not be accepted without other confirming studies. The barriers are seen to be much larger than would be expected from the 0.06 eV electron affinity difference believed to exist between Ge and GaAs.

In a continuation of these studies, Howard *et al.* (1968) have applied high pressures to n–n Ge–GaAs to move the conduction band edges. They conclude that the n–n junction barrier variations are fairly compatible with known conduction-band-edge movements.

4.5.3 Ge Interface Mobility Studies in n–n Ge–GaAs Junctions

In an n–n Ge–GaAs heterojunction an accumulation layer exists on the Ge side because the Ge has a greater electron affinity than the GaAs. If the Ge side is provided with two ohmic contacts and is etched down to a few tenths of a micron, the surface conductance of this accumulation layer (for various heterojunction bias conditions) may be measured.

The mobility observed by Esaki *et al.* (1964a,b) was a few hundred cm^2 V^{-1} sec^{-1} at 77°K and was reasonably constant over a range of interface fields from 2×10^4 to 8×10^4 V cm^{-1} for 4×10^{15} donors cm^{-3} GaAs substrate unit, and from 1.5×10^5 to 5×10^5 V cm^{-1} for 4×10^{16} donors cm^{-3} GaAs substrate unit, with both dc and ac techniques.

The same type of measurements were also made at ice temperature for a heterojunction of lightly doped GaAs with $\sim 4 \times 10^{14}$ donors cm^{-3}. The

TABLE 4.3

Barrier Heights for n–p Ge–GaAs Junctions for Various Interface Orientations

Ge	GaAs	Interface orientations	V_D		
			From $j_{s0}(T)$ (eV)	Measured from (dV/dj) (T) $qV \gg kT$ (eV)	From $C(V)$ (eV)
Sb doped $N_d - N_a = 1.1 \times 10^{17}$ cm^{-3}	Te doped $N_d - N_a = 3 \times 10^{17}$ cm^{-3}	(111)Ga	0.31	0.29	0.57
		(111)As	0.26	0.25	0.37
		(110)	0.21	0.19	0.30
	Te doped $N_d - N_a = 5.9 \times 10^{16}$ cm^{-3}	(111)Ga	0.50	0.49	0.59
		(111)As	0.37	0.36	0.36
		(110)	0.36	0.36	0.45
	Undoped $N_d - N_a = 9.4 \times 10^{14}$ cm^{-3}	(111)Ga	0.52	0.52	0.62
		(111)As	0.39	0.39	0.49

result indicated again, a constant mobility (~ 1000 cm^2 V^{-1} sec^{-1}) over a range of interface fields from 10^4 to 2×10^4 V cm^{-1}. The direct observation of the accumulation layer gives additional evidence for the Anderson model of the n–n heterojunction in which the barrier space charge is of simple character determined by the bulk properties of the two semiconductors rather than the type found in Ge–Si n–n junctions where interface charges are dominant.

Interface conductance studies of a similar nature have also been reported by L. L. Chang (1965a,b) up to field strengths of 3×10^5 V cm^{-1} for GaAs$_{0.9}$P$_{0.1}$ substrates. These studies indicate that the electron mobility in the Ge at the interface depends on the field but that the dependence is weaker than that which would be expected for a purely diffuse scattering process.

4.6 Other Studies of Isotype Heterojunctions

Other n–n heterojunctions that have been studied include the pairs Ge–GaAs$_{1-x}$P$_x$, Ge–GaP, Ge–CdSe, Si–GaP, GaSb–GaAs, GaAs–InSb, and InP–GaAs.

Switching studies have been made of n–n heterojunctions by many who have fabricated these structures. The studies usually showed switching times of less than 10^{-9} sec, and the observations were usually equipment or circuit limited. Brownson's study (1965) of n–n Ge–Si diode switching is one of the most extensive for heterojunctions. Since this time, however, high performance Schottky metal–semiconductor junctions have become available as high-speed diodes and n–n heterojunctions appear to offer no advantages compared with these.

Hampshire *et al.* (1970b) have made use of the bias-dependent "optical" properties of a CdSe–Ge n–n heterojunction. At a particular photon wavelength the photo-emf is zero and the wavelength for cross-over of the sign of the photo-emf may be adjusted by the applied bias voltage. This is used as the basis of a null detector, which responds to photon energies between 0.73 and 0.86 eV with a sensitivity of 0.43 V μ^{-1} for null point in the linear region. This detector has applications in pyrometry (Hampshire *et al.*, 1970a).

Chapter 5 | Optical Properties of Heterojunctions and Heterojunction Lasers

5.1 Introduction

Heterojunctions have interesting optical properties. An important one is the window effect in which light of energy between E_{g2} and E_{g1} passes through the wide-gap material and is absorbed in the vicinity of the junction. The wide-gap material may be of low resistivity and substantial thickness within a limit set by free carrier absorption of photons. Therefore, the sheet resistance of the photocell is low, which is advantageous as far as power output is concerned. Other potentially interesting optical properties include (a) sensitive phototransistor action and (b) the possibility of photodiodes between materials such as the II–VI semiconductors that are difficult to dope in p–n homojunction form.

As discussed in Chapter 1, there are also several concepts by which heterojunction structures may lead to up-conversion of light energy, for example, from infra-red to visible, with useful efficiency.

Let us begin with a brief discussion of conventional (p–n homojunction) photocell action. The differences that exist for heterojunctions are then developed. The results of optical measurements on several kinds of heterojunctions (Ge–GaAs, Ge–Si, p–n, n–n) are presented and interpreted in terms of absorption, collection, and band-structure effects. Finally heterojunctions in solar cells and lasers are considered.

5.2 Photocell Action in (p–n) Homojunctions

For a simple model of a p–n homojunction diode under illumination, the current density is

$$J = J_L - J_0 \exp\{(qV/kT) - 1\} \tag{5.1}$$

where J_0 is the Shockley-model leakage current of the diode without illumination, and J_L is the maximum current density that corresponds to the applied illumination. As shown in Fig. 5.1(a), the electric field of the junction depletion causes the photoinduced holes to move to the p-side of the junction and the electrons to move to the n-side. The p-side therefore tends to become positive with respect to the n-side, and the J–V characteristic with light applied is as shown by the full line in Fig. 5.1(b). The open-circuit voltage is obtained from (5.1) by setting J equal to zero:

$$V_{OC} = (kT/q)\ \ln\{(J_L/J_0) + 1\} \qquad (5.2)$$

According to this simple model the maximum open-circuit voltage that could be achieved under very high illumination conditions would correspond to leveling of the conduction band edges in Fig. 5.1(a) and therefore is a value somewhat less than the energy gap of the semiconductor.

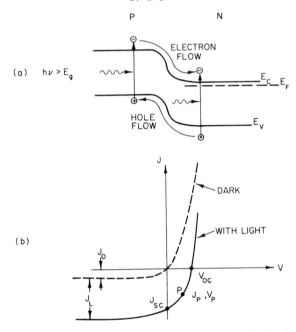

Fig. 5.1. *Conventional photodiode (p–n homojunction) under illumination.* (a) Short-circuit current conditions with light; (b) current–voltage characteristics with and without light.

The power output (per unit area) of the photocell is

$$P = JV = [J_L - J_0\ \exp\{(qV/kT) - 1\}]V \qquad (5.3)$$

This is a maximum at the point P in Fig. 5.1(b) where dP/dV is zero. Hence

$$\{(qV_P/kT) + 1\}\ \exp(qV_P/kT) = J_L/J_0 + 1 \qquad (5.4)$$

and

$$J_P = J_L - J_0\ \exp(q(V_P/kT) - 1) \qquad (5.5)$$

From the simple Shockley model for generation in a diode

$$J_0 = q(D_n/\tau_n)^{1/2}(n_i^2/N_A) + q(D_p/\tau_p)^{1/2}n_i^2/N_D \qquad (5.6)$$

where

$$n_i^2 = N_c N_v \exp(E_g/kT) \qquad (5.7)$$

For a given J_L it is apparent that both J_P and V_P depend upon J_0, which in turn depends upon the energy gap of the semiconductor if other factors such as the diffusion coefficients, lifetimes, and doping densities are relatively unchanged. Following Wolf (1960), J_0 is shown as an exponential function of E_g in Fig. 5.2(a) for typical assumed parameter values for D, τ, and the doping densities. Characteristic factors which relate to the shape of the

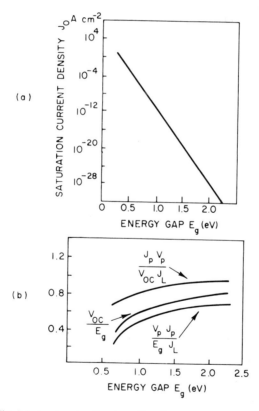

Fig. 5.2. *p–n homojunction photocell performance versus energy gap.*
(a) Assumed saturation current density versus width of the energy gap, (b) voltage factor V_{OC}/E_g, curve factor $V_P J_P/V_{OC} J_L$, and characteristic factor $V_P J_P/E_g J_L$, versus width of energy gap. (After Wolf, 1960.)

curve between V_{OC} and J_{SC} are then as shown in Fig. 5.2(b). The term $V_{P}J_{P}/E_{g}J_{L}$ represents efficiency if the photons are all of energy E_{g}. The performance is seen to increase as semiconductors of larger energy gap are considered.

5.3 Heterojunction *p–n* Photocells

Heterojunction *p–n* photocell concepts have been studied by Anderson (1962), Perlman (1964), Donnelly and Milnes (1966a), and others.

In general, in heterojunctions the smaller of the two energy gaps determines the voltage output that can be achieved from the photocell. Figure 5.3 shows a *p–n* heterojunction under illumination for short-circuit and open-circuit conditions. From Fig. 5.3(b) the open-circuit output voltage is less than the energy gap of the small-gap material. The short-circuit current that can be collected corresponds to the number of photons in the energy range $E_{g2}-E_{g1}$. The principal advantage of a heterojunction photocell, therefore, is not basically in the voltage or current performance but rather in the extent to which the surface-recombination losses and sheet-resistance losses are reduced

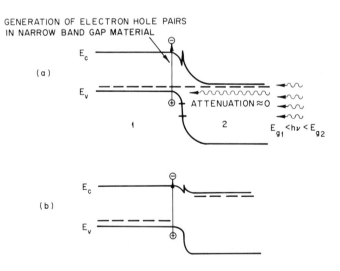

Fig. 5.3. *p–n heterojunction under illumination.*
(a) Under low illumination and short-circuit current conditions. Since the output voltage is zero, the Fermi levels are equal on either side of the junction. (b) Under high illumination and open-circuit voltage conditions. The conduction band is seen to be almost in a flat-band condition and the output voltage is somewhat less than the energy gap of the small-gap material.

because the window effect allows the junction to be placed deeper from the surface than in a homojunction cell. The magnitude of this advantage depends on the particular design geometry and other practical matters of fabrication. The performance is adversely affected, of course, if there is any tendency for the interface to have a high recombination rate.

The window effect creates a band-pass region of sensitivity to photons of energy between E_{g2} and E_{g1}. This is shown in Fig. 5.4, which is for a p–n Ge–GaAs heterojunction.

Although heterojunction optical properties are of considerable potential

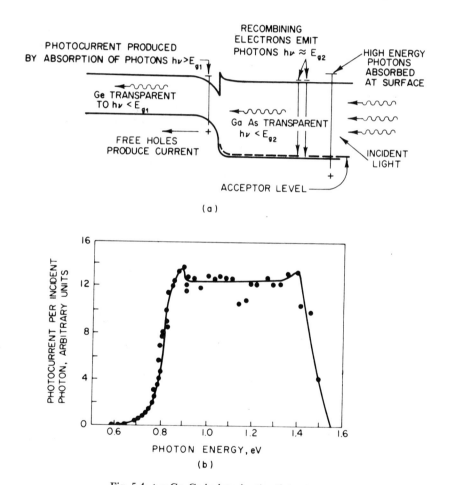

(a)

(b)

Fig. 5.4. *p–n* Ge–GaAs *heterojunction photoresponse.*
(a) Schematic showing photon absorption; (b) short-circuit photocurrent per incident photon. (After Anderson, 1962.)

interest, relatively few studies have been made of quantum efficiencies (electrons collected for a number of photons absorbed) of heterojunction photostructures. Fabrication and contact problems are factors that have made such measurements difficult. It seems likely that photocells of n–GaP/p–Si, n–ZnSe/p–Ge, n–ZnSe/p–GaAs, $Ga_xAl_{1-x}As$/GaAs, GaN_xAs_{1-x}/GaAs, and $Ga_xAl_{1-x}P$/GaP would be of interest since the energy gap ranges are good. Also most of these pairs are close in lattice match and the interface-state effects therefore may be small. Some indication that interface recombination effects need not be serious may be inferred from heterojunction transistor

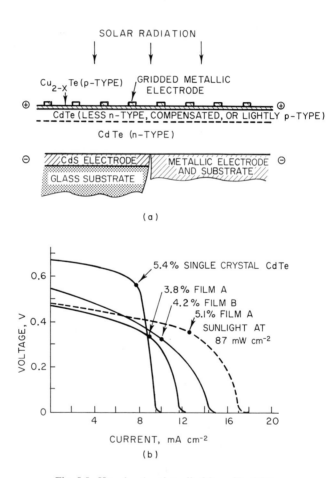

Fig. 5.5. *Heterojunction photocell of* $Cu_{2-x}Te$–CdTe.
(a) Structure of two film versions of the cell, (b) voltage–current characteristic under 80 mW cm^{-2} artificial illumination. (After Cusano, 1963.)

studies where current gains in excess of 20 are commonly seen. There are also a few direct quantum efficiency measurements available such as 75% reported for heterojunctions of $GaAs_{0.8}P_{0.2}/GaAs$ by Ramachandran and Moroney (1964).

As mentioned earlier, a special feature of a heterojunction is that it allows the formation of junctions between semiconductor materials that cannot be doped both *p*-type and *n*-type. Cusano (1963), Aven and Garwacki (1963), Dutton and Muller (1968), and others have worked on this problem with II–VI compound semiconductors, most of which are difficult to achieve with heavy *p*-type doping because of compensation mechanisms. Figure 5.5(a) shows some typical polycrystalline CdTe film structures of Cusano in which the processing creates a *p*-type $Cu_{2-x}Te$ front layer. The rectification and photodiode action is therefore attributed to the heterojunction between the

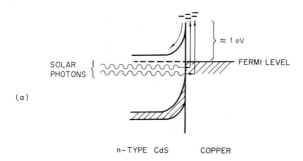

(a)

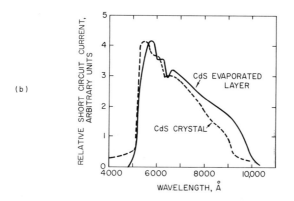

(b)

Fig. 5.6. CdS–*copper photocell.*
(a) Energy diagram showing barrier of about 1 eV. Photons excite electrons from the metal over this barrier into the semiconductor. (b) Spectral response of CdS–Cu photovoltaic cells. (After Moss, H.I., 1961.)

p-type CuTe and the *n*–CdTe. Figure 5.5(b) shows the voltage–current characteristics obtained for several structures under illumination with the efficiencies of the optimum power points indicated. Solar conversion efficiencies of up to 6% were obtained with polycrystalline film cells and 7.5% for single-crystal cells.

Photoresponse can also be obtained from Schottky-barrier type hetero-junctions. Figure 5.6(a) shows the energy band diagram for a cell formed between *n*–CdS and a copper layer (which was not heat-treated to limit the possibility that a CuS layer might exist). The photoresponse, Fig. 5.6(b), shows that photons in the energy range from 2.4 eV (the band gap of CdS) to about 1 eV (the Schottky-barrier height) lift electrons over the barrier from the metal into the semiconductor to provide the output current. Efficiencies in excess of 4% have been obtained from such cells, fabrication costs are low, and the cells can be deposited on flexible substrates.

Studies have been made of the light emitted from heterojunction cells under forward current flow between II–VI semiconductors (such as *n–p* ZnSe–ZnTe) by Kot *et al.* (1965), Aven and Garwacki (1967b), Tsujimoto *et al.* (1967), and others. In general, the efficiencies in terms of photons emitted per electron of current flow have been quite low $(10^{-4}–10^{-5})$ at room temperature.

5.4 Optical Properties of Isotype *n–n* Heterojunctions

The optical properties of *n–n* Ge–Si heterodiodes have been studied by Donnelly and Milnes (1966d). The band diagrams, the *I–V* characteristics, and the open-circuit photoresponse spectra for three diodes of different doping levels are shown in Fig. 5.7.

The band diagrams, Fig. 5.7(a), are only rough approximations. They were obtained by assuming a constant electron affinity difference of 0.18 eV which is in agreement with the photoelectric measurements of Allen and Gobeli (1962). The density of interface states was assumed to be 8% of the surface states found by Allen and Gobeli on cleaved silicon surfaces. This value was chosen since the relative lattice constants of Ge and Si result in the germanium atoms bonding with all but approximately 8% of the atoms expected at a silicon surface. In Fig. 5.7(a), l_1 and l_2 are the depletion widths.

For the photoelectric response studies, the monochromatic light was incident on the Si perpendicular to the junction. During these measurements, the open-circuit photovoltage was kept below 2 mV so the results shown in Fig. 5.7(c) represent the small-signal case where the open-circuit voltage is directly proportional to the short-circuit current. The sign of the response

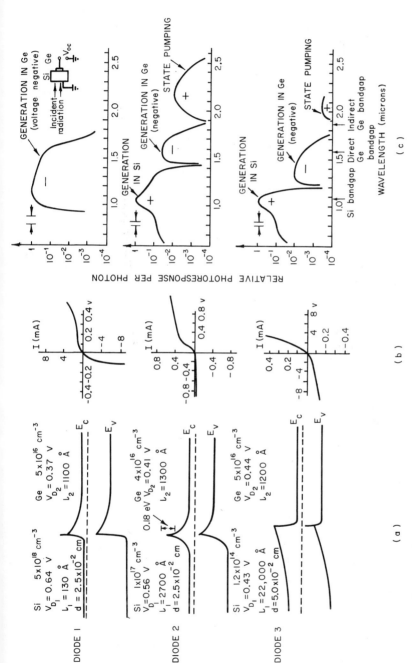

Fig. 5.7. *Characteristics of n–n Ge–Si diodes (3 doping types) at room temperature. (After Donnelly and Milnes, 1966d.)*

refers to the polarity of the Ge with the Si grounded. The wavelengths corresponding to the energies of the direct and indirect Ge bandgaps and the Si bandgap are indicated in the diagram.

Since a high degree of carrier recombination may be expected at the interface states in Ge–Si junctions, the minority photocarriers may be assumed to recombine at the interface. At wavelengths of sufficient energy to excite carriers in the Ge but not in the Si, the light is transmitted to the Ge where absorption occurs producing a photocurrent density J_{p2} and therefore the photoresponse is negative. From 1.85 μ (0.67 eV) to 1.54 μ (0.805 eV) indirect or phonon assisted absorption occurs and the response is expected to increase with decreasing wavelength. At 1.54 μ direct absorption begins and the photoresponse is expected to be relatively constant.

At energies above the Si band gap, photons are absorbed in the Si. Because of the very gradual absorption edge and long diffusion length in the Si, a photocurrent J_{p1} is generated, while J_{p2} is now attenuated. It is then possible for the photocurrent to reverse sign and become proportional to J_{p1}.

There is general similarity between the expected results and the observed photoresponse curves of Fig. 5.7. In diode 1 there is high probability of tunneling through the extremely thin silicon barrier. Therefore only the Ge photoresponse is observed. In diodes 2 and 3, three distinct regions are observed. At wavelengths greater than the effective Ge band gap wavelength (1.85 μ), a positive photoresponse is noted. This region may be caused by photoexcitation of the electrons from the Ge valence band to the Si conduction band or to excitation of electrons from interface states or from traps in the bulk silicon bandgap. The other two regions correspond to the negative Ge photoresponse and the positive Si photoresponse as predicted. The change of sign of the photoresponse, however, does not always correspond to the silicon bandgap as expected. The predominate reason for this is that the positive photoresponse due to electron excitation is still present at wavelengths of sufficient energy to generate carrier pairs across the band gap in the Ge. This positive photoresponse tends to cancel the negative Ge response, and may actually do so causing the photoresponse to change sign before the Si bandgap is reached.

Van Opdorp and Vrakking (1967) have also studied photoeffects in n–n Ge–Si heterojunctions and have explained their results in terms of a model of two Schottky diodes connected back to back. They find evidence exists for the photoemission of electrons from the Ge valence band through the interface to the Si conduction band. They conclude that this is the most probable interpretation for the sign-reversal point (previously discussed) in the photoresponse versus wavelength curves lying considerably below the Si energy gap. Similar effects have been seen in n–n Ge–GaP heterojunctions (van Ruyven *et al.*, 1965).

The optical properties of *n–n* Ge–Si heterojunctions have also been studied by Yawata and Anderson (1965). In particular they describe an effect by which photoexcited holes become trapped in the valence band energy notch and so modulate the energy barrier in the conduction band. It is claimed that this modulation effect is quite large and that the device makes a practical photodetector.

5.5 Heteroface Solar Cell Considerations

Heterojunctions have not been developed to a high performance level for solar cells. Only limited attention has been given to this possibility because silicon solar cell technology has been capable of meeting most of the present performance requirements.

The performance factors that are usually desired in solar cells include

(a) high efficiency for the ambient spectrum,
(b) sustained performance in the environment used (e.g., under irradiation conditions in space),
(c) sustained performance over a wide temperature range or to a certain upper temperature,
(d) high output power per unit weight,
(e) high output voltage and a fairly flat voltage-regulation versus load curve, and
(f) low cost and ease of manufacture.

Solar cells with efficiencies of about 12% are available in silicon. These represent the outcome of many years of development and are an effective match to most of the needs. Si has the advantages of being

(a) a light-weight elemental semiconductor of highly developed technology,
(b) an indirect-gap material with a fairly gradual absorption edge which allows the photons to create carriers well below the surface,
(c) a high lifetime semiconductor with fairly good mobilities and therefore large diffusion lengths which aid collection,
(d) a material with fairly low surface recombination velocity, 10^3 cm sec^{-1}, so that recombination loss at the surface is small, and
(e) an adequate material in radiation resistance if suitably protected by transparent covers.

A typical solar spectrum is shown in Fig. 5.8 with shaded sections of the curve that represent the number of carrier pairs that would be generated in semiconductors of various energy gaps. A low energy-gap material can use more of the photons in the spectrum but the photovoltage generated is low, and the low gap semiconductor does not efficiently use the high energy photons.

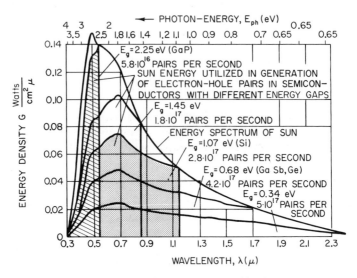

Fig. 5.8. *Solar cell energy spectrum considerations.*
The energy spectrum of the sun on a bright, clear day at sea level, and the parts of this spectrum utilizable in the generation of electron–hole pairs in semiconductors with energy gaps of 2.25, 1.45, 1.07, 0.68, and 0.34 eV, respectively. Listed is the number of electron–hole pairs generated, assuming an abrupt absorption edge with complete absorption and zero reflection on its high energy side. (After Wolf, 1960.)

At present GaAs is the only semiconductor that can be considered a serious competitor to Si for solar cells. The following comments may be made in comparing the two materials:

(a) GaAs has a larger energy gap (1.45 versus 1.1 eV) which represents a better power match to the solar spectrum [see Fig. 5.9(a)] and a better temperature performance. At 0°C the η_{max}-values (excluding losses) are 28.5 and 24% for GaAs and Si, respectively. At 200°C the figures are 12 and 4% according to the calculations of Wysocki and Rappaport (1960). When a recombination component is added, Fig. 5.9(b), GaAs still maintains its temperature superiority.

(b) GaAs, however, suffers from the problem that it has a steep absorption edge because it is a direct gap material, and this creates carriers near the surface. As GaAs tends to have a high surface recombination velocity and low diffusion length the losses are greater than for silicon. The surface-recombination velocity problem is discussed by Kleinman (1961).

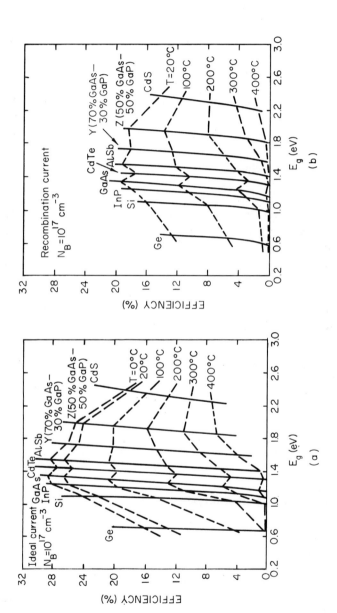

Fig. 5.9. (a) *Conversion efficiency as a function of energy gap for ideal current–voltage characteristics. (After Wysocki and Rappaport, 1960.) (b) Conversion efficiency as a function of band gap for solar cells with recombination currents.*

(c) Efficiencies that have been achieved for GaAs solar cells are nevertheless quite high. Wysocki *et al.* (1966) report on GaAs cells in the range 6.7–10.3% and Si $n–p$ cells in the range 9.2–11.8%. This was in a study of irradiation by 185–530 keV protons. According to Loferski (1963) the highest efficiency reported for GaAs is 13% and for silicon 15%. Recently Ellis and Moss (1970) have suggested that the surface recombination problem in GaAs could be overcome by a built-in field, and the efficiency of GaAs solar cells raised to 20%.

The addition of a heterosemiconductor face to a homojunction GaAs solar cell should offer the possibility of some improvement in efficiency. Studies need to be made to show whether a wide-gap semiconductor with suitable metallurgical and electrooptical properties is available.

Conceptually, the wide band-gap heterosemiconductor would act as a window that admits photons of energy less than E_{g1}. Those of energy between E_{g1} and E_{g2} (the GaAs band gap) would create carriers in the GaAs homojunction. The wide-gap layer could be quite thick to provide a lower internal cell resistance than can be achieved with homojunction cells. If the layer were to be several mils thick, it might provide inherent radiation resistance.

The lower internal cell resistance made possible by the heteroface contact should allow the adjacent GaAs layer to be redesigned to be more lightly doped and therefore perhaps higher in diffusion length and so a better collector of the photoinduced carriers. The role of the built-in fields in the junction region would also have to be studied in the course of the redesign.

The $n–n$ or $p–p$ interface between the heterosemiconductor and the GaAs might have a recombination velocity substantially below the value of 10^5 cm sec^{-1} that seems usual for a GaAs surface. If this expectation were realized, calculations suggest that the efficiency of a GaAs cell might be raised several percent.

Such solar cells have not been constructed, but one can speculate that the heteroface might be provided by wide-gap semiconductors such as GaN (3.2 eV), $GaAs_xN_{1-x}$, or $Ga_xAl_{1-x}P$.

5.6 Heterojunction Solar Cell Analysis

Although one approach to solar cell improvement is with heteroface structures covering homojunction cells, the performance possibilities of true heterojunction cells also need consideration.

The expected performance of heterojunction solar cells may be studied by computing the efficiency for various possible design parameters. Most of the loss terms such as reflection and bulk-recombination losses and spreading resistance losses may be allowed for in the calculations. However, it is usual to neglect interface recombination since, in general, its magnitude is not known.

The heterojunction pairs for which such calculations are discussed below are limited to those of close lattice match and reasonably close thermal coefficients of expansion. Hopefully, therefore, there is a chance that such heterojunctions may be made without the fabrication processes introducing high densities of interface states at the junctions and therefore significant recombination through these states.

All the solar cells considered are assumed to be of the same area (1 $\times$ 2 cm) and to have a surface contact made of the grid structure shown in Fig. 5.10. This grid structure was selected because it is commonly used for Si solar cells (Handy, 1967). In this configuration, approximately 12.8% of the surface area is covered by the contact.

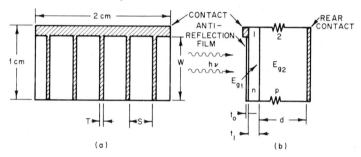

Fig. 5.10. *Solar-cell geometry considered.*
(a) View of input face showing contact fingers: $W = 0.9$ cm, $S = 0.4$ cm, $T = 0.0127$ cm.
(b) Cross section.

The parameters used in the calculations by Sahai and Milnes (1970) are given in Table 5.1. The solar spectrum values were taken from Moon (1940) and are very close to the values (Johnson, 1954) now used as standard for above-atmosphere solar radiation.

From Table 5.1 it is seen that the heterojunction cells were assumed to have n-type doping for the wide-gap material. For most heterojunction pairs ΔE_c is considerably less than ΔE_v. By selecting for consideration n-window p-base heterojunctions, we have the advantage that the electrons which flow into the n-region from the p-base, where almost all the photon absorption takes place, are not seriously impeded by the relatively small ΔE_c spike. In the other class of heterojunction, p large-gap, n small-gap, the relevant spike is ΔE_v and is large, and it is usual to observe a great deal of interface recombination.

A second reason for the n-window p-base choice is that most of the current is caused by electrons collected as minority carriers from the narrow band-

TABLE 5.1

Values Assumed in Heterojunction Solar Cell Design Calculations

Solar cell	Cell	t_0 (Å)	t_1 (μ)	d (mils)	S (cm sec^{-1})	L_n (μ)	L_p (μ)	D_n (cm^2 sec^{-1})	D_p (cm^2 sec^{-1})	N_D (cm^{-3})	N_A (cm^{-3})	τ_n (sec)	τ_p (sec)
n–p ZnSe–Ge	A-1	950	7.5	20	10^5	707	0.2	50	0.5	10^{18}	10^{17}	10^{-4}	10^{-9}
	A-2	950	28	20	10^5	707	0.3	50	1	2×10^{17}	10^{17}	10^{-4}	10^{-9}
n–p ZnSe–GaAs	B-1	825	7.5	5	10^5	2.8	0.2	80	0.5	10^{18}	5×10^{17}	10^{-9}	10^{-9}
	B-2	825	28	5	10^5	2.8	0.3	80	1	2×10^{17}	5×10^{17}	10^{-9}	10^{-9}
n–p GaP–Si	C-1	925	5	20	10^5	215	0.3	23	1	10^{18}	2×10^{16}	2×10^{-5}	10^{-9}
	C-2	925	250	20	10^5	215	0.3	23	1	10^{18}	2×10^{16}	2×10^{-5}	10^{-9}
n–p Si–Si	D-1	900	0.5	20	10^3	215	1.7	23	3	10^{19}	2×10^{16}	2×10^{-5}	10^{-8}
	D-2	900	1	20	10^3	215	1.7	23	3	10^{19}	2×10^{16}	2×10^{-5}	10^{-8}
n–p GaAs–Ge	E-1	1300	0.5	20	10^5	848	2	72	4	10^{18}	2×10^{16}	10^{-4}	10^{-8}
	E-2	1300	1	20	10^5	848	2	72	4	10^{18}	2×10^{16}	10^{-4}	10^{-8}
n–p GaAs–GaAs	F-1	750	0.5	5	10^5	3.5	2	125	4	10^{18}	10^{16}	10^{-9}	10^{-8}
	F-2	750	1	5	10^5	3.5	2	125	4	10^{18}	10^{16}	10^{-9}	10^{-8}
n–p GaAs–GaAs	G-1	750	0.5	5	10^5	0.7	2.2	4.5	5	5×10^{16}	10^{19}	10^{-9}	10^{-8}
	G-2	750	1	5	10^5	0.7	2.2	4.5	5	5×10^{16}	10^{19}	10^{-9}	10^{-8}

gap semiconductor, and the diffusion length of electrons tends to be appreciably more than that of holes in the semiconductors of interest to us.

The diode equation was assumed to be of the type

$$J = J_{d0}[\exp(qV/kT) - 1] + J_{rg0}[\exp(qV/2kT) - 1] \qquad (5.8)$$

where J_{d0} and J_{rg0} are the diffusion and depletion layer recombination–generation components of the reverse saturation current density. Their values are given by

$$J_{d0} = q\left(\frac{N_D D_n \exp[-q(V_D - \Delta E_c)/kT]}{L_n \tanh(d/L_n)}\right.$$
$$\left. + \frac{N_A D_p \exp[-q(V_D + \Delta E_v)/kT]}{L_p \coth(t_1/L_p + \phi)}\right) \qquad (5.9)$$

$$J_{rg0} = q\left(\frac{n_{i1} l_1}{2K_1 \tau_{01}} + \frac{n_{i2} l_2}{2K_2 \tau_{02}}\right)\frac{1}{q[V_D(V_D - V)]^{1/2}/kT} \qquad (5.10)$$

where N_D is the donor concentration in the n-type material; N_A is the acceptor concentration in the p-type material; V_D is the built-in potential of the diode; ΔE_c is the conduction band discontinuity expressed in volts; ΔE_v is the valence band discontinuity expressed in volts; L_n is the electron diffusion length in the p-type material; L_p is the hole diffusion length in the n-type material; t_1 is the thickness of the n-type material (surface layer); d is the thickness of the p-type material (base region); τ_{01}, τ_{02} are the minority carrier lifetimes in the depletion regions on the two sides of the junction; n_{i1}, n_{i2} are the intrinsic carrier concentrations in the two semiconductors; l_1, l_2 are the depletion layer widths at no bias ($V = 0$) in the two semiconductors; K_1, K_2 are given by $K_1 = (1 + N_1 \epsilon_1/N_2 \epsilon_2)^{-1}$ and $K_2 = 1 - K_1$, where N_1, N_2, ϵ_1, and ϵ_2 are the impurity concentrations and dielectric constants of the two semiconductors; ϕ is given by $\tanh \phi = SL_p/D_p$, where S is the surface recombination velocity for the window semiconductor (assumed to be n-type) and D_p is the hole diffusion constant.

It should be noted that J_{rg0} in Eq. (5.8) is dependent on applied voltage V as given by Eq. (5.10) because the widths of the depletion regions, as well as the electric field strength in these regions, depend on applied voltage. For Eq. (5.10) it was assumed that the junction was of a step type.

5.6.1 *Computation of Reflection Loss*

Since the refractive index of most semiconductors is high (for Si, $n \sim 3.5$), the reflection loss from the surface of the photocell is prohibitive unless some kind of antireflection film is used. The refractive index is represented in complex notation as $n - ik$, where n and k are related to the reflection coefficient ρ and absorption coefficient α by

$$k = \alpha\lambda_0/4\pi \tag{5.11}$$

$$n = (1 + \rho)/(1 - \rho) + \{[(1 + \rho)/(1 - \rho)]^2 - (1 + k^2)\}^{1/2} \tag{5.12}$$

where λ_0 is the wavelength of light in free space. For a heterojunction cell, one has to consider in principle an additional reflecting surface at the n–p junction. In practice, however, this complex situation may be simplified by assuming that the first semiconductor layer is thick for optical computations. This may be justified (Vasicek, 1960) by showing that the path difference in the first semiconductor $2n_1t_1$ is greater than $5\lambda_0$. This allows us to first compute the composite reflection coefficient of the heterojunction cell and then take into account the reduction in reflection caused by a transparent ($k = 0$) antireflection film. The formulae to be used for computing the overall reflection (R) and the transmission (T_r) into the second semiconductor material are:

$$R = \frac{R_0 + R_{12} + 2(R_0R_{12})^{1/2}\cos(4\pi n_0 t_0/\lambda_0)}{1 + R_0R_{12} + 2(R_0R_{12})^{1/2}\cos(4\pi n_0 t_0/\lambda_0)} \tag{5.13}$$

$$T_r = \frac{(1 - R_0)(1 - R_1)(1 - R_2)\exp(-4\pi k_1 t_1/\lambda_0)}{[1 - R_1R_2\exp(-8\pi k_1 t_1/\lambda_0][1 + R_0R_{12} + 2(R_0R_{12})^{1/2}\cos(4\pi n_0 t_0/\lambda_0)]} \tag{5.14}$$

where

$$R_0 = \left(\frac{n_0 - 1}{n_0 + 1}\right)^2, \qquad R_1 = \frac{(n_1 - n_0)^2 + k_1^2}{(n_1 + n_0)^2 + k_1^2}$$

$$R_2 = \frac{(n_1 - n_2)^2 + (k_1 - k_2)^2}{(n_1 + n_2)^2 + (k_1 + k_2)^2} \tag{5.15}$$

$$R_{12} = \frac{R_1 + (R_2 - 2R_1R_2)\exp(-8\pi k_1 t_1/\lambda_0)}{1 - R_1R_2\exp(-8\pi k_1 t_1/\lambda_0)}$$

The subscripts 0, 1, and 2 refer to the antireflection film, first semiconductor, and second semiconductor, respectively; R_{12} denotes the composite reflection value for regions 1 and 2.

Since n and k for all the materials are functions of the wavelength of the light, the design of an antireflection film for a heterojunction solar cell should be based on maximizing the overall usable number of solar photons transmitted into the second semiconductor. For the homojunction cells considered, the optimum antireflection layer thickness, t_0, was selected by maximizing the total usable photons transmitted into the cell. Figure 5.11 shows typical variation of transmission as a function of antireflection film thickness for ZnSe–GaAs and GaAs–GaAs cells. The optimum thickness of the SiO antireflection film is seen to be about 800 Å for the ZnSe–GaAs cell and about 700 Å for the GaAs–GaAs cell. Table 5.2 lists the optimum values for SiO and SiO$_2$ films for other heterojunction pairs considered and SiO

TABLE 5.2

Optimum Values for Antireflection Film Thickness

Solar cell material	SiO film		SiO$_2$ film		Reflection loss[a] without any film (%)
	Optimum thickness (Å)	Reflection loss[a] (%)	Optimum thickness (Å)	Reflection loss[a] (%)	
ZnSe–Ge	950	11.22	1275	12.24	32.13
ZnSe–GaAs	800	4.92	1075	6.32	25.88
GaP–Si	900	8.62	1200	12.5	30.35
GaAs–Ge	1300	14.35	1500	18.07	34.22
GaAs–GaAs	700	7.87	950	14.9	36.6
Si–Si	800	10.12	1050	16.07	35.45

[a] Reflection loss values are for the complete usable energy range of the sun's spectrum for the solar cell, viz. for $h\nu > E_{g \text{ small}}$.

is seen to be preferable for all pairs. In the studies that follow, all the cells being compared were therefore assumed to be coated with SiO of optimum thickness. (Figures 5.13 and 5.14 include curves of the spectral response of the antireflection coating.)

5.6.2 *Collection Efficiency and Spectral Response*

5.6.2.1 *Holes in the n-Layer.* The minority carrier (hole) diffusion equation under constant monoenergetic illumination is

$$D_p \, dp^2/dx^2 - (p - p_0)/\tau_p + \alpha_1 N \exp(-\alpha_1 x) = 0 \qquad (5.16)$$

where p_0 is the equilibrium hole concentration, τ_p is the hole lifetime, α_1 is the absorption coefficient (a function of wavelength), N is the number of photons cm^{-2} sec^{-1} entering the surface layer, and x is measured from the n-semiconductor surface as shown in Fig. 5.12(a). If N_s represents the photon density in the solar spectrum at that wavelength, then

$$N = N_s(1 - R) \qquad (5.17)$$

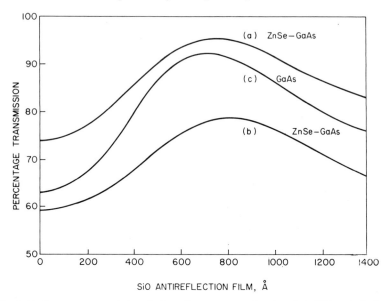

Fig. 5.11. *Percentage transmission of solar-cell spectrum as a function of the* SiO *antireflection film thickness.*
(a) Transmission into the surface layer (ZnSe) of a ZnSe–GaAs solar cell. The percentage represents the number of photons relative to the number in the full solar spectrum with $h\nu > 1.4$ eV. (b) Transmission into the base layer of a ZnSe–GaAs cell. Curve (b) is lower than curve (a) because photons of energy greater than 2.67 eV do not reach the junction and there is an additional reflection at the junction interface. (c) Transmission into the surface layer of a GaAs–GaAs solar cell (photons of energy greater than 1.4 eV) for comparison with curve (a).

where R is given by Eq. (5.13). The boundary conditions for solving Eq. (5.16) are

$$D_p \, dp/dx = S(p - p_0) \qquad \text{at} \quad x = 0$$

and

$$p = p_0 \exp(qV/kT) \qquad \text{at} \quad x = t_1 - l_1 = t'$$

where S is surface-recombination velocity and V is the bias voltage across the diode. Then the solution for the photocurrent density collected from the undepleted n-layer by diffusion, under short-circuit voltage conditions, is

$$\begin{aligned}
J_S{}^p = q\Bigg[&\frac{N\alpha_1{}^2 L_p{}^2}{\alpha_1{}^2 L_p{}^2 - 1}\left(\frac{D_p + S/\alpha_1}{D_p + SL_p}\exp(-t'/L_p) - \exp(-\alpha_1 t')\right) \\
&+ \frac{N\alpha_1 L_p \tanh(t'/L_p + \phi)}{\alpha_1{}^2 L_p{}^2 - 1} \\
&\times \left(\frac{\alpha_1 L_p(D_p + S/\alpha_1)}{D_p + SL_p}\exp(-t'/L_p) - \exp(-\alpha_1 t')\right)\Bigg]
\end{aligned} \qquad (5.18)$$

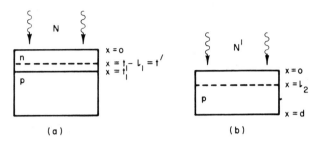

Fig. 5.12. *Representation of heterojunction coordinates for computation of spectral response of (a) surface layer and (b) base region.*

Lines at $x = t'$ and $x = l_2$ represent the two depletion edges of the junction.

Each photon absorbed in the depletion region is assumed to contribute one current carrier, and therefore the photocurrent density due to the depletion region is given by

$$J_{DR}^p = qN(\exp(-\alpha_1 t') - \exp(-\alpha_1 t)) \tag{5.19}$$

5.6.2.2 *Electrons in the p-Region.* The diffusion equation for this region, following the coordinate system of Fig. 5.12(b), is

$$D_n \, d^2 n/dx^2 - (n - n_0)/\tau_n + \alpha_2 N^1 \exp(-\alpha_2 x) = 0 \tag{5.20}$$

where α_2 is the absorption coefficient of the p-base region and N^1 is the photon density transmitted into the base region, as given by

$$N^1 = N_s T_r \tag{5.21}$$

where T_r is given by Eq. (5.14). The boundary conditions are

$$n = n_0 \exp(qV/kT) \quad \text{at} \quad x = l_2 \quad \text{and} \quad n = n_0 \quad \text{at} \quad x = d$$

Then the solution for the photocurrent density due to absorption and diffusion in the undepleted base layer is

$$J_B^n = q\left[\frac{\alpha_2 N^1 \exp(-\alpha_2 l_2)}{\alpha_2 + 1/L_n} - \frac{2\alpha_2 N^1 \exp(-\alpha_2 l_2)[\exp(-d/L_n) - \exp(-\alpha_2 d)]}{L_n(\alpha_2{}^2 - 1/L_n{}^2)[\exp(d/L_n) - \exp(-d/L_n)]} \right] \tag{5.22}$$

The photocurrent density due to absorption in the base depletion region is

$$J_{DR}^n = qN^1[1 - \exp(-\alpha_2 l_2)] \tag{5.23}$$

The total photocurrent density of the solar cell is given by

$$J_T = J_{S^p} + J_{DR}^p + J_{B^n} + J_{DR}^n \tag{5.24}$$

The various components of the photocurrent are obtained by integration over the photon energy range of interest ($h\nu > E_g$) for the region under consideration. The spectral response, i.e., the number of electrons collected per incident photon and the total photocurrent collected by the junction were computed for various heterojunction pairs. The spectral response of a ZnSe–GaAs heterojunction (cell B-1) is shown in Fig. 5.13. The contribution of the ZnSe layer is zero since the collection for photon energies greater than the ZnSe band gap is negligible. The contributions of the bulk base region and the base layer depletion region are shown separately in Fig. 5.13. For

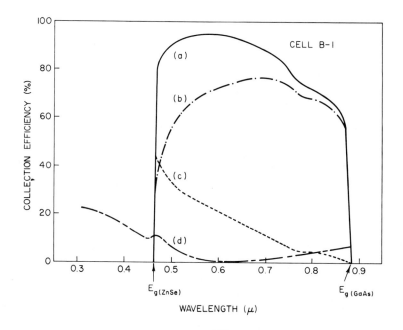

Fig. 5.13. *Spectral response of* ZnSe–GaAs *heterojunction solar cell.*
(a) Computed collection efficiency (total), (b) bulk base region component of collection efficiency, (c) base depletion layer component, (d) reflection loss.

this cell the base depletion region at zero voltage was 458 Å wide. Figure 5.14 shows the spectral response of a comparable GaAs homojunction cell (cell F-1). The contributions of the n–GaAs region and the p–GaAs base region are also shown separately in Fig. 5.14. Comparison of Figs. 5.13 and 5.14 shows that the collection in the p–GaAs is higher for the heterojunction cell, as might be expected, since there is no absorption in the window region. However, the heterojunction cell suffers a cutoff at the band gap of the ZnSe.

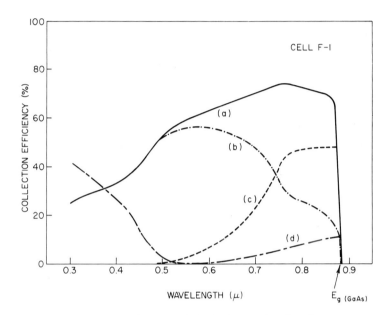

Fig. 5.14. *Spectral response of a GaAs homojunction solar cell.*
(a) Computed collection efficiency (total), (b) total base region response, p–GaAs, (c) response of the n–GaAs layer, (d) reflection loss.

The surface layer, base layer, and total collection efficiencies of various solar cells are given in Table 5.3. The "window effect" is quite beneficial for the ZnSe–GaAs cell relative to GaAs–GaAs solar cells. This is readily seen from Fig. 5.15 which also includes curves for Si–Si and GaP–Si cells. For the GaP–Si heterojunction cells relative to Si–Si cells, the collection efficiency is low primarily because the optical absorption edge of Si is not sharp and the surface recombination velocity of Si can be controlled to low values (10^3 cm sec^{-1} as compared to 10^5 cm sec^{-1} assumed for GaAs). However, when series resistance effects are taken into account (Section 5.6.3), the window effect is found to lower the series resistance of the GaP–Si cell and bring up the power efficiency to a level comparable to that of Si–Si homojunction cells.

Table 5.3 also includes data on n–p ZnSe–Ge and n–p GaAs–Ge solar cells. These pairs have good lattice-match conditions and minority carrier collection has been demonstrated in them. However, Ge is not wide enough in band-gap to make these efficient solar cells (Table 5.4).

TABLE 5.3

Collection Efficiencies of Various Regions of Solar Cells

Solar cell	Cell	Surface layer collection efficiency[a] (%)	Base layer collection efficiency[b] (%)	Total collection efficiency[c] (%)
n–p ZnSe–Ge	A-1	0	97.3	87.8
	A-2	0	97.3	84.2
n–p ZnSe–GaAs	B-1	0	87.3	72.3
	B-2	0	86.7	69.6
n–p GaP–Si	C-1	1.4	84.9	74.8
	C-2	0.007	83.2	65.9
n–p Si–Si	D-1	96.0	83.9	86.2
	D-2	87.1	82.0	83.4
n–p GaAs–Ge	E-1	58.6	96.1	83.2
	E-2	41.8	95.7	74.2
n–p GaAs–GaAs	F-1	58.5	84.4	64.3
	F-2	41.0	77.8	45.1
p–n GaAs–GaAs	G-1	53.6	75.8	58.5
	G-2	29.5	68.1	33.8

[a] Relative to the number of photons of the solar spectrum absorbed in the surface layer with $h\nu > E_{g2}$.

[b] Relative to the solar photons entering the base layer with $h\nu > E_{g2}$.

[c] Relative to the number of photons of the solar spectrum entering the surface layer with $h\nu > E_{g2}$.

5.6.3 Series Resistance Calculations

The major component of the series resistance of a solar cell is usually the resistance of the thin surface layer. In heterojunction cells the window effect allows one to increase the thickness of the surface layer and thus reduce the series resistance. In order to make comparisons between different cells feasible, it is assumed that good quality ohmic contacts can be made to the semiconductors so that the contact resistance is negligible. Internal resistance will thus be the sum of base region resistance and surface layer resistance. The base region resistance R_B is given by

$$R_B = \rho_B d/A$$

where ρ_B is base region resistivity and A is the area in square centimeters of the cell.

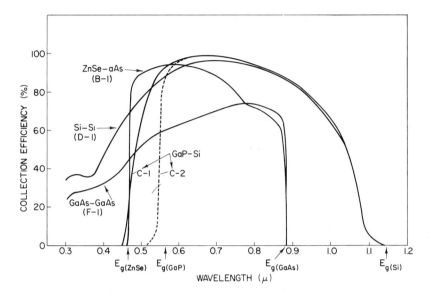

Fig. 5.15. *Collection efficiencies of heterojunction and homojunction solar cells.*

Surface layer resistance was computed by the method described by Handy (1967). For the geometry shown in Fig. 5.10 the area A is 2 cm² and the surface layer resistance R_S is, in Handy's symbols,

$$R_S = \frac{R_c}{1 + R_c/R_p} \qquad (5.25)$$

where

$$R_c = R_4\left(1 + \frac{R_1}{R_3 + \frac{1}{2}R_5} + \frac{R_1}{R_1 + R_4}\right)\bigg/\left(2 + \frac{R_1 + R_4}{R_3 + \frac{1}{2}R_5}\right)$$

$$R_p = \frac{R_c(R_c + R_1)}{2(2R_c + R_1)}, \qquad R_4 = \frac{\rho_s r_3}{t_1 s}, \qquad R_5 = R_4 \frac{r_3}{2W - r_3}$$

and r_3 is given by

$$(2r_3/s)^2 = 2W/r_3 - 1 - 2(W/r_3 - 1)^2 \ln[W/(W - r_3)] \qquad (5.26)$$

and ρ_s is the resistivity of the surface layer. For s equal to 0.4 cm and W as 0.9 cm, solution of Eq. (5.26) gives 0.267 for the value of r_3.

Values of R_1, R_3, and R_7 which are resistances of contact strip, grid strip, and contact resistance of bottom electrode to bulk, were taken from Handy as

0.002, 0.4, and 0.08 Ω, respectively. The total series resistance R_T was computed as

$$R_T = R_B + R_S + R_1 + R_7 \qquad (5.27)$$

Values of R_T for the cells in Table 5.4 range from 0.183 to 1.752 Ω depending on the dopings and mobilities assumed and on the thickness of the window layer in the heterojunction cells. For the Si homojunction cells considered, the values were 0.561 and 0.958 Ω for two different cell proportions. This compares with a value of 0.72 Ω measured for high efficiency n^+–p cells of this geometry (Handy, 1967).

5.6.4 *Solar Cell Efficiency Calculations*

The I–V relationship of the solar cell can be represented by

$$I = A_{\text{TOT}}(J_{d0} + J_{rg0})\{\exp[q(V - IR)kT] - 1\} - J_T A_{\text{ACT}} \qquad (5.28)$$

where A_{TOT} and A_{ACT} are the total area and active area of the solar cell and J_T is the photocurrent density collected by the junction under short-circuit conditions. A schematic representation of a solar cell delivering power to a load is given in Fig. 5.16. No simple analytical relationship between V

Fig. 5.16. *Schematic representation of solar cell delivering power to a load.*

and I can describe the maximum power point of the I–V relationship given by Eq. (5.28). Therefore, for each cell the I–V characteristic was computed on a point by point basis and the maximum power point determined by examination. The current and voltage at this point, denoted by I_P and V_P, along with the overall solar power conversion efficiency η_0 are listed in Table 5.4. This overall efficiency (%), which includes the major loss terms, is given by

$$\eta_0 = \frac{V_P I_P}{0.139 \times 2} \times 100 \qquad (5.29)$$

where 0.139 is the solar energy density (W cm^{-2}) incident on the cell, and 2 cm^2 is the total area of the cell.

Included also in Table 5.4 are calculated efficiencies for each cell with relaxation of various loss terms. These higher efficiencies are given to compare with prior treatments where such simplifications have been made, and because they represent certain upper bounds. However, they are not achievable in practice and attention should be centered mainly on the η_0 values.

TABLE 5.4

Solar Energy Conversion Efficiencies of Solar Cells

Solar cell	Cell	R_T (Ω)	Solar cell efficiency[a] (%)				V_P (V)	I_P (mA)	(V_P/V_{OC})
			Practical η_0	η_1	η_2	η_3			
n–p ZnSe–GaAs	B-1	0.403	13.35	13.66	16.76	19.34	0.814	45.63	0.88
	B-2	0.422	12.8	13.09	16.71	19.34	0.809	44.0	0.88
n–p GaAs–GaAs	F-1	0.51	10.32	10.6	18.28	21.11	0.733	39.14	0.87
p–n GaAs–GaAs	G-1	0.933	10.48	10.9	20.73	24.01	0.82	35.52	0.87
n–p GaP–Si	C-1	1.752	10.15	12.5	13.40	15.47	0.471	59.93	0.73
	C-2	0.183	10.74	10.94	13.40	15.47	0.553	54.0	0.86
n–p Si–Si	D-1	0.958	11.69	13.33	17.54	20.25	0.477	68.2	0.78
	D-2	0.561	11.98	12.89	17.54	20.25	0.499	66.8	0.81
n–p ZnSe–Ge	A-1	0.404	8.18	9.48	11.32	13.15	0.244	93.0	0.72
	A-2	0.423	7.84	9.11	11.32	13.15	0.241	90.6	0.72
n–p GaAs–Ge	E-1	0.501	6.77	8.1	11.75[b]	13.66[b]	0.223	84.5	0.70
	E-2	0.321	6.45	7.14	11.75[b]	13.66[b]	0.233	76.9	0.74

[a] η_0 includes series resistance loss and all other losses, except the interface recombination factors which are presently unknown for heterojunctions, η_1 assumes the series resistance loss to be negligible ($R_T = 0$), η_2 assumes $R_T = 0$ and no reflection loss and perfect collection in the window region, η_3 as for η_2 but without loss due to surface contact area, i.e., A_{TOT} and A_{ACT} assumed to be both 2 cm².

[b] These η-values for GaAs–Ge are not comparable with the values above them. They are for complete collection h$\nu > E_{g2}$ rather than taking account of the narrow window effect. This supposes that collection from the surface layer is possible, which may be only partially true.

The results show that heterojunction solar cells can be hoped for with efficiencies, η_0, comparable to or even greater than the efficiencies obtainable with Si or GaAs homojunction cells. The treatment given, however, neglects junction interface recombination. This is a relatively unknown factor in heterojunction solar cells since few acceptable structures have been made for study.

The theoretical results, with this proviso, show that n–p ZnSe–GaAs heterojunction solar cells should be capable of greater than 13% efficiency. This compares with about 10–12% for GaAs and 12% for Si homojunction cells calculated on a similar basis. These values do not include the effects of built-in drift fields on cell performance. In practical Si cells this raises the measured performance to above 13%. A comparable improvement in n–p ZnSe–GaAs cells might be expected by the provision of built-in fields.

The output voltage of a ZnSe–GaAs cell at optimum load power is calculated to be in excess of 0.8 V and the voltage decline between zero and full load power is 12%. Both of these values are better than for Si homojunction cells, for which the load voltage may be about 0.5 V and the voltage regulation about 20%. Furthermore the ZnSe–GaAs cell should have an advantage over Si cells in performance at high temperatures since the energy gap of GaAs is greater than that of Si.

GaP–Si heterojunction cells are also seen to be interesting from Table 5.4, although the efficiencies expected are a little lower than for Si cells.

Another potential advantage of heterojunction cells is the possibility of low radiation damage in outer space conditions. Data are not readily available for effects of radiation damage in ZnSe, GaP, and GaAs. However, in heterojunctions when the surface layer is made thick enough and the window-region is utilized to create photocarriers in the base region, the effect of radiation damage may be small if the damage is confined mainly to the surface layer which does not contribute to the carrier collection. In Si homojunction cells, transparent covers are used to protect the cell from radiation damage. It is possible that these covers could be dispensed with in heterojunction cells having wide window regions.

5.7 GaAs–Al$_x$Ga$_{1-x}$As Heterojunction Lasers

Laser action was first observed in Cr doped Al$_2$O$_3$ in 1960. This was followed by the development of gas lasers in 1961 and of GaAs diode lasers in 1962. The evolution of GaAs injection lasers up to a few years ago is reviewed in the book "Gallium Arsenide Lasers," edited by Gooch (1969). Other useful reviews are those of Nathan (1963), Rediker (1965), and Stern (1966), to mention only a few such sources. This extensive literature is mentioned since some familiarity with basic injection laser concepts, not at a very detailed level, is assumed in the discussion that follows.

Injection diode lasers have been slow to find useful applications in competition with other laser systems, primarily because of low power capability and the need for operation below room temperature. With the development of heterojunction confinement lasers, continuous-wave (cw) operation has been achieved at room temperature and above (Hayashi, Panish and coworkers). The threshold current density for lasing action has been lowered by a factor of at least twenty. The advantages of injection lasers include small size and ease of modulation by signals superimposed on the injected current. Factors, however, that continue to limit the application of injection lasers are that they do not have the narrow spectral line width, large exit aperture size, narrow beam angle, and coherence of competing lasers. Some typical values for these and other parameters have been given by Vallese (1970).

The concept of a heterojunction diode laser began to develop in 1963. Toward the end of the year, Kroemer (1963) published a discussion of the various possibilities that might exist.* The most important of Kroemer's suggestions was the concept that injected carrier pile-up, or confinement, in the narrow active region would make the achievement of population inversion possible at lower current densities. Kroemer also speculated that heterojunctions might open the way to the achievement of laser action in indirect-gap semiconductors. This was further examined by Wang (1963) and Wang and Tseng (1964) who proposed that a GaAs–Ge heterojunction might be used to achieve direct injection into the $k = (000)$ valley of Ge. However, studies show that the residence time in this valley is extremely short before scattering to the indirect-gap valley, and laser action from indirect-gap materials such as Ge and Si is not likely to be achieved in this way.

Physical realization of the heterojunction laser concept in relation to GaAs depended on finding a wider-gap semiconductor with an excellent lattice match (to minimize interface state recombination) and with suitable barriers and refractive index conditions to provide both carrier and photon confinement. Such a material exists, Al$_x$Ga$_{1-x}$As, but some years of work were necessary with this alloy system by Alferov and co-workers in the USSR, and Panish and Hayashi, and Kressel and co-workers in the USA before striking results were obtained. The properties of Al$_x$Ga$_{1-x}$As–GaAs heterojunctions are discussed briefly in the next few pages. This is followed by a review of the laser properties that have been achieved.

5.7.1 Al$_x$Ga$_{1-x}$As *and Its Interface with* GaAs

AlAs is a semiconductor with a band gap greater than that of GaAs, and a very close lattice match. AlAs has been prepared by only a few investigators. The surface of the material, unless carefully protected after growth, tends not to be very stable, although recently better behavior in this respect has been reported. The high reactivity of the Al also tended to result in high doping densities due to contamination during preparation. However, AlAs

* The filing of an "authors' certificate" on heterojunction laser concepts by Alferov and Kazarinov (1963) in the USSR has been reported (Alferov, 1970a).

was found to have a lattice constant of 5.661 Å which is almost the same as the lattice constant of 5.654 Å for GaAs. The AlAs band gap (indirect) was found to be about 2.1 eV.

Since AlAs itself was not very stable, studies were undertaken of the alloy system $Al_xGa_{1-x}As$, and it was found that this gave acceptable material stability provided x was less than 0.8. Furthermore, it was found that layers of $Al_xGa_{1-x}As$ could be grown on GaAs by liquid-phase epitaxy (Rupprecht *et al.*, 1967). Control of this process improved when the ternary phase diagram for Al–Ga–As was established by Panish and Sumski (1969). The

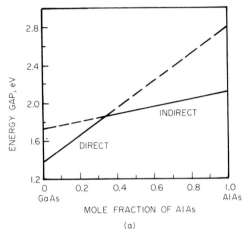

(a)

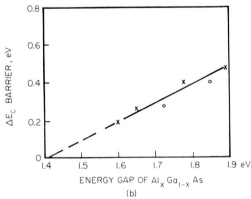

(b)

Fig. 5.17. AlGaAs/GaAs *properties.*

(a) Composition dependence of $Al_xGa_{1-x}As$ direct and indirect energy gaps from photo-response measurements. (After Casey and Panish, 1969.) (b) Discontinuity in conduction band ΔE_c of $Al_xGa_{1-x}As$–GaAs heterojunction versus energy gap of the $Al_xGa_{1-x}As$: ×, *n–p* $Al_xGa_{1-x}As$–GaAs; ○, *p–n* $Al_xGa_{1-x}As$–GaAs. (After Alferov, 1970b.)

variation of the direct and indirect energy gaps as a function of the molecular fraction of AlAs was found to be as in Fig. 5.17(a). This suggested that an Al content in the range 0.2–0.4 would be adequate to give a suitable band-gap difference for heterojunction studies. However, the form of the energy barriers between Al$_x$Ga$_{1-x}$As and GaAs needed to be established. By electrical measurements of *p–n* and *n–p* structures in this heterojunction system, Alferov and co-workers concluded that the discontinuity in the valence band ΔE_v was close to zero, whereas that in the conduction band ΔE_c was equal to the difference in E_g of the heterojunction components, Fig. 5.17(b).

For Al$_{0.3}$Ga$_{0.7}$As–GaAs, the value of ΔE_c is about 0.4 eV and the band gap of the Al$_{0.3}$Ga$_{0.7}$As is 1.8 eV. The energy band diagrams that may be expected therefore are as shown in Fig. 5.18. For an *n–p* Al$_x$Ga$_{1-x}$As–GaAs junction the band diagram with zero voltage applied is Fig. 5.18(a). The bending of the bands shown is caused by electrons moving from the *n*–Al$_x$Ga$_{1-x}$As into the *p*–GaAs to provide alignment of the Fermi levels on the two sides of the junction. Interface state effects are assumed negligible

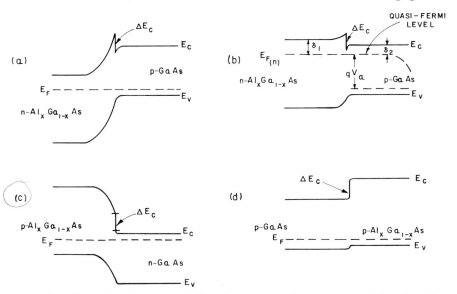

Fig. 5.18. *Energy band diagrams of* Al$_x$Ga$_{1-x}$As–GaAs *heterojunctions assuming* $\Delta E_c \sim 0.4$ eV *and* ΔE_v *is negligibly small.*

(a) For an *n–p* Al$_x$Ga$_{1-x}$As–GaAs junction, with no voltage applied. Grading of the Al content over a few hundred Å at the interface would reduce the effect of ΔE_c. (b) Forward voltage V_a applied to junction (a). The quasi-Fermi level in the *p*–GaAs represents the injected electron density, which may be greater than the electron density of the emitter. (c) Energy band diagram for a *p–n* Al$_x$Ga$_{1-x}$As–GaAs junction, with no voltage applied. (d) Energy band diagram for *p–p* heterojunction, neglecting interface state effects.

in this diagram, which is probably permissible because of the excellent lattice match conditions. Application of a forward voltage, the $Al_xGa_{1-x}As$ being made negative, results in the band diagram of Fig. 5.18(b). Here the Fermi levels are separated by an energy corresponding to an applied voltage V_a, and a quasi-Fermi level is shown in the p–GaAs to represent the density of injected electrons before they recombine with holes farther from the interface. Since in Fig. 5.18(b) the Fermi level spacing δ_2 is less than δ_1, it may be inferred that the density of injected electrons in the GaAs is greater than the electron density in the n–$Al_xGa_{1-x}As$ that is providing the injection (the densities of states in the $Al_xGa_{1-x}As$ and in the GaAs are assumed to be not very different). This effect is known as "superinjection" and is a special feature of heterojunctions. Such action has been examined extensively by Alferov and co-workers (Alferov et al., 1969a; Alferov, 1970a).

The energy band diagram for p–n rather than n–p junctions of $Al_xGa_{1-x}As$– GaAs is shown in Fig. 5.18(c). From the barriers that exist, it is apparent that current flow in this junction is predominantly by injection of holes into the n–GaAs. The energy diagram for a p–GaAs/p–$Al_xGa_{1-x}As$ structure is shown in Fig. 5.18(d) and ΔE_c is seen to be the main barrier in the conduction band.

If a double heterojunction structure n–$Al_xGa_{1-x}As/p$–GaAs/p–$Al_xGa_{1-x}As$ is considered, the resulting energy band structure may be envisaged by imagining Figs. 5.18(b) and (d) placed side by side. Hence ΔE_c in Fig. 5.18(d) will act as a confinement barrier for injected electrons in the GaAs and the valence band barrier of Fig. 5.18(b) will provide confinement of holes.

The interface between $Al_xGa_{1-x}As$ and GaAs, grown by liquid phase epitaxy, appears to be relatively free of defects that behave as nonradiative recombination centers. This has been studied by examination of the photoluminescence of a GaAs surface excited by He–Ne laser radiation (6328 Å) through an $Al_{0.6}Ga_{0.4}As$ layer (Hayashi and Panish, 1970). At room temperature the external efficiency of the photoluminescence (8800 Å) was much higher if measured from a region where the GaAs surface was covered with the $Al_xGa_{1-x}As$ layer as compared to the same GaAs without the layer. This indicates that the interface has many fewer nonradiative recombination centers than a bare GaAs surface. The shape of the photoluminescence spectrum, Fig. 5.19, is the same whether the AlGaAs is present or not. The absence of a high-energy band tail suggests that the change from GaAs to $Al_xGa_{1-x}As$ occurs in less than the penetration depth ($\frac{1}{4}\ \mu$) of the He–Ne radiation. Another indication of the low density of nonradiative recombination centers at the interface is the high external efficiency of spontaneous emission from single heterostructure mesa diodes. This is observed even at current densities as low as 1 A cm^{-2} (about 3×10^{13} cm^{-3} excited carriers).

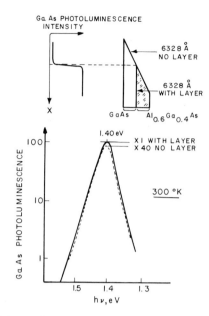

Fig. 5.19. *Photoluminescence spectra of a* GaAs *substrate surface with* (————) *and without* (- - - - -) *a covering layer of* Al$_x$Ga$_{1-x}$As.

As illustrated, the two spectra were obtained on the same angle-lapped substrate. The substrate and layer were both *p*-type. (After Hayashi and Panish, 1970.)

Therefore at much larger current densities, a few times 10^3 A cm^{-2}, which correspond to laser operation, the interface nonradiative recombination effects may be expected to be negligible.

Another interesting experiment that has been carried out is the study of the diffusion length of electrons created by He–Ne laser illumination within a small penetration distance ($\frac{1}{4}\,\mu$) of the GaAs–Al$_x$Ga$_{1-x}$As interface, as indicated in Fig. 5.20. With the *p*–GaAs of thickness 1.8 μ, 95% of the created electrons reached the *n–p* GaAs–GaAs interface, while for a thickness of 4.7 μ the collection was 80%. From these values, the electron diffusion length in the *p*–GaAs was concluded to be 6–7 μ, or perhaps 50% less if allowance is made for the probable effect of the Zn concentration gradient in the *d* region. A diffusion length of this magnitude points up the need for electron confinement in laser structures, where the active region is normally considerably less than this in width.

The details of the behavior of Al$_x$Ga$_{1-x}$As–GaAs heterojunctions have been explored by Alferov and co-workers (1970a–i). If the GaAs is lightly or moderately doped ($<2 \times 10^{17}$ cm^{-3}), injection over-the-barrier is observed. The current–voltage characteristics of such heterojunctions are described by the expression

$$I = I_{01} \exp(qV/2kT) + I_{02} \exp(qV/kT) \qquad (5.30)$$

where the temperature dependences of the factors I_{01} and I_{02} agree with the Sah–Noyce–Shockley model for carrier recombination in the space-charge

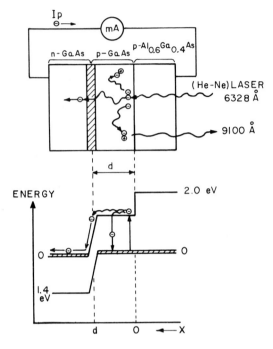

Fig. 5.20. *A photoinjection technique used for determination of the diffusion length of electrons in the active region of a single confinement heterostructure laser.*
(After Hayashi and Panish, 1970.)

region and in the bulk of the semiconductor. For more heavily doped junctions tunneling effects become dominant. From a study of the electroluminescence spectra and of the polarization of the light as a function of bias voltage, thermoinjection filling of tails and "diagonal" tunneling recombination transitions are found to occur. The details of these matters will not be examined here.

5.7.2 *Single and Double Heterojunction Confinement* $Al_xGa_{1-x}As$–$GaAs$ *Lasers*

In a heterojunction laser structure, two confinement actions have to be considered. There is the confinement of the injected carriers by the energy barriers in the conduction and valence bands. Also there is the waveguide confinement of the photons caused by the refractive index changes at the $GaAs$–$Al_xGa_{1-x}As$ interfaces. Both confinement effects contribute to the lowering of the threshold current density for laser action. The carrier confinement predominantly controls the population inversion and therefore the gain β of the laser cavity, and allows the width of the GaAs active region to be made narrower. The refractive index confinement reduces the photon losses from the laser cavity and relates therefore mainly to the loss term α in laser analysis. Waveguide and resonant mode action in lasers has been discussed by McWhorter (1963), by Nelson and McKenna (1967), by Zachos and Ripper (1969), and by Adams and Cross (1970).

A comparison of homojunction, single-heterojunction, and double-heterojunction structures is shown in Fig. 5.21. The confinement barriers

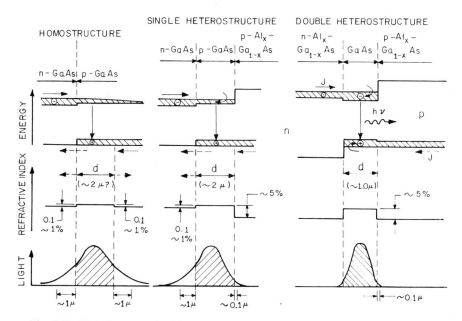

Fig. 5.21. *Physical structure, bandshapes under operating voltage, refractive index steps, and optical power distribution in homostructure, single heterostructure, and double heterostructure laser diodes.* (After Panish and Hayashi, 1970.)

in a homojunction structure are seen to be quite small and are primarily a consequence of the doping differences. The spread of the light on either side of the active region is seen to be quite large. In the single-heterostructure laser there is an electron confinement barrier of about 0.4 eV and roughly a 5% decrease in refractive index in passing from the p–GaAs to the p–Al$_x$Ga$_{1-x}$As. This confines the spreading of the light at this interface. However at the n–p GaAs interface there is only a small change of refractive index and considerable photon loss occurs into the n–GaAs. At high bias conditions, with thin widths of the p–GaAs region, injection of holes may occur into the n–GaAs and laser action may be affected. For double heterojunction structures, as shown in Fig. 5.21, carrier and photon confinement can be expected for both sides of the active region.

A comparison is shown in Fig. 5.22 of the power output as a function of peak current for the three kinds of laser structures. The threshold current density

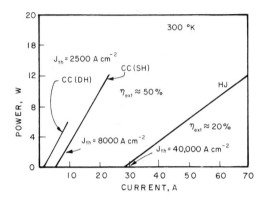

Fig. 5.22. *Comparison of power output as a function of peak current for close confinement, double heterojunction* (DH); *close confinement, single heterojunction* (SH); *and homojunction* (HJ) *lasers having the same area.*
(After Kressel, 1970.)

for laser action is seen to have been reduced from 40,000 to 8000 A cm^{-2} by going to a single-heterojunction structure. For a double-heterojunction laser, the threshold current density is in the range of 1000–2500 A cm^{-2}, depending upon the dimensions of the structure and on other details of the fabrication. The dependence of the threshold current density upon the cavity length L may be expected to follow the relationship:

$$J_{th} = (1/L\beta) \, [\alpha L + \ln(1/R)] \qquad (5.31)$$

where α is the internal loss per unit length, β is the gain factor per unit length and per unit current density, and R is the reflectivity coefficient (Pilkuhn and Rupprecht, 1967). The observed variations of J_{th} with $1/L$ for the three kinds of structures is shown in Fig. 5.23.

For general use, laser lengths are typically in the range 250–500 μ. From the α and β values given, it is seen that the successive degrees of confinement decrease the loss from 60 to 10 cm^{-1} and increase the gain factor from 1.7 to 20 cm^{-1} kA^{-1}. Further study of double-heterojunction lasers suggests that the data may be fitted by a straight line corresponding to J_{th}^2 proportional to $1/L$. The gain is found to be an increasing function of the operating current density according to a power law J^m, where m is 2 or somewhat higher

[Goodwin and Selway (1970) (single-heterostructure lasers); Hayashi *et al.* (1971) (double-heterostructure lasers)].

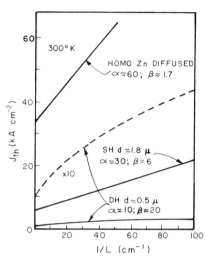

Fig. 5.23. *Threshold current density J$_{th}$ versus 1/L for homostructure,* SH, *and* DH *lasers.*
The units for the loss term α are cm^{-1} and for the gain term β are cm^{-1} kA^{-1}. (After Panish and Hayashi, 1970.)

The threshold current density in a double heterostructure is almost inversely proportional to the thickness, d, of the active GaAs region, as shown in Fig. 5.24. The smaller the active thickness that has to sustain population inversion, the smaller the current and the current density needed to provide it. For a single-heterostructure laser however, loss of holes occurs at 300°K if the active thickness d is made smaller than about 2 μ, because of imperfect confinement at the n–p GaAs interface, and therefore the threshold current density passes through a minimum as shown by the upper curve in Fig. 5.24.

The far field pattern of the beam of light from an injection laser depends on the resonant modes and diffraction limits imposed by the rectangular symmetry of the active region (Zachos and Dyment, 1970). However, under certain conditions, the distribution can be Gaussian which is convenient for subsequent optical processing. For a homostructure or a single-heterostructure laser the beam spread perpendicular to the junction plane may correspond to a half-angle of 10–15°. For effective collimation, an optical system with an aperture of about $f/2$ must be used (Gooch, 1969). Single-heterostructure lasers, with spectral line widths ∼20 Å, have been shown to be capable of retrieving holographic information with about 200 × 200 resolvable lines (Firester and Heller, 1970). In double confinement structures with very narrow active-region thicknesses the half-angle for the beam spread is 40° or more and the collimation problem is correspondingly more difficult.

The lasing light output from a DH diode is generally polarized with the optical electric field vector of the radiation parallel with the plane of the

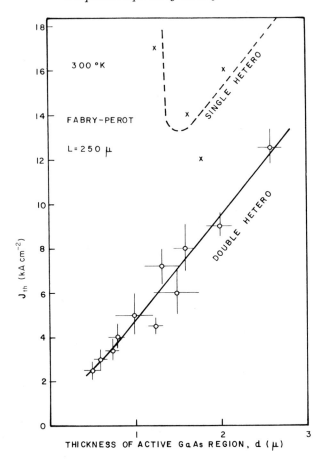

Fig. 5.24. *Dependence of the threshold current density on the thickness of the* GaAs *active layer in double confinement heterostructure lasers.*
(After Panish and Hayashi, 1970.)

heterojunction. This corresponds to TE modes within the laser cavity and is in contrast to SH or homostructure lasers in which the field distribution may be approximated by TEM modes and the emission generally does not have a well defined polarization.

The room temperature dc continuous wave performance of a double-confinement laser structure is shown in Fig. 5.25. The rapid increase in light output beyond the threshold current is apparent and the differential efficiency is about 5%. At 1.3 J_{th} (1.0 A) the total light output for the two ends of the diode is 20 mW and the overall quantum efficiency is 1.6%. Under pulsed conditions the threshold current density is about 20 to 30% lower than that for cw operation.

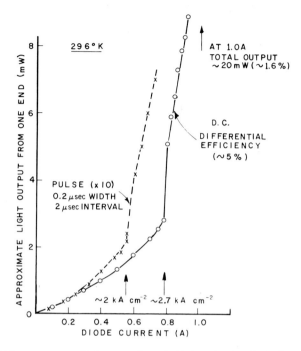

Fig. 5.25. *Light output versus diode current for a double heterojunction* GaAs *laser diode at room ambient temperature for a* 10% *pulsed duty cycle and for dc (continuous wave) operation.* Diode length 370 μ, width 80 μ, and area 3×10^{-4} cm^2. (After Panish and Hayashi, 1970.)

The threshold current density for heterostructure lasers decreases as the temperature is lowered below 300°K. For example a DH laser, with an active region thickness of 1 μ, and a 300°K threshold current density of 4×10^3 A cm^{-2} may have thresholds of 2×10^3 A cm^{-2} at 200°K and less than 0.6×10^3 A cm^{-2} at 100°K.

If the active region of a confinement laser is Al$_x$Ga$_{1-x}$As instead of GaAs, the wavelength of the emitted light is shortened and enters the visible range of the spectrum. For the spectral range 9000–8000 Å the quantum efficiency is relatively constant, but for shorter wavelengths the efficiency begins to drop rapidly and is down by a factor of 10 at 7000 Å which is obtained for x about 0.30 (Kressel *et al.*, 1970a).

5.7.3 Fabrication and Mounting Techniques

Confinement-type GaAs lasers are fabricated by liquid-phase epitaxy from Ga solutions. The boat used typically has four separate solutions suitably doped so that by movement of the substrate crystal four successive layers may be grown. Such a system is shown in Fig. 9.28, p. 265.

The substrate crystal is n–GaAs with Te or Si doping in the range $1\text{–}4 \times 10^{18}$ cm^{-3} and is oriented for growth on a $(\overline{1}\overline{1}\overline{1})$ or (100) surface. The first layer grown is $n\text{–}Al_xGa_{1-x}As$ usually 2–5 μ thick, typically Sn doped, with x between 0.2 and 0.4. The second layer is the p–GaAs active region (0.4–2 μ thick) doped with Si and probably with some Zn by diffusion from layer 3. Layer 2 may also contain a small amount of Al, either deliberately provided or carried over from layer 1. Layer 3 is $p\text{–}Al_xGa_{1-x}As$, $3\text{–}8 \times 10^{18}$ cm^{-3}, with x in the range 0.2–0.4, and thickness 1–2 μ. The fourth and final layer is p–GaAs, $3\text{–}5 \times 10^{18}$ cm^{-3}, doped with any convenient p-type dopant, often Ge. Its function is to provide for a better contact than would be possible directly to $Al_xGa_{1-x}As$. Layers 3 and 4 are both kept quite thin since the main heat sinking for the laser is normally provided on layer 4. For improved thermal performance, the heat sink may be a type II diamond which has up to five times the thermal conductivity of Cu. Further details of fabrication have been given by Hayashi *et al.* (1971).

One form of fabrication that merits special mention is the stripe contact geometry shown in Fig. 5.26. The stripe provides a convenient way of

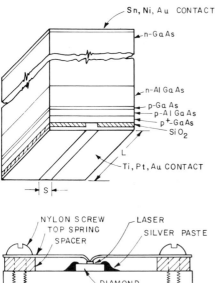

Fig. 5.26. GaAs–AlGaAs *double hetero-junction laser with stripe geometry contact.*

The laser is mounted with the stripe side down on a metallized diamond heat sink having five times the thermal conductivity of copper. The p–GaAs layer is the light-emitting region.

obtaining a small device area with some lateral heat flow that allows a high cw operating temperature. The optimum stripe width S is between 10 and 15 μ for typical laser structures (Dyment and D'Asaro, 1967; Dyment *et al.*, 1969b). The lasing region of the junction is somewhat wider than the stripe width because of current spreading.

Two failure modes are found in GaAs homojunction lasers. These are catastrophic failure in which the facet of the laser is damaged due to excessive optical flux density, and also a gradual degradation that is related to the high current densities involved. With the development of heterojunction lasers of low operating-current density, the degradation effect has been considerably reduced. Diodes subject to a cw injection-current density of 8×10^3 A cm^{-2} show very little degradation after 1000 hr of operation at room temperature (Ripper, 1970b). In other studies at 300°K, single-heterojunction lasers operating at 1 W mil^{-1} of facet length for a pulse current of 50×10^3 A cm^{-2} with a duty cycle of 0.04% have shown less than 20% degradation of light output after 1000 hr (Kressel, 1970).

Kressel *et al.* (1970e) have recently shown that for a given pulse length and shape the power level at which damage occurs is lower for double-heterostructure than for single-heterostructure lasers. For double hetero-structures with threshold current densities of 2000–3000 A cm^{-2}, catastrophic failure occurs at optical power levels 2 to 3 times lower than for single hetero-structures. This is presumably a result of the high optical flux densities due to the narrow active regions for double heterostructures. In order to reduce this problem, Lockwood *et al.* (1970) and Kressel *et al.* (1971) have developed a structure they term a large optical cavity (LOC) injection laser. This is a double heterostructure in which the recombination region and optical cavity are independent of one another. The structure consists of a p–Al$_x$Ga$_{1-x}$As region (width d_1), a thin p–GaAs region (d_2) which is the recombination light-emitting layer, an n–GaAs region (d_3) which is the optical cavity, a n–Al$_x$Ga$_{1-x}$As region (d_4), and an n–GaAs substrate. Radiation is produced in region d_2 by the injection of electrons from d_3. This radiation excites the higher order LOC modes of the large cavity consisting of regions d_2 and d_3. Typically, d_2 is 0.5–1 μ in width and d_3 1–6 μ. As d_3 is increased, the threshold current density goes from 1200 to 8600 A cm^{-2} but the structures have differential quantum efficiencies above 40%. At room temperature for a 3% duty cycle, the power efficiency was 20%, which is the highest room tem-perature value reported to date. The threshold for catastrophic damage for these devices is considerably higher than for conventional single-hetero-structure lasers similarly operated.

Chapter 6 | Metal–Semiconductor Barriers

6.1 The Schottky Model

Barriers at metal–semiconductor contacts, if they follow the simple Schottky model, are determined by the difference in the work function, ϕ_m, of the metal and the electron affinity, χ_s, or work function, ϕ_s, of the semiconductor. The energy diagrams that are expected are shown in Figs. 6.1 and 6.2 for metal contacts on n- and p-type semiconductors. In Fig. 6.1(b), for instance, the barrier to the movement of electrons from the n-semiconductor into the metal is $(\phi_m - \phi_s)$, and the barrier to the reverse flow of electrons from the

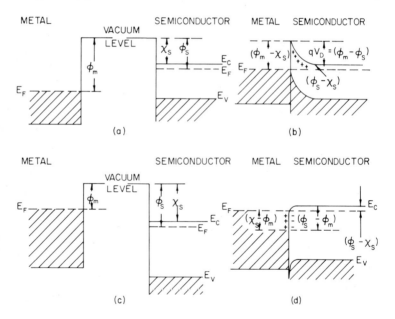

Fig. 6.1. *Energy level diagrams of metal contacts to n-type semiconductors.*
(a) and (b) with $\phi_m > \phi_s$, (c) and (d) with $\phi_m < \phi_s$. Contact (b) acts as a rectifier, since a barrier $(\phi_m - \phi_s)$ exists in the conduction band of the semiconductor. The forward bias direction is for the semiconductor negative with respect to the metal and electrons flow from the semiconductor into the metal. Contact (d) is ohmic since virtually no barrier exists in the conduction band. (After van der Ziel, "Solid State Physical Electronics," 2nd ed., © 1968. By permission of Prentice-Hall, Inc., Englewood Cliffs, New Jersey.)

metal to the semiconductor is $(\phi_m - \chi_s)$. If the junction is forward biased with an external voltage source, V_a, so that the semiconductor is negative with respect to the metal, the forward barrier becomes $q(V_D - V_a)$. However, the reverse barrier $(\phi_m - \chi_s)$ remains, in a first-order model, relatively unaffected by applied voltage or by the doping level of the semiconductor. Therefore $(\phi_m - \chi_s)$ will be defined as the barrier of the metal–semiconductor pair.

Comparison of Fig. 6.1(b) and (d) shows that the junction is rectifying for an *n*-type semiconductor if $\phi_m > \phi_s$, and ohmic if $\phi_m < \phi_s$. For a *p*-type semiconductor (Fig. 6.2) the converse is expected to be true. The experimental evidence for most semiconductors, whether *n*- or *p*-type,

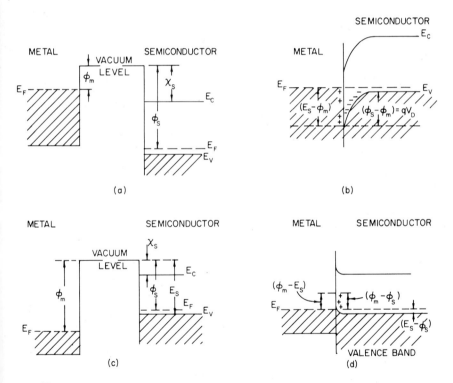

Fig. 6.2. *Energy level diagrams of metal contacts to p-type semiconductors.*
(a) and (b) are for $\phi_m < \phi_s$. Contact (b) therefore acts as a rectifier since a barrier $(\phi_s - \phi_m)$ exists to the flow of holes. This barrier to hole flow is lowered if the semiconductor is made positive with respect to the metal. E_s is $\chi_s + E_g$. In (c) and (d) since $\phi_m > \phi_s$ there is no barrier to current flow and the contact is ohmic. (After van der Ziel, "Solid State Physical Electronics," 2nd ed., © 1968. By permission of Prentice-Hall, Inc., Englewood Cliffs, New Jersey.)

however, does not support this simple model which neglects the effects of interface states.

For a while, however, let us continue with the simple Schottky model. In attempting to verify the $(\phi_m - \chi_s)$ barrier concept, difficulties arise. One is that the experimental determinations of the work functions of metals are widely spread in value. For instance the range of ϕ_m values that have been measured for Al is from 2.98 to 4.36 eV. Even for a nonreactive metal like Au the range was 4.0 to 4.92 until recent determinations under high-vacuum conditions (free of Hg or oil film contamination from vacuum pumps) suggested the value 5.2 eV (Huber, 1966; Riviere, 1966).

One must come to terms with this difficulty. This may be done by accepting mean values (as Michaelson, 1950), by placing more weight on recent determinations, or by showing the range graphically. All three methods will be used here since the treatments discussed reflect the attitudes of various workers. Table 6.1 shows typical values used. Also included are values of

TABLE 6.1

Work Function and Electronegativity Values

Metal	Mean value[a] ϕ_m(eV)	Recent determinations (eV)	Full range (eV)	Electronegativity E_N(eV)
Mg	3.46	—	2.74–3.79	1.2
Al	3.74	4.2	2.98–4.36	1.5
Si	($\chi_s = 4.01$)	—	—	1.8
Cu	4.47	4.4–4.5	3.85–5.61	1.9
Zn	(3.86)	—	3.08–4.65	1.6
Ge	($\chi_s = 4.13$)	—	—	1.7
Ni	4.84	4.6–5.1	3.67–5.24	1.8
Ag	4.28	4.2–4.4	3.09–4.81	1.9
Cd	(4.08)	—	3.68–4.49	1.7
Sn	4.11	—	3.12–4.64	1.7
Mo	4.28	—	4.08–4.48	—
Au	4.58	4.7–5.2	4.0–5.2	2.4
W	4.63	—	4.25–5.01	2.3
Pt	5.29	5.48	4.09–6.35	2.2

[a] From Michaelson (1950), or for Zn and Cd, from the mean of the full range.

electronegativity. Electronegativity (Pauling, 1960; Gordy and Thomas, 1956) represents the ability of different atoms to attract electrons when the atoms are in close proximity as in chemical bonding. Thus the effective transfer of electronic charge between a metal and a semiconductor in almost atomic contact may perhaps be represented by the difference of electronegativity values rather than the difference in work functions. Some workers (for instance Mead and Spitzer, 1963a, b) have adopted this approach and use the electronegativity for the metal rather than the work function. A feature in favor of the use of the Pauling electronegativity scale is that the values are well defined to within about ± 0.1 eV. The electronegativity of a compound semiconductor such as GaAs may perhaps be treated as the value for the higher electronegativity component, since it is assumed that these provide the sites for the most intimate contact with the metal atoms.

The results of the work-function approach, $\phi_m - \chi_s$, and the electronegativity approach are shown in Table 6.2 for contacts of Au and Al on a number of n-type semiconductors. Neither approach is adequate to predict barrier heights. Many barriers are predicted to be negative, representing ohmic behavior, although the experimental contacts are rectifying.

TABLE 6.2

Contacts between Metals and Semiconductors

Semiconductor (n-type)	Metal	Experimental barrier height and/or type of contact (eV)	Theoretical barrier height work function approach $\phi_m - \chi_s$ (eV)	Theoretical barrier electronegativity approach $E_{Nm} - E_{Ns}$ (eV)
Si	Au	0.81	0.57	0.60
	Al	0.70–0.77	− 0.26 ohmic	− 0.30 ohmic
Ge	Au	0.45	0.45	0.70
	Al	0.48	− 0.38 ohmic	− 0.20 ohmic
GaP	Au	1.30	0.55	0.3
	Al	1.05	− 0.29 ohmic	− 0.6 ohmic
GaAs	Au	0.90	0.51	0.4
	Al	0.80	− 0.33 ohmic	− 0.5 ohmic
CdS	Au	0.78	0.08	− 0.1 ohmic
	Al	ohmic	− 0.76 ohmic	− 1.0 ohmic

In a development of the work-function approach, Geppert *et al.* (1966) by an averaging procedure have attempted to find effective values for the work functions of metals and compatible values for the electron affinities of semiconductors in contact with them. However, for an effective approach to the problem of metal–semiconductor barrier heights, specific consideration must be given to surface state effects. The role that interface state effects play in the band structure was first considered quantitatively by Bardeen (1947). More recently Mead and co-workers have added greatly to our understanding of the problem for a wide range of semiconductors.

6.2 Determination of Barrier Heights

The techniques for measuring barrier heights between metals and semi-conductors are worth considering before interpreting the results in terms of an interface state model.

The Schottky equation for current flow over a barrier (Chapter 7) is of the form

$$J = C \exp(-\phi_B/kT)[\exp(qV_a/kT)-1]$$

where C is a constant (although somewhat temperature dependent) which depends on the model of carrier flow (emission or diffusion) considered. The characteristic of $\ln J$ versus the applied voltage, V_a, is therefore as shown in Fig. 6.3(a). Hence the barrier can be determined from the activation energy plot for the metal–semiconductor junction for a fixed forward bias voltage, V_a, as shown in Fig. 6.3(b).

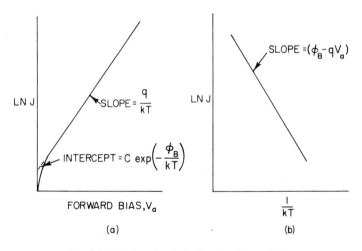

Fig. 6.3. *Metal–semiconductor junction characteristics.*
(a) *I–V* characteristic of forward biased metal–semiconductor contact. (b) Activation energy plot of forward biased metal–semiconductor contact. (After Mead, 1966b.)

Spectral response measurements of photoexcitation in metal–semiconductor junctions provide another method of determining barrier heights. As shown in Fig. 6.4, light may be applied to the front wall of the photocell, if the metal is very thin, or through the semiconductor to the junction. If the photon energy exceeds the barrier height but is less than the semiconductor band gap, photoemission of electrons from the metal into the semiconductor is observed [process (a) in Fig. 6.4(b)]. If the photon energy exceeds the band gap of the semiconductor, direct band-to-band excitation occurs, as shown in Fig. 6.4(c) by the sharp increase in response.

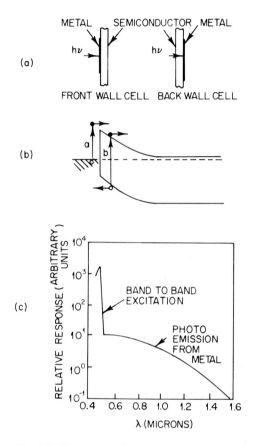

Fig. 6.4. *Photoresponse of metal–semiconductor junctions.*
(a) Photocells illuminated from front or rear, (b) photoexcitation processes depending on photon energy, (c) photoemission of electrons from the metal into the semiconductor develops into band to band excitation as the photon energy is increased. (After Mead, 1966b.)

The short-circuit photocurrent expected for excitation over a barrier is proportional to $(h\nu - \phi_B)^2$ (R. A. Smith, 1959) provided $(h\nu - \phi_B)$ is more than a few kT. Therefore, the square root of the photocurrent response, $R^{1/2}$, plotted against $h\nu$ should give a straight line, and extrapolation of this line to intercept the energy axis gives the barrier height ϕ_B.* This is shown in Fig. 6.5(a)

* Although $R^{1/2}$ is suggested by the Fowler model, better linearity for certain ranges of semiconductor doping may be obtained by plotting $R^{1/3}$ or $(Rh\nu)^{1/2}$ versus $h\nu$. Models taking account of the density of states function have been suggested for both of these approaches.

and (b) for Al on $n-$ and p–GaAs. From Fig. 6.1(b) the barrier height ϕ_{Bn} is expected to be $(\phi_m - \chi_s)$, and from Fig. 6.2(b) the barrier ϕ_{Bp} is $(E_s - \phi_m)$ where E_s is $(\chi_s + E_g)$. The sum of the two barrier heights $(\phi_{Bn} + \phi_{Bp})$ should therefore be equal to E_g. This is confirmed by the results in Fig. 6.5, where ϕ_{Bn} and ϕ_{Bp} add to 1.35 eV, which is reasonably close to the band gap of GaAs.

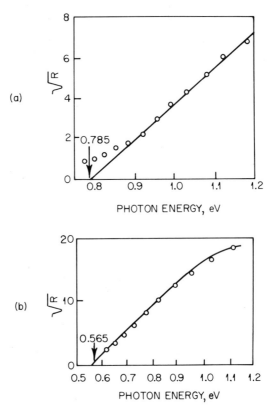

Fig. 6.5. *Barrier height determinations from the photoresponse of Al–GaAs metal–semiconductor photojunctions.*

(a) n-type GaAs, (b) p-type GaAs. The vertical scale is arbitrary. (After Spitzer and Mead, 1963.)

The barrier height is also determinable from capacitance versus reverse voltage measurements. For a uniformly doped semiconductor, C^{-2} versus voltage is a straight line and its voltage intercept provides the diffusion barrier height.

Table 6.3 presents observed barrier heights for a number of metals on some of the most commonly used semiconductors.

TABLE 6.3

Barrier Heights for Metals on n-Type Semiconductors

Metal	Observed barrier heights (eV) 300°K													
	Si	Ge	SiC	GaP	GaAs	GaSb	InP	InAs	InSb	ZnS	ZnSe	CdS	CdSe	CdTe
Al	0.50–0.77	0.48	2.0	1.05	0.80			Ohmic		0.8		Ohmic		0.76
Ag	0.56–0.79			1.20	0.88		0.54	Ohmic	0.18 (77°K)	1.65		0.35–0.56	0.43	0.66
Au	0.81	0.45	1.95	1.30	0.90	0.60	0.49	Ohmic	0.17 (77°K)	2.0	1.36	0.68–0.78	0.49	0.60
Ca	0.40													
Cr	0.57–0.59													
Cu	0.69–0.79	0.48		1.20	0.82					1.75	1.10	0.36–0.50	0.33	
Mg	0.56–0.68			1.04						0.82	0.70			
Mo														
Na	0.43											0.45		
Ni	0.67–0.70				0.86									
Pb	0.40–0.79													
Pd	0.71									1.87		0.62		
Pt	0.90			1.45						1.84	1.40	0.85–1.1		
PtSi	0.85													
W	0.66	0.48											0.37	0.58
WSi$_2$	0.86													

Considerable care must be given to preparing the semiconductor surface before the evaporation of the metal, if fundamental and reproducible barrier heights are to be achieved. Most chemical preparation methods fail to remove completely layers on the surface of the semiconductor. For instance, ellipsometer measurements on Si show that the final HF etching step of most chemical preparation procedures fails to remove 10–20 Å of oxide on Si. Turner and Rhoderick (1968) have studied this problem for metal–n-type Si Schottky barriers and find substantial differences between cleaved and chemically prepared surfaces. Their results from C^{-2} plots for various metals on the (111) Si face (for 1 to 10 Ω cm material) are given in Table 6.4.

TABLE 6.4

Heights of Metal–Silicon Schottky Barriers

		Barrier height (eV)		
		Cleaved surfaces		Chemically prepared surfaces after aging
Metal ϕ_m	(eV)	Turner and Rhoderick (1968)	Archer and Atalla (1963)	Turner and Rhoderick (1968)
Pb	4.20	0.79	—	0.41
Al	4.20	0.76	0.77	0.50
Ag	4.31	0.79	0.76	0.56
Cu	4.52	0.79	0.77	0.69
Au	4.70	0.82	0.81	0.81
Ni	4.74	0.70	0.68	0.67

In general some aging is observed in chemically prepared barriers if left in air for several weeks, or heated to 100°C for about 30 min. The equilibrium barrier obtained is then rather stable and fairly independent of the details of the final chemical preparation steps. The aging seems to be associated with a slow change in the film that already exists before evaporation. The effect did not suggest a subsequent growth of oxide by diffusion of oxygen through the metal since there was no evidence of aging of the cleaved diodes and no dependence of aging on the thickness of the metal in the case of the chemically prepared ones.

The time scale of the aging suggests an ionic activity within the oxide film rather than an electronic activity. Turner and Rhoderick suggest that the aging process is similar in nature to the aging of MOS transistors, and that it can be attributed to the migration of charged ions through the oxide. If the ions are positively charged, as is normally supposed, they would be attracted towards the metal by the built-in field. During this migration their charge distribution together with the compensating negative charges on the surface of the metal would give rise to a dipole which would modify the barrier height, and the dipole would disappear when the ions reach the metal and their charge is neutralized. Such an explanation of the aging is consistent with the experimental facts and would explain why the aging is accelerated by heating. In particular, the presence of charged ions acting as "slow" states could give rise to the nonlinearity in the C^{-2}–V_a plots which was observed immediately after evaporation if the occupation of the states varied with applied voltage. The dominating factor in determining the barrier heights of metal–silicon junctions evaporated on "vacuum cleaved" surfaces seems to be the high surface state density. The barrier heights of these junctions are substantially independent of the work function of the metal. However, the barrier heights of the same metal–silicon junctions prepared by evaporation of the metal on chemically prepared surfaces do depend on the metal, showing that the effect of surface states on these diodes is less important. The barrier heights show an overall variation which is less than the range of work functions of the metals employed. Estimates from a model for this condition give a state density of 2×10^{12} states eV^{-1} cm^{-2}.

Effects of pre- and postannealing treatments on Si Schottky barrier diodes have been studied by Saltich and Terry (1970). They also conclude that intervening oxide layers may be present and affect the barrier heights observed.

6.3 The Barrier Heights Observed for Various Semiconductors

Measured barrier heights for several metals on cleaved n-type Si are given in Fig. 6.6. The Schottky theory line $\phi_B = (\phi_m - 4.01)$ is seen to be too steep to represent the observed barriers.

The barrier heights of various metals on GaAs are almost independent (Fig. 6.7) of the electronegativities of the metals. On the other hand the ZnS results in Fig. 6.7 show that the barriers on this semiconductor change by amounts equal to the electronegativity differences. Clearly GaAs is surface-state controlled, but ZnS is not and conforms with the simple Schottky model. For a semiconductor such as CdS the results are somewhat between, with the barrier change being roughly one-half of the electronegativity difference.

Further measurements show that most semiconductors conform to the pattern represented by the results for Si and GaAs. Namely, the energy barrier ϕ_B does not depend at all strongly on the work function or electronegativity of the metal applied. Mead (1966a) has supplied the explanation. Many covalent-bonded semiconductors have a large density of surface states that form a narrow band centered about one-third of the way up the forbidden gap from the valence band edge. Since this large density of interface states

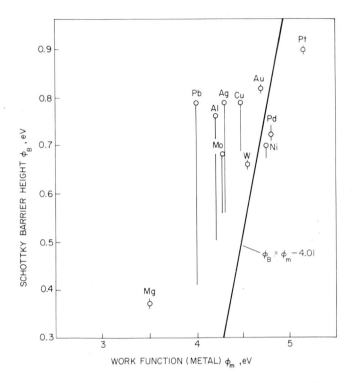

Fig. 6.6. *Barrier height versus work function for metals on n-type silicon.*
The circles are barrier heights obtained with cleaved Si and the vertical lines cover the range of values observed with barriers on chemically prepared surfaces after aging. The diagonal line is a plot of $\phi_B = \phi_m - \chi$ using 4.01 eV for the electron affinity of Si.

can supply or accept charge with relatively little shift of the state Fermi level, the Fermi level of the metal (irrespective of which metal is used) tends to lock into position about a third of the way up the semiconductor band gap. The barrier ϕ_{Bn} from the metal Fermi level to the conduction band edge of the semiconductor is therefore almost independent of the metal. This suggests that the barrier for various semiconductors might be proportional to the energy gap of the semiconductor. This has been shown to be so (Fig. 6.8) for a wide range of semiconductors.

Earlier we concluded that $\phi_{Bn} + \phi_{Bp} = E_g$ from simple reasoning based on Figs. 6.1 and 6.2 for a semiconductor without surface states. For a surface-state-controlled semiconductor, this conclusion is still valid provided the states are localized in the same region of the band gap for both n- and p-type doping (see Fig. 6.5).

For materials of a strongly ionic nature such as SiO_2 or ZnS the barrier

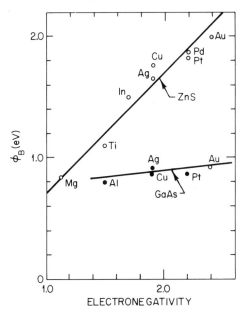

Fig. 6.7. *Barrier heights versus electronegativity for various metals on* ZnS *and* GaAs *(surface state controlled).* (After Mead, 1966b.)

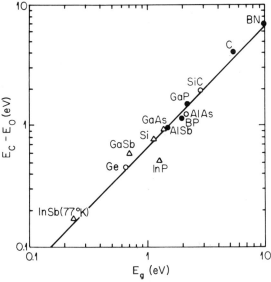

Fig. 6.8. *Location of interface Fermi level relative to conduction band edge for gold contacts on various surface state controlled materials.*

The line is $\phi_{Bn} = E_c - E_0 = \frac{2}{3}E_g$. ($\bigcirc$) *n*-type, ($\bullet$) *p*-type, ($\triangle$) both. (After Mead, 1966b.)

heights are directly related to the electronegativity of the metal. Figure 6.9 shows barrier energy diagrams for Al and Au on SiO_2 (Deal et al., 1966). Experimental results similar to these are shown plotted in Fig. 6.10, as barrier height versus metal electronegativity. The experimental points are represented reasonably well by a slope $(d\phi_B/dE_N)$ of unity. For the semiconductor GaSe, which is only partially ionic in its bonding, the variation

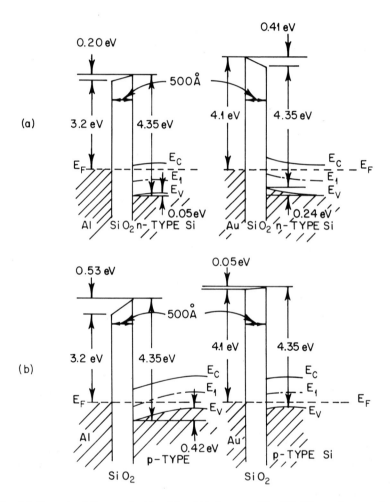

Fig. 6.9. *Energy band diagrams for Al and Au field plates on oxidized* Si. (*Zero surface-state charge density is assumed.*)

(a) n-type silicon, (b) p-type silicon. By appropriate choice of the electrode material, the underlying n-type Si surface can be varied from accumulation to depletion and the p-type surface can be varied from almost flat band to inversion. (After Deal et al., 1966.)

of ϕ_{Bn} with E_N is seen to have a slope of 0.6. On the other hand, for vacuum-cleaved Si, which is covalent in its bonding, the variation of barrier height with metal, from Table 6.4, is very slight.

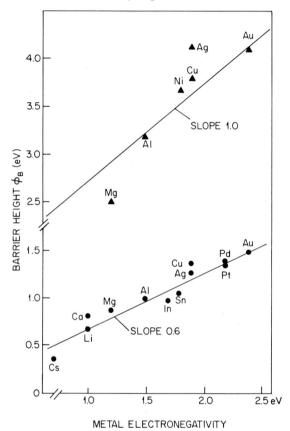

Fig. 6.10. *Barrier heights of various metals on* SiO$_2$, *and on GaSe, versus the electronegativity E_N of the metal.*

(▲) SiO$_2$, (●) GaSe. (After Kurtin *et al.*, 1969.)

Over the range of materials of immediate interest, the difference in electronegativity ΔE_N of the two constituents of a binary compound gives a crude but monotonic measure of the ionicity of the compound. This suggests that the role of the ionic-covalent transition may be illustrated by plotting the slope factor $(d\phi_B/d\chi_m)$ versus electronegativity difference ΔE_N of the compound. The results obtained by Mead and co-workers are given in Fig. 6.11. There is evidently a well-defined transition between the interface properties corresponding to "ionic" materials and those corresponding to "covalent" materials. Apparently in ionic semiconductors the wavefunctions of electrons associated with the cations and anions overlap insufficiently to create interface states with energies near the center of the band gap that if present would control the barrier heights by Fermi-level stabilization at the metal–semiconductor interface.

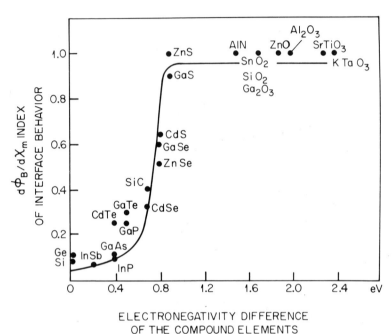

Fig. 6.11. *Variation of* $d\phi_B/d\chi_m$ *versus the electronegativity difference, which is large for a high degree of ionic bonding.*
(After Kurtin *et al.*, 1969.)

In conclusion, there has been very considerable progress in understanding metal–semiconductor barriers in recent years, and the models that now exist of the situation are acceptable. However, as in most subjects, we can hope for continued improvement from further studies.

Chapter 7 | Metal–Semiconductor Junction Behavior

7.1 Characteristics Expected from Emission over the Barrier

The first-order approach to the modeling of current in a metal–semiconductor junction is to assume that the carrier flow is caused by thermionic emission over the barrier and to neglect all tunnel effects and image-force barrier lowering effects. Figure 7.1 illustrates the effect of forward and reverse voltage bias on the barrier between a metal and an n-type semiconductor. At zero bias the electron flux from the semiconductor into the metal is given by the number of electrons having an energy of qV_D, or greater, that are directed per second towards unit area of the interface. For a Maxwellian distribution the result is

$$J_0 = qN_D(kT/2\pi m^*)^{1/2} \exp(-qV_D/kT) \qquad (7.1)$$

where m^* is the effective electron mass.

At zero voltage the net flow over the barrier must of course be zero, and J_0 therefore may also be written in terms of the electron flux from the metal into the semiconductor over the barrier ϕ_B. From Fig. 7.1(a) the barrier height is the diffusion barrier plus the doping step $E_c - E_F$,

$$\phi_B = qV_D + \delta \qquad (7.2)$$

From simple semiconductor theory, with N_c the effective density of states at E_c,

$$\exp(-\delta/kT) = N_D/N_c = N_D/2(2\pi m^* kT/h^2)^{3/2} \qquad (7.3)$$

With the aid of Eq. (7.3), the expression for J_0 from (7.1) may be rewritten as

$$J_0 = \frac{4\pi}{h^3}qm^*k^2T^2 \exp(-\phi_B/kT) = AT^2 \exp(-\phi_B/kT) \qquad (7.4)$$

This is similar to the expression obtained for the emission of electrons from a metal into vacuum over a barrier ϕ_B (see, e.g., van der Ziel, 1968), and A would be the Richardson constant if the free electron mass could be used.

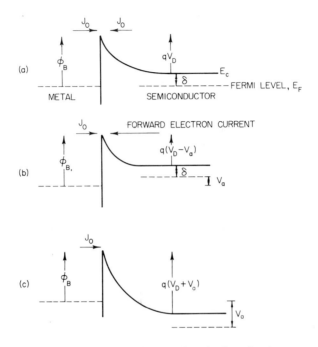

Fig. 7.1. *Energy diagrams for a metal–semiconductor junction.*
(a) Zero-bias voltage. (b) Forward-bias voltage causes electron flow from the semi-conductor over the barrier $(V_D - V_a)$. (c) Reverse-bias voltage exposes the small electron flow J_0 from the metal to the semiconductor.

When a forward bias voltage V_a is applied to the junction, as in Fig. 7.1(b), the effective barrier in the semiconductor becomes $q(V_D - V_a)$ and the electron flow from the semiconductor into the metal is enhanced by a factor $\exp(qV_a/kT)$. The current–voltage relationship for the junction is therefore

$$J = J_0[\exp(qV_a/kT) - 1] \tag{7.5}$$

This emission model supposes that electrons emitted from the metal into the semiconductor have no difficulty moving into the bulk of the semi-conductor. However, the semiconductor depletion region is usually suffi-ciently thick that the electron flow in the region of the junction is controlled by the field and diffusion equations. The analysis is then more involved.

Before discussion of these matters, it is necessary to examine image force lowering that may be expected to affect the barrier height and shape in a metal–semiconductor junction.

7.2 Field Lowering of the Image-Force Barrier

7.2.1 *Schottky Effect at Metal–Vacuum Surface*

Electrons being emitted from a metal cathode into vacuum see a barrier which depends upon the field strength at the cathode surface. The Schottky equations for this effect will be developed with respect to emission into vacuum and then transferred to emission of electrons from a metal into a semiconductor.

The Fermi level of the metal in Fig. 7.2(b) is at the work function energy ϕ_{WF} below the vacuum level. An electron at a distance x from the metal surface experiences an attractive force to the metal. At the interface the lines of field must be perpendicular to the metal surface since it is assumed the surface is a perfectly conducting sheet. The field lines are therefore as though the electron of charge $-q$ induces an image charge $+q$ at a distance $-x$ inside the metal [Fig. 7.2(a)]. By Coulomb's equation the force attracting the electron to the metal is therefore

$$F = q^2/4\pi\epsilon_0(2x)^2 \qquad (7.6)$$

where ϵ_0 is the dielectric constant of free space, 8.85×10^{-14} F cm^{-1}. Integration of Eq. (7.6) from $x = \infty$ to a finite x provides the expression

$$\phi(x) = -q/(16\pi\epsilon_0 x) \qquad (7.7)$$

for the electron energy (in units of electron volts) near the metal. [In Eq. (7.7) at $x = 0$, $\phi(x)$ goes to minus infinity instead of to $-\phi_{WF}$. This, however, is not physically significant since substituting x equal to say 3Å in (7.7) shows that $\phi(x)$ is then still only about -1 eV, and Eq. (7.7) is mostly applied at distances of some hundreds of angstroms or more.]

If an electric field $\mathscr{E}$ V cm^{-1} is now applied in the vicinity of the metal–vacuum interface, the energy of an electron at a distance x becomes

$$\phi(x) = -(q/16\pi\epsilon_0 x) - \mathscr{E}x \qquad (7.8)$$

This has a maximum value at a value of x given by $(q/16\pi\epsilon_0\mathscr{E})^{1/2}$. Hence the barrier lowering produced is $\Delta\phi$ eV [Fig. 7.2(c)] where

$$\Delta\phi = (q\mathscr{E}/4\pi\epsilon_0)^{1/2} \qquad (7.9)$$

If $\mathscr{E}$ is 10^4 V cm^{-1}, then $\Delta\phi$ is about 0.039 eV and occurs at a distance 235 Å from the metal surface.

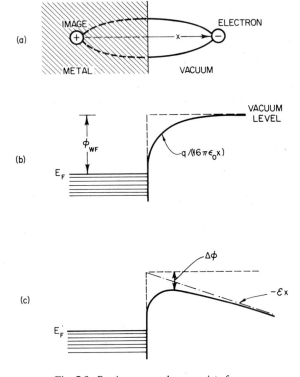

Fig. 7.2. *Barrier at a metal–vacuum interface.*
(a) Electron in vacuum with image charge in the metal. (b) Electron energy barrier in the absence of applied field. (c) External applied field, $\mathscr{E}$, reduces the barrier height by $\Delta\phi$.

7.2.2 *Barrier Lowering at Metal–Semiconductor Interfaces*

At a metal–semiconductor interface a Schottky barrier lowering effect, comparable to that just presented for a metal–vacuum surface, must be expected. The current density J_0 may therefore be written as

$$J_0 = AT^2 \exp[-(\phi_B - \Delta\phi)/kT]$$
$$= AT^2 \exp(-\phi_B/kT)\,\exp[(q\,\mathscr{E}/4\pi\epsilon\epsilon_0)^{1/2}/kT] \qquad (7.10)$$

In the depletion region the maximum field strength at the junction, from Poisson's equation, is

$$\mathscr{E} = [2qN_D(V_D - V_a)/\epsilon\epsilon_0]^{1/2} \qquad (7.11)$$

The field decreases linearly with distance from the junction on the semi-conductor side, but as an approximation we need not take account of this since the distance for the barrier lowering effect is small.

From (7.10) and (7.11) the reverse bias current I_R of a metal–semi-conductor junction should vary as $\ln(I_R)$ proportional to $\mathscr{E}^{1/2}$, or $\ln(I_R)$ proportional to $(V_D - V_a)^{1/4}$. Experimental results that verify this barrier lowering dependence are presented later in the chapter.

7.3 Carrier Flow in the Barrier Region

Consider electron transport in the barrier region of a metal n-type semi-conductor junction where tunneling effects are neglected. Much of our recent understanding of this is due to C. R. Crowell and co-workers.

The electron potential energy $(q\psi)$ variation with distance is shown in Fig. 7.3. The barrier is high enough that the charge density between the metal surface and $x = w$ is essentially that of the ionized donors, where w is the edge of the electron depletion layer. The rounding of ψ near the metal–semiconductor interface is due to the superimposed effects of the electric field associated with the ionized donors (which is shown by the dotted extrapolation of ψ) and the attractive image force experienced by an electron when it approaches the metal. As drawn, Fig. 7.3 shows an applied voltage difference between the metal and the semiconductor bulk that would give rise to a flow of electrons into the metal. The quasi-Fermi level, or imref, ϕ_n eV associated with the electron current density J in the barrier is shown

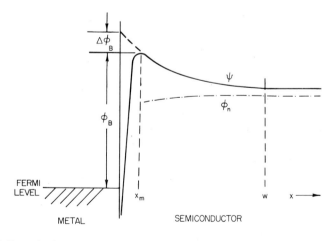

Fig. 7.3. *Energy barrier versus distance, including image force lowering* $\Delta\phi_B$, *in a metal–semiconductor junction.*

schematically as a function of distance in Fig. 7.3. Throughout the region between x_m and w,

$$J = -q\mu n \, d\phi_n/dx \qquad (7.12)$$

where μ is the electron mobility and n the electron density at the point x.

$$n = N_c \exp(\phi_n - \psi)/kT \qquad (7.13)$$

where N_c is the effective density of states in the conduction band and T is the electron temperature. For the region between x_m and w it is assumed that the electron temperature is equal to the lattice temperature. Equations (7.12) and (7.13) will not be applicable between x_m and the metal surface since there the potential energy changes rapidly in distances comparable to the electron mean free path. In this region the distribution of carriers cannot be described thus by an imref nor be associated with an effective density of states. If this portion of the barrier acts as a sink for electrons, however, the current flow may be described in terms of an effective recombination velocity v_R at the potential energy maximum (Crowell and Sze, 1966b,c). Hence

$$J = q(n_m - n_0)v_R \qquad (7.14)$$

where n_m is the electron density at x_m when the current is flowing; n_0 is a quasi-equilibrium electron density at x_m, i.e., the density which would occur if it were possible to reach equilibrium without altering the magnitude or position of the potential energy maximum. It is convenient to measure both ϕ_n and ψ with respect to the Fermi level in the metal. Then

$$\phi_n(w) = V_a$$

$$n_0 = N_c \exp(-q\phi_B/kT)$$

and

$$n_m = N_c \exp[q(\phi_n(x_m) - \phi_B)kT] \qquad (7.15)$$

where ϕ_B is the barrier height and V_a is the applied voltage.

If n is eliminated from Eq. (7.12) and (7.13) and the resulting expression for ϕ_n integrated between x_m and w, then

$$J = \frac{qN_c v_R}{1 + v_R/v_d} \exp\left(-q\frac{\phi_B}{kT}\right)\left(\exp\frac{qV_a}{kT} - 1\right) \qquad (7.16)$$

where

$$v_d = \left[\int_{x_m}^{\omega} \frac{q}{\mu kT} \exp\left(-\frac{q}{kT}(\phi_B - \psi)\right) dx\right]^{-1} \qquad (7.17)$$

is an effective diffusion velocity associated with the transport of electrons from the edge of the depletion layer at w to the potential energy maximum. If the electron distribution is Maxwellian for $x \geq x_m$, and if no electrons return from the metal other than those associated with the current density qn_0v_R, the semiconductor acts as a thermionic emitter.

Then

$$v_R = A^* T^2/qN_c \tag{7.18}$$

where A^* is the effective Richardson constant for the semiconductor surface orientation. Crowell (1965) has determined A^* for a variety of semiconductor tensor effective masses. At 300°K, v_R is 7.0×10^6, 5.2×10^5, and 1.0×10^7 cm sec^{-1} for $\langle 111 \rangle$ oriented n-type Ge, $\langle 111 \rangle$ n-type Si, and $\langle 111 \rangle$ n-type GaAs, respectively. If $v_d \gg v_R$, the factor of the exponential terms in Eq. (7.16) is dominated by v_R and the thermionic theory is the most applicable. If, however, $v_d \ll v_R$, the diffusion process is dominant. If we were to neglect image force effects and if the electron mobility were independent of the electric field, v_d would be equal to $\mu \mathscr{E}$, where $\mathscr{E}$ is the electric field in the semiconductor near the boundary. The standard Schottky result would then be obtained. To include image force effects in the calculation of v_d, the appropriate expression for ψ in Eq. (7.17) is

$$\psi = \phi_B + \Delta\phi_B - \mathscr{E}x - q^2(16\pi\epsilon x) \tag{7.19}$$

where ϵ is the image force permittivity and $\Delta\phi_B$ is the image force lowering of the barrier. This is given by

$$\Delta\phi_B = (q\mathscr{E}/4\pi\epsilon)^{1/2} \tag{7.20}$$

if $\mathscr{E}$ is constant for $x < x_m$. If the electron mobility is constant, and the electric field is constant for $x < x_m(1 + 2kT/q\Delta\phi_B)$, then from Eq. (7.17),

$$v_d \approx \mu\mathscr{E}/\beta \tag{7.21}$$

where

$$\beta \equiv q\frac{\Delta\phi_B}{2kT} \int_0^\infty \exp\left[-q\frac{\Delta\phi_B}{2kT}\frac{\gamma^2}{(1+\gamma)}\right] d\gamma \tag{7.22}$$

and $\gamma \equiv (x - x_m)/x_m$; β is the approximate factor by which v_d is reduced due to the modification of the barrier by the image force (Crowell and Sze, 1966c). When the Schottky lowering is comparable to or less than kT/q, β is virtually unity. This is typically the case for Schottky barriers near room temperature. Even at high fields the effect of the image force should not appreciably affect the current flow because with materials of moderately high mobility the condition $v_d > v_R$ can be satisfied in spite of the fact that β is greater than unity.

In summary, Eq. (7.16) gives a result which is a synthesis of Schottky's diffusion theory and Bethe's thermionic emission theory, and which predicts currents in essential agreement with the thermionic emission theory if $\mu \mathscr{E}(x_m) > v_R$.

7.4 Experimental Characteristics

The current–voltage characteristics of Au–n-type Si and Au–n-type GaAs junctions (Kahng, 1963, 1964) are shown in Figs. 7.4 and 7.5. As the slopes of the experimental curves are usually slightly greater than unity, an empirical factor η is customarily introduced into Eq. (7.5) so that it becomes

$$J = J_0[\exp(qV_a/\eta kT) - 1] \qquad (7.23)$$

where

$$J_0 = AT^2 \exp(-\phi_B/kT) \qquad (7.24)$$

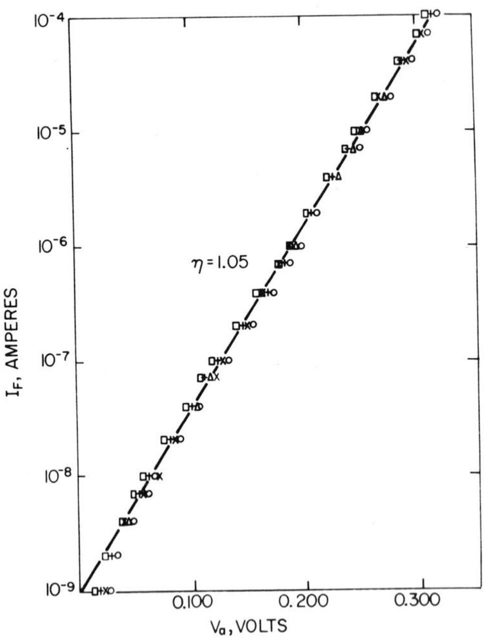

Fig. 7.4. *Forward characteristics for* Au–Si *metal–semiconductor junctions.* Room temperature, n-type Si, 1.1 Ω cm; ($\bigcirc$), ($\times$), ($\triangle$), ($+$), and ($\square$) represent the five diodes. (After Kahng, 1963.)

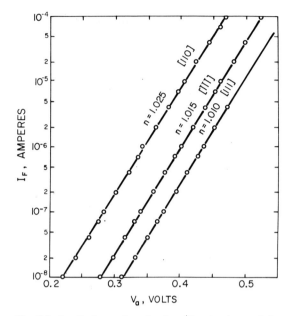

Fig. 7.5. Au–GaAs *metal–semiconductor junction characteristics.*
The current is different for the three crystal orientations, because the barrier height ϕ_B is dependent on orientation. [After Kahng (1964), *Bell Syst. Tech. J.*, **43**, 215, Fig. 1. Copyright 1964, American Telephone and Telegraph Co.; reprinted by permission.]

This form, however, is not consistent with thermodynamics ($J_f = J_r$ at $V_a = 0$) or time reversal invariance. Crowell and Rideout (1969a,b) have shown that the form should be

$$J = J_0\{\exp(qV_a/\eta kT) - \exp[(-1 + 1/\eta)qV_a/kT]\} \qquad (7.23a)$$

For the forward bias condition, Eq. (7.23a) becomes

$$J \approx J_0 \exp(qV_a/\eta kT)$$

whence

$$\ln[J/\exp(qV_a/\eta kT)] \approx \ln J_0 \qquad (7.25)$$

This suggests plotting as shown in Fig. 7.6, from which J_0 may be determined for temperatures at which the measurements were made. An activation energy plot may then be made for J_0 versus $10^3 \, T^{-1}$ from which the barrier height for the diode is inferred to be 0.690 eV and the quasi-Richardson constant, A, is 9.1 A cm^{-2} °K^{-2}. Equation (7.25), as presented, does not include any term for bulk series resistance drop; this should be allowed for.

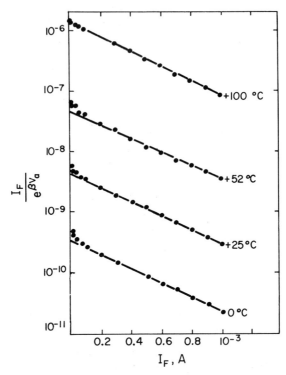

Fig. 7.6. *Forward current curves from which J_0 may be determined for a* Au–Si *diode.* 1.1 Ω cm, $a = 1.8 \times 10^{-3}$ cm², β is $q/\eta kT$ from Eq. (7.25). (After Kahng, 1963.)

(The procedure is only meaningful, of course, if η is close to unity.*) The activation plots for four different resistivities of the *n*-type Si are given in Fig. 7.7. The variation of ϕ_B and A with doping may be related to many problems. One problem, as discussed in Chapter 6, is that a thin oxide layer usually exists at the metal–semiconductor interface. Although perhaps only about 20 Å thick and thin enough for tunneling, it affects barrier heights. It may also be nonuniform in thickness to an extent that influences the effective area of the junction. A second important problem is that surface leakage and perimeter-edge field effects influence the characteristics of Schottky barriers that are not provided with guard rings.

One technique resulting in more nearly ideal characteristics is that of Yu and Mead (1970). This is illustrated in Fig. 7.8 for an Al on *n*–Si junction.

---

* For a few metal–semiconductor junctions (such as Pt on GaP, or Ta on ZnTe) the energy barrier may be greater than $0.5 E_g$. For such a condition, R. Williams [*R.C.A. Rev.* (1969) **30**, 306] has proposed a recombination model for the current flow and this results in a slope η equal to $\phi_B/0.5 E_g$. For Ta on *p*–ZnTe the barrier ϕ_B is 1.51 eV and E_g is 2.26 eV, hence Williams' model predicts $\eta = 1.34$, which is the value observed in experiments.

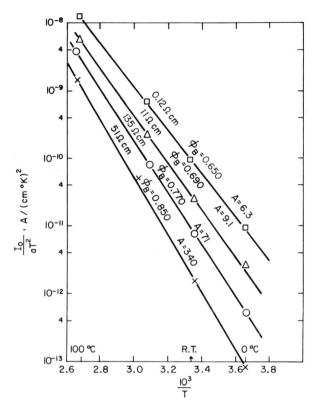

Fig. 7.7. *Activation energy plots of I_0 determined from forward characteristics to obtain ϕ_B and A. a is the diode area. (After Kahng, 1963.)*

The n–n^+ Si (111) wafer is oxidized and provided with windows. Aluminum is then electron-beam evaporated in a diffusion-pump vacuum system. The area defined by subsequent photoresistant masking and etching is such that the Al overlaps the oxide to act as a field plate. When the diode is reverse biased, this field plate keeps the underlying surface in depletion (see the cross-hatched area in the insert sketch in Fig. 7.8) thus reducing soft breakdown due to accumulation at the corner. The junction is then subjected to a brief low-temperature heat treatment, typically in the 400–500°C range. This step, presumably, allows the Al to reduce the thin SiO$_2$ layer at the interface and make intimate contact to the Si.

The I–V characteristic in Fig. 7.8 obeys the theoretical thermionic emission equation over about six decades of current with η very close to unity. The intercept of the forward current at zero applied bias, from Eq. (7.4), gives a barrier height of 0.70 eV if A is taken as the free electron value

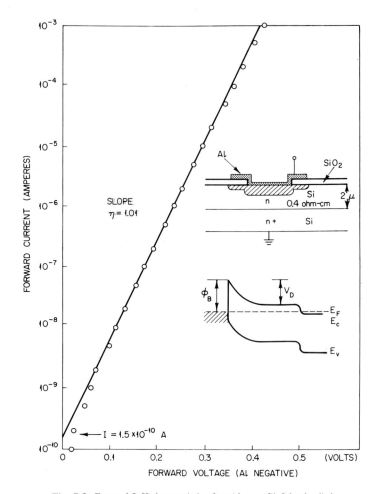

Fig. 7.8. *Forward I–V characteristic of an* Al *on n–*Si *Schottky diode.*
Insert shows the Al overlapping the SiO_2 for control of the depletion at the diode peri-
meter. Area of the diode is 0.85×10^{-5} cm². (After Yu and Mead, 1970.)

of 120 A cm^{-2} °K^{-2}. [The detailed studies of Crowell and Sze (1966c) and
Lepselter and Andrews (1969) suggest that for n-type Si the constant A
should be only slightly less than this in value.] From the activation energy
plot of $\ln(I_F)$ versus $1/T$ at a fixed forward bias, the barrier is found to be
0.69 eV. Photoemission excitation of electrons from the metal into the semiconductor also confirmed this barrier height.

The reverse characteristics of the Al on n–Si junctions made by Yu and
Mead are shown in Fig. 7.9. The breakdown voltages are about 60% of those

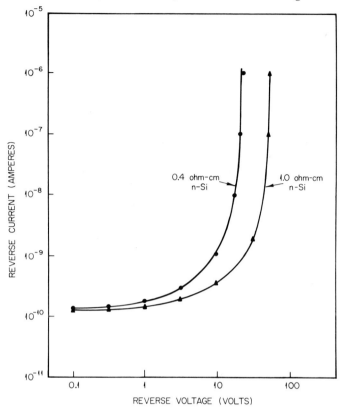

Fig. 7.9. *Reverse characteristics of* Al *on n–Si Schottky diodes constructed with overlap of the* Al *onto* SiO$_2$ *to control the edge field.*
(After Yu and Mead, 1970.)

of a plane abrupt diode and are presumably limited by the depletion region curvature.

7.5 Schottky Diodes with *p-n* Junction Guard Rings

The work of Lepselter and Sze (1968b), Zettler and Cowley (1969a,b), and others, has led to Schottky diodes with nearly ideal characteristics achieved by guard ring structures. The essential features of such a diode are shown in Fig. 7.10(a). The *p*-type diffused guard ring extends in normal planar fashion under the oxide. The Schottky barrier, which is formed in the interior of the ring, is electrically contacting the *p-n* junction so that the depletion region of the total reverse-biased structure has the general shape indicated in Fig. 7.10(b). In this case, the regions of highest electric field will depend upon the depth and profile of the diffused junction. For an abrupt diffused junction of small radius of curvature, the highest fields will be at the periphery of the junction and the breakdown will be lower than that of a plane parallel junction. On the other hand, for an ideal linearly graded junction, the breakdown voltage is higher than that for a plane junction, so that a hybrid diode with a linearly graded *p–n* junction would be expected to have its breakdown limited first by the abrupt junction Schottky barrier. This approach has been taken by Lepselter and Sze and control of the avalanche breakdown voltage and location by tailoring the diffused junction profile has been demonstrated. Hybrid diodes may have voltage ratings in excess of 100 V, compared with a few tens of volts for conventional diodes.

The equivalent circuit of a hybrid structure is shown in Fig. 7.10(c) together with a sketch of the anticipated *I–V* characteristics of the two diode components alone and the total *I–V* characteristic. It is seen that the composite characteristic will be dominated by the *p–n* junction at voltages at

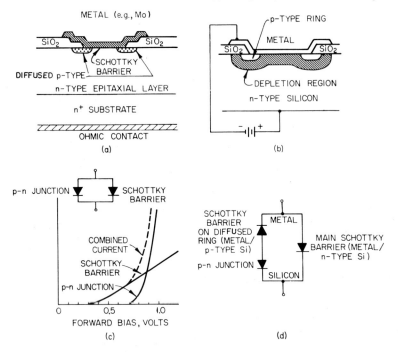

Fig. 7.10. *Schottky diode with p–n junction guard ring.*
(a) Structure for "hybrid" diode. (b) Reverse-biased hybrid diode, showing depletion region shape. (c) Hybrid-diode equivalent circuit, assuming contact between metal and diffused ring is ohmic. Qualitative *I–V* characteristic expected from the structure. (d) Equivalent circuit of hybrid diode, assuming rectifying contact between metal and diffused ring. (After Zettler and Cowley, 1969a.)

which the *p–n* junction injects appreciable minority carrier charge and hence modulates the conductivity of the epitaxial layer. Injection by the *p–n* junction will limit the high-speed switching operation at high-current levels due to minority carrier charge storage. At low levels, the *I–V* characteristics is essentially that of the Schottky barrier and the switching speed will not be limited by storage effects.

The model shown in Fig. 7.10(c) assumes that the metal used to form the Schottky barrier to the *n*-type Si forms an ohmic contact to the diffused *p* region. This is effectively the case for a p^+ diffusion with virtually any metal since the current flow can proceed by a tunneling mechanism. The "ohmic" contact can also exist independently of the diffused region surface concentration with the use of a metal (or silicide) which has a high barrier height on *n*-type Si and therefore a low barrier on *p*-type Si. For instance, the barrier for PtSi is 0.85 eV on *n*–Si and 0.25 eV on *p*–Si (Lepselter and Andrews, 1969). For such conditions there is a very high saturation current for the (reverse polarity) Schottky barrier diode formed on the diffused *p* region.

In Chapter 6 it was shown that the sum of the barrier heights of a given metal on *n*- and *p*-type samples of a given semiconductor equals the band gap of the semiconductor. Hence, a metal having a barrier height of, say, 0.6 eV on *n*-type silicon will, to a good approximation, have a barrier height of about $1.1 - 0.6 = 0.5$ eV on *p*-type Si. Clearly, then, it is possible to form opposite polarity rectifying Schottky barriers over a diffused *p–n* junction provided that the concentration at the surface of the diffused region is not so high as to permit large tunneling currents. The equivalent circuit of Fig. 7.10(c) should then be modified to the three-diode model shown in Fig. 7.10(d). The essential feature of this modification is that the metal *p*-type Schottky barrier in series with the *p–n* junction is reverse biased when the main Schottky barrier and *p–n* junction are forward biased so that most of the applied forward voltage for this branch will appear across the reverse-biased Schottky barrier. Hence, the current which the *p–n* junction can inject is limited by the reverse current of the metal *p*-type Schottky barrier which, in turn, is determined by the effective barrier height of the metal on the diffused ring. Thus, the rectifying contact to the *p* region can be used to eliminate partially or completely the charge storage under heavy forward bias.

Figure 7.11 shows the forward and reverse characteristics of a Mo on *n*–Si Schottky diode with a guard ring construction as in 7.10. The barrier of the Mo on the *n*–Si is 0.68 eV and the barrier for the Mo on the *p*–Si ($N_A = 5 \times 10^{17}$ cm^{-2}) guard ring region is 0.42 eV. With this barrier of 0.42 eV and an area of 10^{-5} cm^2 the leakage current of the guard ring Schottky barrier is only 160 μA. Thus the Mo on *p*–Si guard ring junction prevents the *p–n* junction, Fig. 7.10(d), from contributing significantly to charge storage under forward bias conditions for the main Mo on n–Si Schottky junction.

Since the barrier for Au on p-type Si is about 0.30 eV compared with 0.42 eV for Mo, guard-ring hybrid diodes based on Au are found to exhibit more charge storage than those based on Mo. In a comparative study, diodes

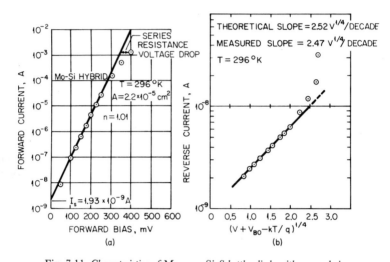

Fig. 7.11. *Characteristics of* Mo *on* n–Si *Schottky diode with a guard ring.*
(a) Forward I–V characteristic. (b) Reverse characteristic of typical Mo–Si hybrid diode, shows essentially exact agreement with image force-lowering theory. (After Zettler and Cowley, 1969a.)

based on Au were found to be storing greater than 100 pC of charge for 15 mA forward-current, whereas Mo based diodes at the same current had only 1.4 pC of stored charge.

The noise theory of Schottky barrier diodes has been discussed by Cowley and Zettler (1968). The hybrid Mo–Si diodes are found to have low f^{-1} noise with corner frequencies (at which the f^{-1} noise value falls below the value for thermal and shot noise) of a few hundred hertz.

Trapping and transient-capacitance effects in Schottky barrier diodes have been studied by Furukawa and Ishibashi (1967a,b), Senechal and Basinski (1968), and Roberts and Crowell (1970).

7.6 Field Effect Barrier-Lowering Studies

As discussed in Section 7.2.2 the J_0 current of a metal–semiconductor junction in a simple model should vary as

$$J_0 = AT^2 \exp[-(\phi_B - \Delta\phi)/kT]$$
$$= AT^2 \exp(-\phi_B/kT) \exp[(q\,\mathscr{E}/4\pi\epsilon\epsilon_0)^{1/2}/kT] \qquad (7.10)$$

where the barrier lowering is $\Delta\phi$ and is related to the maximum field strength

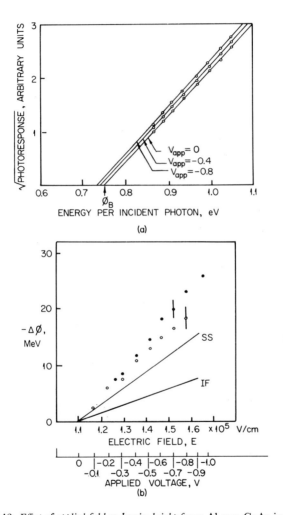

Fig. 7.12. *Effect of applied field on barrier height for an Al on n–GaAs junction.*
(a) Photoresponse versus incident photon energy for various sample bias conditions (sample temperature 300°K). (b) Shift in barrier energy versus maximum electric field in the barrier depletion region. (—) theoretical; (●) experimental, constant $h\nu$; (○) experimental, photocurrent extrapolation. IF is the barrier shift expected from standard image force theory with $\epsilon = 11.5$ for GaAs. (After Parker *et al.*, 1968.)

$\mathscr{E}$ at the junction. Since the field strength is given by

$$\mathscr{E} = [2qN_D(V_D - V_a)/\epsilon\epsilon_0]^{1/2} \qquad (7.11)$$

the reverse current I_R should plot as $\ln(I_R)$ proportional to $(V_D - V_a)^{1/4}$.

Early studies of this relationship tended to show the desired $(V_D - V_a)^{1/4}$ relationship, but the agreement depended on the relative dielectric constant term ϵ being taken as unity. More recently, with the reduction of excess surface leakage, the relationship has been satisfactorily confirmed for silicon with Mo barriers [see Fig. 7.11(b)] and with Al barriers (Yu and Mead, 1970) with ϵ taken as the normal relative dielectric constant value of 12. The problem of barrier lowering with applied field has also been examined by Parker *et al.* (1968) for Al on n–GaAs junctions. These investigators measured the barrier from the photoresponse. As shown in Fig. 7.12(a), the change with applied reverse voltage is observable although fairly small. A second approach was also used, based on the equation for the photocurrent I_p

$$I_p = C(h\nu - \phi_B)^2 \qquad (7.26)$$

The photon energy was held constant and the photocurrent was measured as a function of reverse bias voltage. If it is assumed that C is a constant independent of the bias, then $\Delta\phi$ can be inferred from this experiment. The barrier changes given by both experiments were very similar, as shown by the points in Fig. 7.12(b). The line marked IF corresponds to the barrier lowering expected from Eq. (7.10) with ϵ taken as 11.5 relative to ϵ_0. Correction of the slope by Parker and his co-workers is in terms of an interface state model that results in the line marked SS. Study of this approach is best left to the individual reader.

7.7 Thermionic Tunneling and Other Concepts

Depending on the doping of the semiconductor, and interface fabrication care, the factor η in Eq. (7.23)

$$J = J_0 \exp[(qV_a/\eta kT) - 1]$$

may or may not be independent of temperature. For example, Padovani and Sumner (1965) have observed η as high as 30 at low temperatures for Au–nGaAs junctions. Their data could be fitted by modifying (7.23) with the introduction of a factor T_0 in the temperature, as follows:

$$J = AT^2 \exp[-\phi_B/k(T + T_0)] \exp[(qV_a/k(T + T_0)) - 1] \qquad (7.27)$$

For Au–GaAs the excess temperature T_0 is about 50°K, and for Au–Si T_0 is 35°K (Padovani, 1967). Equation (7.27) may be rearranged as

$$\ln(J/T^2) = \ln A - (\phi_B - qV_a)/k(T + T_0) \tag{7.28}$$

In an attempt to fit the Au–GaAs results to this equation, Padovani (1966) concludes that the effective barrier is a function of temperature

$$\phi_B = \phi_{B0} - \beta_T \Delta T \tag{7.29}$$

where β_T is 2.1×10^{-4} eV °K^{-1}.

From the Mead and Spitzer studies (1964) GaAs has a high density of states that fixes the Fermi level of the metal at about two-thirds of the band gap from the conduction band edge. Since the band gap of GaAs decreases with increasing temperature, at about 4.3×10^{-4} eV °K^{-1}, one might expect the barrier ϕ_B also to decrease with temperature and the β observed by Padovani is not unreasonable.

Let us turn now to various other factors that have been neglected and should be considered. In particular, for fairly heavily doped junctions further progress in the understanding of the characteristics of metal–semiconductor junctions depends on recognizing that electron flow through the barrier by field emission tunneling can be a significant factor where the barrier is thin. The electron flow may take place either from the Fermi level of the semiconductor or over some range of energy levels part way up the barrier. The latter effect has been termed thermionic field emission by Padovani and Stratton (1966a). For forward bias the two effects are as shown in Fig. 7.13(a). For reverse bias the emission is from the metal into the semiconductor, and this is shown in Fig. 7.13(b).

The expression obtained for the current density is

$$J = \frac{A T^2 \pi E_{00} \exp[-2E_B^{3/2}/3E_{00}(E_B - E)^{1/2}]}{kT[E_B/(E_B - E)]^{1/2} \sin\{\pi k T[E_B/(E_B - E)]^{1/2}/E_{00}\}} \tag{7.30}$$

for low temperature conditions in which F-emission dominates. The term E_{00} is

$$E_{00} = q[N_D/2\epsilon]^{1/2}/[(2m^*)^{1/2}/\hbar] \tag{7.31}$$

These equations are not tractable enough for further development here. Padovani and Stratton with such equations have been able to explain the forward and reverse characteristics of Schottky barriers on GaAs doped

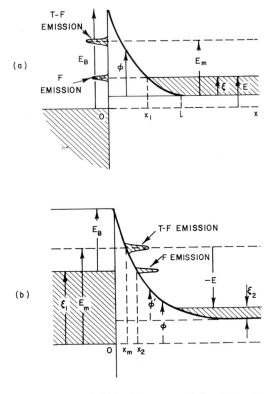

Fig. 7.13. *Field emission and thermionic-field emission currents in heavily-doped metal–semiconductor junctions.*
(a) Energy diagram for forward bias. (b) Energy diagram for reverse bias. (After Padovani and Stratton, 1966a.)

5×10^{16} atoms cm^{-3} and above. Crowell and Rideout (1969a,b) in continued analysis of thermionic-field emission currents have further developed the approach and have been successful in explaining various discrepancies between barrier heights deduced from photothreshold, capacitance–voltage and current–voltage characteristics.

In recent years Schottky barrier studies have been made on a wide range of semiconductors and references to many of these papers may be found by consulting the bibliography through the index provided.

7.8 Minority Carrier Flow in Metal–Semiconductor Junctions

The minority carrier current in a Schottky barrier diode is generally small. Scharfetter (1965), however, has shown that the electric field resulting from the majority carrier current produces a minority carrier current many orders of magnitude above that predicted by diffusion theory. The ratio of the minority carrier current to the total current is defined as the ratio (γ). The current ratio increases with forward current with the relationship

$$\gamma = n_i{}^2 J / b N_D{}^2 J_{ns} \tag{7.32}$$

where n_i and N_D are the intrinsic and doping concentrations, b the mobility ratio, J_{ns} the Schottky diode saturation current density, and J the diode forward current density. As an example, a 5 Ω cm n-type Si–Au diode will have a current ratio of 5% at a current density of 350 A cm^{-2}. The hole flow for the metal n–n^+ diode shown in the energy diagram of Fig. 7.14 is, of course, from left to right and therefore is affected by the energy barrier in the

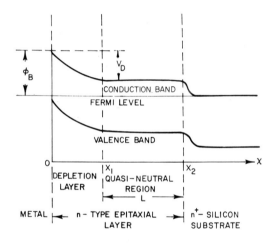

Fig. 7.14. *Energy diagram of a n–n⁺ metal–semiconductor Schottky barrier diode.*
(After Scharfetter, 1965.)

valence band at the n–n^+ interface. The minority carrier stored charge per unit area (Q), for Schottky diodes made on thin epitaxial layers, depends upon the characteristics of the epitaxy–substrate interface and can become very significant when this interface is highly reflecting (i.e., has a low value of surface recombination velocity). For large applied bias and negligible bulk recombination the stored charge is given by

$$Q = q n_i{}^2 D_p J / N_D J_{ns} S \tag{7.33}$$

where q is the electronic charge, D_p the diffusion constant, and S the surface recombination velocity. In measurements on experimental epitaxial diodes the interface was not found to be highly reflecting but was characterized by a recombination velocity of about 2000 cm sec^{-1}. This value applied to the 5 Ω cm silicon–gold diode yields a storage time (Q/J) of about $\frac{1}{3}$ nsec.

Charge storage effects in guard-ring (hybrid) diode structures have been mentioned briefly in Section 7.5, and have been discussed more extensively by Zettler and Cowley (1969a,b).

7.9 Metal–Semiconductor Junction Applications

7.9.1 Mixers and Related Applications

Metal–semiconductor diodes, or hot carrier diodes as they are sometimes called, are widely used as low-level detecting and mixing diodes at microwave frequencies.

In the metal–semiconductor diode expression

$$J = J_0 \exp[(qV_a/\eta kT) - 1]$$

the term η is close to unity, and this is a favorable factor in mixer performance. The burn-out resistance and noise characteristics tend also to be superior to those of point-contact mixer diodes (Anand and Howell, 1968).

Their frequency response is high because they are RC limited rather than minority-carrier-limited diodes. With n–n^+ epitaxy techniques, the barrier capacitance can be low (say <0.1 pF at zero bias on a 10^{17} cm^{-3} n-doped region), and the series resistance low (say <10 Ω) because the n layer is thin and the n^+ region contributes little resistance. The barrier dot may have a radius of 5–10 μ. A typical geometry for incorporation in a stripline configuration is shown in Fig. 7.15. Such diodes are capable of good mixer performance at carrier frequencies of tens of GHz, either as discrete components in waveguides or as balanced-mixers in striplines.

Consider an n–n^+ structure with a barrier of radius, a, which is large compared with the thickness, w, of the epitaxial layer. The series resistance is given by

$$R_s = w\rho/\pi a^2 \propto w/a^2 N_D\mu_n$$

and the capacitance by

$$C_B \propto a^2(N_D\epsilon_r\epsilon_0)^{1/2}$$

where N_D is the doping density of the n-region. Hence the parasitic time constant is

$$R_sC_B \propto w(\epsilon_r\epsilon_0)^{1/2}/N_D^{1/2}\mu_n$$

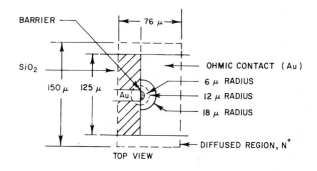

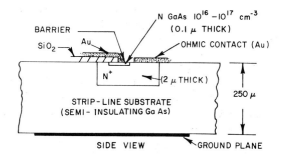

Fig. 7.15. *Schottky barrier mixer diode for incorporation in stripline at* 10 GHz.

For good high-frequency performance the epitaxial layer should be thin, say 1 μ; and the doping density should be chosen to give a large value for the term $N_D^{1/2}\mu_n$ and an acceptable value for the reverse breakdown voltage of the diode. For n-type Ge the chosen N_D value may be 10^{18} carriers cm^{-3} and μ_n may be 1500 cm^2 V^{-1} sec^{-1}; for n–GaAs 2 $\times$ 10^{17} carriers cm^{-3} and 4500 cm^2 V^{-1} sec^{-1}. The conversion loss performance as a function of frequency is then as shown in Fig. 7.16 for n–GaAs, n–Ge, and p–Si Schottky barrier mixer diodes. The advantage of the high mobility of the GaAs is apparent. However, this is only a simplified approach to the problem with many effects neglected. For more information reference should be made to Oxley and Summers (1966) and Leighton (1970), and journals such as the *IEEE Transactions on Microwave Theory and Techniques.*

Metal–semiconductor diodes are also in widespread use for all manner of high-speed low-power switching applications including computer logic uses. The very exact logarithmic relationship that they display over many decades of current has led to their use in logarithmic converter circuits.

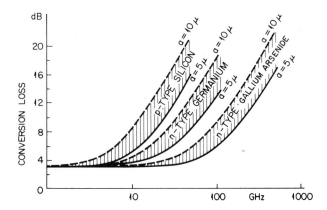

Fig. 7.16. *Ideal mixer conversion loss as a function of carrier frequency for Schottky barrier diodes.* The epitaxial layer is assumed to be 1 μ thick, a is the radius of the diode and the assumed doping levels are 2×10^{17} cm^{-3} for n-type GaAs, 10^{18} carriers cm^{-3} for n-type Ge, and 5×10^{18} carriers cm^{-3} for p-type Si. (After Oxley and Summers, 1966.)

Metal–semiconductor diodes are also used as variable capacitances in parametric circuits for frequency multiplication. References to mixer and parametric uses of metal–semiconductor diodes are found in the papers by Rusch and Burrus (1968), Cowley and Sorensen (1966), and Cerniglia *et al.* (1968).

7.9.2 *Photodiode Applications*

Silicon surface-barrier photocells in which the light passes through a thin (150 Å) gold layer have been studied by Ahlstrom and Gartner (1962).

More recently in the design of photodiodes as high-speed detectors of modulated light, an avalanche mode of operation has been used to give sensitivity and speed. Schematically this is shown for homojunction cells in Fig. 7.17. In Fig. 7.17(a), x_1 and x_2 represent the boundaries of the depletion region. Such structures have been studied by Biard and Shaunfield (1967) and others. Melchior *et al.* (1968) have recently described a Pt–Si Schottky barrier avalanche photodiode, Fig. 7.17(c). The n-type Si was 0.5 Ω cm, and the Pt layer was about 100 Å and formed PtSi of reasonably low sheet resistance. A diffused guard ring was used to reduce edge breakdown effects. The avalanche gain at the He–Ne laser wavelength was 35 and the speed of response was less than 10^{-9} sec.

A Schottky barrier photodiode with Pt on GaAs, and a guard-ring formed by proton-irradiation, has also been described recently (Lindley *et al.*, 1969). Operating at a gain of 100 these photodiodes exhibit gain-bandwidth

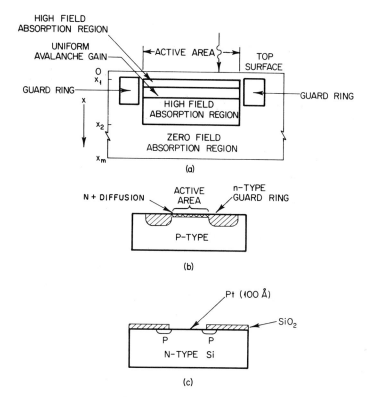

Fig. 7.17. *Avalanche photodiodes.*
(a) Schematic model for homojunction diode. (b) n^+-p Si homojunction diode. (c) Schottky barrier avalanche photodiode (Pt–Si). (After Biard and Shaunfield, 1967.)

products greater than 50 GHz and enhanced signal-to-noise ratio in excess of 30 dB.

In another application of the photoproperties of surface barrier diodes, Yamamoto and Ota (1968) have used the dependence of the capacitance of a Au on n–GaAs diode on illumination for FM light detection and for control of parametric oscillations.

7.9.3 *Transistor Applications of Schottky Barriers*

In saturated switching of conventional $n-p-n$ or $p-n-p$ junction bipolar transistors the turn-off speed is limited by storage-time. The conventional way to reduce storage time is to control minority-lifetime by gold-doping. But gold-doping also decreases the current gain β, which is proportional to the minority lifetime, hence cannot eliminate the problem. G.A. May (1968) has shown that the storage time can be practically eliminated by replacing the conventional $p–n$ collector junction by a Schottky barrier with the metallic side forming the collector region. This eliminates the injection of minority-carriers from the collector into the base region, as well as minority charge storage in the collector region. Theoretically the storage time is approximately β times less than that of the best possible gold-doped transistor with similar geometry and β.

The energy band diagram of a Schottky barrier-collector transistor (SBCT) and a typical experimental structure are shown in Fig. 7.18.

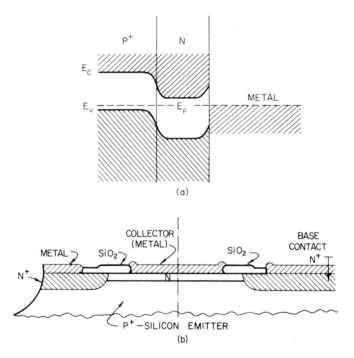

Fig. 7.18. *Schottky barrier-collector transistor.*
(a) Energy diagram for zero bias condition; (b) typical experimental structure. (After G.A. May, 1968.)

The SBCT requires only one $p–n$ junction, hence fabrication is possible with materials otherwise difficult to fabricate into bipolar transistors. For example, an SiC bipolar transistor might be feasible using the structure of Fig. 7.18(b).

The advantages of the SBCT over the conventional bipolar transistor are (a) practically zero inverse β, and therefore reduced storage time, (b) practically zero collector series resistance, (c) zero collector transit time, there being no collector depletion layer (advantage over planar type only), (d) the collector is formed in the normal contacting step, hence simpler fabrication and a wider choice in fabrication methods and materials, (e) no minority-charge lifetime-control processing and associated problems.

The principal disadvantage of this transistor is one shared with alloyed junction transistors, namely for high-frequency transistors where the base-width must be small, the maximum operating collector to emitter voltage is limited simultaneously by base reach-through and collector junction break-down. For a base-width of 0.5 μ the highest possible estimated operating voltage for silicon is approximately 12 V, where the base doping level is adjusted so that reach-through occurs as breakdown voltage is approached.

The SBCT shows promise as a switching element. The fabrication technique can be adapted to integrated circuits readily. Very high-speed logic may be achieved with saturated-switching circuitry using the SBCT.

Another promising application for metal–semiconductor junctions is as the gate of junction field effect transistors, Mead (1966a). The metal gate can in some fabrication processes be used as the mask to provide autoregistration of the source and drain regions with respect to the gate.* Kurtin and Mead (1968b) have also pointed out that the Schottky barrier gate is useful for the construction of field-effect devices since it avoids the difficulties of p–n junction formation in wide band-gap materials, and the Schottky barrier depletion layer is not affected by the presence of surface states. A properly formed Schottky barrier has a low reverse current and therefore a Schottky-barrier gate FET may have an acceptable input impedance, although it will not have the very high input impedance of an MOS structure.

7.9.4 *Metal-Base Transistors*

Transistors with metal bases, in which the carrier transport from emitter to collector is by hot electron flow, are of potential interest for several reasons. The frequency response might be high because of the absence of minority carrier storage effects. Furthermore the base region could be quite thin (a few hundred angstroms) and yet still be of reasonably low resistance, so that the RC time-constant limit might be high. Another reason for interest is that simple thin film procedures of metal evaporation and oxide growth might lead to the fabrication of very low cost active devices.

---

* Schottky barrier field-effect transistors are the subject of a number of papers in the March 1970 issue of the *IBM Journal of Research and Development*.

Most of these advantages have turned out to be difficult to achieve, and thin-film metal-base transistor studies are not very active these days. It is interesting to review why this is so.

Some metal-base transistor structures that have been proposed from time to time are shown in energy diagram form in Fig. 7.19. The triode in 7.19(a) has metal for the emitter and collector and these are separated from the base by thin insulating barriers. The metal might be Al and the insulating barriers Al_2O_3. Injection would be by tunneling through the very thin emitter-base insulator and the hot electrons injected would have the range (hopefully) to cross the metal-base region and be collected over the barrier between the base and collector metal. However, barriers between metals and insulators tend to be inconveniently high [over 1 eV, see for instance Braunstein *et al.* (1966)] and therefore current collection is difficult (Nelson and Anderson, 1966). Attention therefore turned to the use of a metal–semiconductor collector where the barrier heights are more reasonable. The three structures (b), (c), and (d) in Fig. 7.18 all have semiconductor collectors and metal bases, but differ in the emitter. Their relative merits have been studied by Atalla and Soshea (1963). They concluded that the SMS structure was the most promising in terms of gain–bandwidth product.

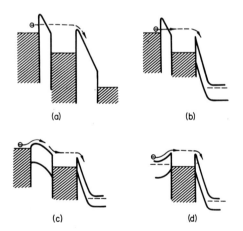

Fig. 7.19. *Hot electron metal-base transistor energy diagrams.*
(a) *m–i–m–i–m* triode, (b) *m–i–m–s* tunnel emitter triode, (c) SCL emitter triode, and (d) semiconductor–metal emitter triode (SMS). (After Atalla and Soshea, 1963.)

For the tunnel-emitter triode Fig. 7.19(b) the insulator thickness between the emitter and the base must be a few tens of Å for useful tunnel current densities. There is considerable difficulty achieving this without pin-hole effects. Indeed it has been suggested that most of the transistor action seen in such structures is associated with direct injection through pin-holes. This kind of injection has been studied by Spratt (1962) and his co-workers, leading them to propose an "edge-effect" transistor in which injection was into a depletion layer. Lavine (1964) has presented a review of depletion-layer and other transistors.

However, considering the SMS transistor once more, it is clear that there are two major problems to be discussed. The first concerns the mean free path of hot electrons in the metal base. The attenuation lengths have been measured by Crowell *et al.* (1962a,b) for Au, and some other metals. For electrons approximately 1 eV in energy above the Fermi level they found mean free paths at room temperature of 740 Å for Au, 440 Å for Ag, 170 Å for Pd, and 200 Å for Cu. Most of the SMS studies have therefore concentrated on Au as the base material and have used thicknesses of a few hundred Å. More recently measurements by Crowell and Sze (1965a–c) of Au films on Ge and Si substrates have given mean free paths of 230 Å for 0.85 eV electrons at room temperature. Therefore the base widths of SMS transistors have to be quite thin, perhaps 100 Å.

Crowell and Sze (1965b,c) have also considered the problem of electron scattering by optical phonons in the emitter and collector barriers of semiconductor–metal–semiconductor structures. The current transfer ratio predicted by their treatment is only 0.68 for Si/Au/Si structures and this neglects base transfer losses and quantum-mechanical reflections at the collector barrier. The quantum reflection problem is treated by Crowell and Sze in another paper (1966b) where they predict the net current transfer ratio to be less than 0.5 exclusive of base transport losses. In experimental studies of SMS structures by Sze *et al.* (1966), Au, Ag, Pd, and Al were used as the base metals; Ge, Si, GaAs, and GaP were used as the emitter semiconductors; and Ge, Si, GaAs, GaP, CdSe, and CdS were used as the collector semiconductors. With about 100 Å base thickness, the highest I_c/I_e

ratios obtained were about 0.3 $\left(\text{for Si} \middle/ \dfrac{\text{Au}}{\text{Ag}} \middle/ \text{Ge and GaP} \middle/ \dfrac{\text{Au}}{\text{Ag}} \middle/ \text{Ge structures} \right)$.

Since the low frequency common-base current gain is so low, power gain if it is to be achieved depends on the attainable impedance ratio. In an appraisal of SMS transistors, Sze and Gummel (1966) conclude that with present technology the impedance ratio appears to be limited to about 100.

For a Si/Au/Ge structure with a collector epitaxial layer 1 μ thick, base-metal thickness of 100 Å, emitter stripe width of 1 μ, and at a current density of 3000 A cm^{-2}, the calculated values at 10 GHz are: 26 dB for the unilateral gain, 29 dB for the maximum stable gain with a resistance ratio of 10^4 and about 10 dB for the transducer gain with a stripe length of 1000 μ and a resistance ratio of 100. No parasitic losses were included in the above figures and optimum adjustments were assumed for reactive components of generator and load immittances.

In conclusion, therefore, the future of SMS thin-film metal base transistors does not look promising because of the low current transfer ratios ($I_c/I_e \sim 0.3$) that seem the best that can be achieved.

Chapter 8 | High Yield Photoemissive Cathodes

8.1 Introduction

Photoemission of electrons from a material is a three stage process involving: (1) absorption of the photon some distance within the bulk of the material with increase of an electron in energy, (2) movement of the excited electron toward the surface of the material, and (3) escape of the electron over the energy barrier at the surface into vacuum. The yield of electrons per incident photon depends upon factors such as the surface reflectivity, the absorption depth, the energy and diffusion length of the electrons with enough energy to pass over the surface energy barrier, the energy loss mechanisms within the escape depth, and the band-bending conditions created by the surface treatment. Performance data for some practical photocathodes are given in Table 8.1.

TABLE 8.1
Properties of Conventional Photocathodes

Cathode	λ_{peak} Å	Maximum quantum efficiency at peak %	Sensitivity (max) $\mu A\ lm^{-1}$	Long wavelength cutoff (1% of peak sensitivity) Å	Dark current A cm^{-2}	Surface type
Ag–O–Cs	8500	1	60	12,000	10^{-12}	S1
Cs$_3$Sb	4000	30	90	6200	10^{-15}	S4/S11
Bi–Ag–O–Cs	4500	10	90	7500	10^{-14}	S10
Na$_2$KSb	3700	30	60	6500	$<10^{-16}$	—
(Cs)Na$_2$KSb	4000	40	230	8500	10^{-16}	S20

The Ag–O–Cs cathode, known as the S1 surface, is seen to be the only material suitable for use in the wavelength range out to 12,000 Å ($\sim$1 eV). However, even at the wavelength for peak sensitivity, the quantum efficiency is only 1% and the sensitivity only about 60 $\mu A\ lm^{-1}$. These values decline rapidly for longer wavelengths. Typically sensitivity values are given relative to incident rather than absorbed radiation. At 5550 Å (2.24 eV), a watt of radiant energy is equivalent to 660 lumens. Hence 1 lumen at this wavelength represents 4.2×10^{15} photons sec^{-1}. If the quantum efficiency were

100% (i.e., 1 electron emitted per incident photon), the current yield would be 680 μA lm^{-1}. The conversion between sensitivity expressed in mA W^{-1} and in electrons per photon depends upon the wavelength as shown in Fig. 8.1. As the sensitivity of the eye decreases for longer wavelengths, more photons are required per lumen and current yields in excess of 680 μA lm^{-1}

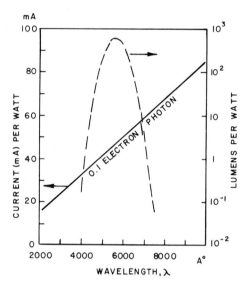

Fig. 8.1. *Relation between sensitivity expressed in* mA W^{-1} *and a quantum yield of* 0.1 *electrons per photon.*
The full line is the current in mA W^{-1} of optical power versus wavelength, for an emission efficiency of 0.1 electron per photon. The dashed line is a curve of lm W^{-1} versus wavelength. It falls off rapidly on either side of 5550 Å, which is the peak of sensitivity of the eye.

become possible without exceeding a quantum yield of 1 electron per photon. (Actually a yield of more than 1 electron per photon contravenes no fundamental laws, since if the photon energy is high enough it can create a high energy electron which in turn impact-ionizes a second electron and both may be emitted. This has been observed only for very high energy ultraviolet photons, as might be expected.)

The properties of conventional photosurfaces have been reviewed by Sommer and Spicer (1965) and by Sommer (1968). The reader is referred to these sources for authoritative discussions on alkali photocathodes.

Electronic processes involved in Ag–O–Cs cathodes have remained somewhat obscure to this day. Cesium has a work function of 1.95 eV and creates lower effective work functions when in monatomic layers on other metals (for instance, 1.54 eV on W). However, as discussed by Sommer (1968), the Ag also plays a role in the performance of Ag–O–Cs surfaces. The performance of the Cs_3Sb photocathode is attributed to the material being a p-type semiconductor with a band gap of 1.6 eV and an electron affinity of $\sim$0.4 eV. Hence the cutoff edge expected is about 2.0 eV: this corresponds to the 6200 Å cutoff wavelength given in Table 8.1.

Studies of Cs_2O showed this to be an n-type semiconductor with a band gap of 2.0 eV and an electron affinity of 0.7–1.0 eV. However, such studies suggested no improvements for Ag–O–Cs surfaces and did not lead to the development of other long wavelength photocathodes. For a considerable number of years therefore, the performance of long wavelength photocathodes remained almost unchanged. Recently, however, the concept of zero or negative work function photocathodes based on III–V semiconductors with Cs or O–Cs coated surfaces has provided a breakthrough to new performance levels. The sensitivity in the region of 1.4 eV has been increased by about two orders of magnitude over that of Ag–O–Cs surfaces, and useful performance (2%) has been obtained at 1.17 eV (1.06 μ).

8.2 Cesiated p^+–GaAs Photocathodes

Although the principal success has been with III–V compounds, the genesis of the technique goes back to studies of the influence of band bending on photoelectric emission from Si single crystals by Scheer (1960) and van Laar and Scheer (1962a,b). In these studies, a coating of Cs was found to create a threshold barrier, E_A, of about 1.4 eV for both p-type and n-type doped Si but the quantum efficiency (Fig. 8.2) was found to be much higher for p-type than for n-type material. The explanation is that the surface of the p–Si becomes n-type and band bending occurs which reduces the effective photoelectric work function to $E_{ph}(p)$ as shown in Fig. 8.3(b). This is much less than the corresponding quantity $E_{ph}(n)$ shown in Fig. 8.3(a).

From an inspection of Fig. 8.3(b), the desirable features of this approach may be inferred. The energy E_A should be as small as the choice of the coating material will allow, and the processing should be such that the band bending is as large as possible. Maximum possible band bending has been obtained if the bottom of the E_A notch approximately lines up with the valence band edge in the bulk. Furthermore the bending should take place in the shortest possible distance so that most of the radiation penetrates to the flat region of the conduction band and the photoinduced electrons have the highest energy. The shorter the band bending distance the smaller the loss of conduction band electrons thermalizing into the notch by scattering processes instead of being emitted. The bulk semiconductor must be heavily doped ($N_A \sim 10^{19}$ cm^{-3}) if the band bending is to occur over a suitably short distance.

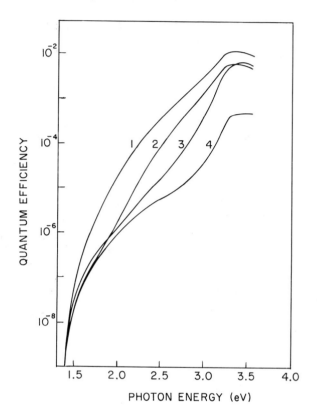

Fig. 8.2. *The influence of bulk doping on the photoemission from* Cs *covered* Si *crystals.*
1. 3×10^{18} B cm^{-3}; 2. 4×10^{17} Ga cm^{-3}; 3. 10^{16} Ga cm^{-3}; 4. 10^{16} P cm^{-3}. (After van Laar and Scheer, 1962a.)

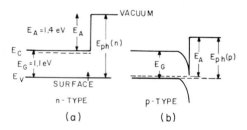

Fig. 8.3. *Energy band models to explain the greater photoemission from p-type* Si(Cs) *than from n-type.*
(After Scheer, 1960.)

From Fig. 8.3(b) the effective barrier for emission of a conduction band electron is $(E_A - E_G)$. Experiments by Scheer and van Laar (1965) showed that E_A was about 1.4 eV for a Cs covered p-type semiconductor such as Si. The quantum efficiency for Si was low because the barrier $E_A - E_G$ (0.3 eV) did not allow direct emission of photoinduced electrons that had thermalized to the conduction band edge. However Scheer and van Laar proposed that the use of a p-type semiconductor with a band gap equal to E_A would allow easy emission of photoinduced conduction band electrons. This was confirmed when they showed that heavily doped p–GaAs ($E_G = 1.4$ eV), vacuum cleaved and coated with Cs, had a quantum efficiency of about 35% at 3.5 eV and a long-wavelength threshold of 9000 Å. The maximum overall sensitivity was about 500 μA lm⁻¹. Quantum efficiency curves versus photon energy and wavelength are shown in Fig. 8.4.

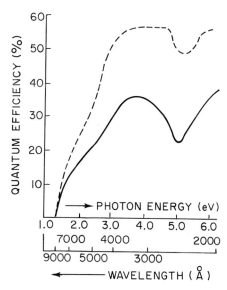

Fig. 8.4. *Spectral distribution of the photo-electric yield for* GaAs–Cs.
The full line curve shows the efficiency in electrons per incident photon and the dashed curve the efficiency per absorbed photon. (After Scheer and van Laar, 1965.)

The generalized energy band diagrams proposed by Scheer and van Laar (1965) are shown in Fig. 8.5. It will be noticed that a thin spike barrier arising from electron affinity differences is included, as in heterojunction band diagrams.

Following the Scheer and van Laar studies, considerable activity developed in the field. This has been reviewed by Bell and Spicer (1970). One objective was to determine procedures by which good yields could be obtained without using vacuum cleaved surfaces, since cleaving is not a convenient manufacturing process. The indications are that these technological problems have been solved. The results of a simple heat-cleaning approach to the

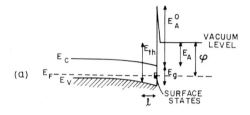

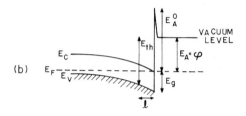

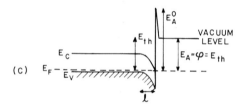

Fig. 8.5. *Band diagrams for a p semiconductor after absorption of electropositive metal atoms.*
(a) Fermi level controlled by surface states; (b) Fermi level not controlled by surface states;
(c) as (b) but heavier p-type doping to give band bending over a shorter distance. $E_A{}^0$ is
the electron affinity of the clean semiconductor surface, E_A the reduced one, φ is the work
function, l is the escape depth, E_{th} is the threshold energy for valence band emission, and
E_g is the band-gap energy. (After Scheer and van Laar, 1965.)

problem are shown in Fig. 8.6. The thermal decomposition of GaAs surfaces
in vacuum occurs at about 670°C (Russell *et al.*, 1966).

Another line of study was the examination of the Cs layer to determine
the influence of O_2. Turnbull and Evans (1968) found that the yield for Cs
on GaAs that had been exposed to the atmosphere was much lower than for
vacuum cleaved GaAs surfaces. However by admitting O_2 to the GaAs(Cs)
system the response could be built up as shown in Fig. 8.7. Yields from the
GaAs–Cs–O structure were usually lower by factors of at least two when the
GaAs was contaminated prior to the Cs–O coating. However, the work
function of Cs–O surface was not greater than 1.4 eV, the value for a clean
GaAs(Cs) structure. The lower yield was attributed to the presence of a

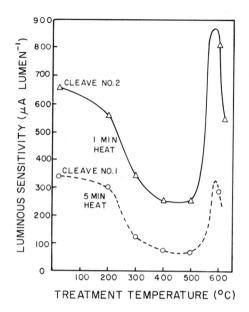

Fig. 8.6. *Effect of the temperature of the heat treatment on the room temperature luminous sensitivity of cleaved p⁺–GaAs, recoated with Cs₂O after each heat treatment.*
(After Bell and Spicer, 1970.)

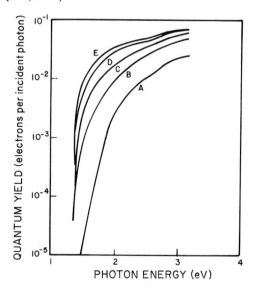

Fig. 8.7. *Spectral response from contaminated* GaAs.
A. GaAs–Cs; B–E. GaAs–Cs–O with progressive increase in Cs–O (B, minimum thickness of Cs–O; E, maximum thickness). (After Turnbull and Evans, 1968.)

barrier at the interface of the GaAs and the Cs–O, rather than to electron losses in diffusing through the Cs–O layer.

In any photocathode structure the simplest model for the probability $P_1(x)$ of an electron created at distance x inside the semiconductor reaching the band bending edge of the junction is given by the diffusion equation

$$P_1(x) = \exp(-x/L) \qquad (8.1)$$

where L is the diffusion length. The number of pairs produced at a depth x per unit time is

$$dn/dt = \alpha(h\nu)I_0 \exp[-\alpha(h\nu)x]\,dx, \qquad (8.2)$$

where $\alpha(h\nu)$ is the absorption coefficient of the semiconductor for a photon of energy $h\nu$ and I_0 is the intensity of the incident light in units of photons sec^{-1}. Under steady state conditions, the number of electrons reaching the junction will be given by the product of Eqs. (8.1) and (8.2) integrated from zero to infinity. The field-induced photoemissive yield will be this quantity multiplied by a factor P_e, the emission efficiency. This is the probability of escape for an electron which reaches the junction. The expression for the photoelectric yield, Y, in electrons per absorbed photon is

$$Y = P_e\,\alpha[h\nu]/(\alpha(h\nu) + (1/L)). \qquad (8.3)$$

This expression appears to have been used first by Simon and Spicer (1960). Eden *et al.* (1967) found that this equation fitted the Scheer and van Laar data if the probability of escape P_e was taken as 0.6 and the diffusion length as $\sim$1500 Å. These workers also found that the photoexcited electrons tend to thermalize in either the Γ or X conduction band minima while diffusing to the surface. This becomes clear from a study of the band structure and of the energy distributions of the photoemitted electrons (Figs. 8.8a and 8.8b).

Further examination of this approach was made by James *et al.* (1968) for GaAs–Cs–O structures. Between the threshold energies 1.4 eV and 2.3 eV almost all the emitted electrons were thermalized into either the Γ or X minima, but for higher photon energies a significant number of unthermalized electrons was seen. Also the absorption coefficient α was so short for photon energies above 2.3 eV that excitation in the band-bending region presented a further complication. Considering only the electrons generated in the bulk region, the photoelectric quantum yields for each minimum are found to be

$$Y_X = \frac{P_X J_X}{qI(1-R)} = \frac{P_X F_X}{1 + (1/\alpha L_X)} \qquad (8.4)$$

$$Y_\Gamma = \frac{P_\Gamma J_\Gamma}{qI(1-R)} = \frac{P_\Gamma}{1 + (1/\alpha L_\Gamma)}\left[F_\Gamma + \frac{F_X}{1 + \alpha L_X}\right] \qquad (8.5)$$

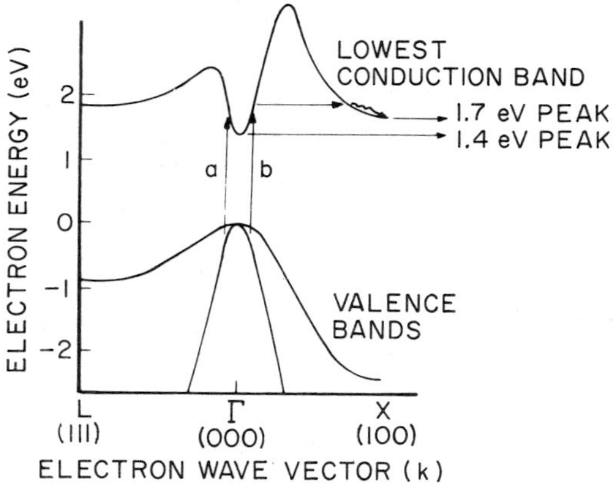

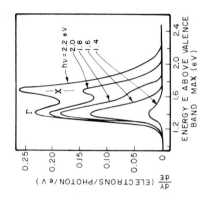

Fig. 8.8b. *Energy distributions for photoemitted electrons GaAs(Cs) for photon energies between 1.4 and 2.2 eV.* The distributions have been normalized to quantum yield. (After Eden *et al.*, 1967.)

Fig. 8.8a. *Schematic of the GaAs band structure near the energy gap showing the relevant excitation–escape processes.*

For photon energies below 1.7 eV, the electrons thermalize in the Γ minimum before escaping into vacuum, while for larger photon energies an increasing number "thermalize" in the X minima.

where F_X is the fraction of the electrons which is excited above 1.7 eV and scatters into the X valley. This can be calculated from the band structure of GaAs and is shown in Fig. 8.9. Thus the unknowns in Eqs. (8.4) and (8.5) are the diffusion lengths L_X and L_Γ and the escape probabilities P_X and P_Γ. The diffusion lengths determine the shape of the yield curves and the escape probabilities determine the magnitude. Therefore L_X and P_X may be determined from the experimental X yield curve and L_Γ and P_Γ from the Γ yield

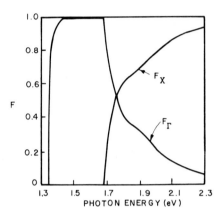

Fig. 8.9. *Fraction of photoexcited electrons which thermalize in each minimum, calculated from the* GaAs *band structure.* (After James et al., 1968.)

curve. The value of L_X, which characterizes the optical phonon scattering from X to Γ, was inferred to be 0.03 ± 0.01 μ at room temperature and was independent of doping. The diffusion length L_Γ was dependent on doping and ranged from 1.6 μ at 1×10^{19} Zn atoms cm^{-3} to 1.0 μ at 4×10^{19} cm^{-3}. The longer the diffusion length the better, particularly for photon energies near the low energy threshold where the absorption depth tends to be long. This suggests the use of the 1×10^{19} cm^{-3} doped GaAs. However the Cs layer work function is 1.4 eV which is not much different from the band gap of GaAs, and so the Fermi level should be kept as low as possible to reduce the effective barrier to photoelectrons. Therefore the yield with 4×10^{19} Zn cm^{-3} is better than the yield with a doping of 1×10^{19} cm^{-3}. The addition of O_2 to a Cs layer on vacuum cleaved GaAs however, changes the situation since it reduces the work function as shown in Fig. 8.10, although as the layer is made thicker the probability of passing through the layer becomes less. Each layer was of monolayer order of thickness but the exact value was unmeasured. For the Γ process and a doping of 1×10^{19} cm^{-3}, a layer thickness of six O–Cs treatments provided the best sensitivity for photons near the threshold energy, as shown in Fig. 8.11a. [Sommer et al. (1970) have noted that Cs_2O is denser than Cs and therefore several O–Cs treatments may be necessary to form a uniform monolayer of Cs_2O.] The performance of the optimized GaAs(O–Cs) structure is shown in Fig. 8.11b. The

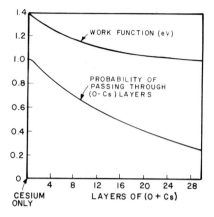

Fig. 8.10. *Effects of additional* O–Cs *layers, showing the work function lowering and the electron absorption as measured experimentally.*
(After James *et al.,* 1968.)

maximum sensitivity is 1000 μA lm^{-1}. At 1.5 eV the sensitivity is almost two orders of magnitude better than an S1 photocathode. The stability of the O–Cs surface was better than that of the Cs surface which deteriorated 15% in two weeks at 10^{-11} Torr at room temperature. Heating the GaAs (Cs) photocathode to 75°C caused loss of Cs from the surface and drastically reduced photosensitivity.

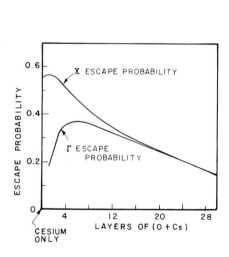

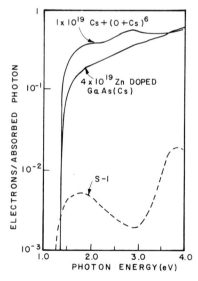

Fig. 8.11a. *Escape probability curves for* X *and* Γ *electrons in GaAs, inferred from the energy distribution of emitted electrons.*
(After James *et al.,* 1968.)

Fig. 8.11b. *Absolute quantum yield curves.*
Shown for optimum Cs only treatment [4 × 10^{19} Zn-doped GaAs(Cs)], optimum O–Cs treatment [1 × 10^{19} GaAs, Cs + (O + Cs)^6], and for comparison, a commercial S1 photocathode. (After James *et al.,* 1968.)

Uebbing (1970) has studied the effect of C contamination on the photo-yield of GaAs(Cs–O) surfaces by Auger electron spectroscopy. This shows that a monolayer of C is sufficient to reduce the yield to zero (probably by preventing the proper sticking of the Cs–O layer). Similar studies are in process for other contaminants.

The studies of James *et al.* (1968) were with vacuum cleaved, Zn-doped GaAs and the electron diffusion length was 1.6 μ for a doping of 1×10^{19} Zn cm^{-3}. Recently Garbe and Frank (1970) investigated the photoemission from Si-doped p-type GaAs layers obtained by liquid-phase epitaxy on (110) or (100) substrates. The carrier concentrations of the samples were between 1.7×10^{19} cm^{-3} and 6.6×10^{17} cm^{-3}. They were heat cleaned and activated by Cs,O or CsF,Cs layers. White light sensitivities between 1120 and 545 μA lm^{-1} were obtained. The electron diffusion lengths were greater than 3 μ (Si as a dopant is normally found to give better diffusion lengths in GaAs than Zn). The quantum yield at 1.4 eV was 20% which is several times better than reported for (Zn)GaAs–Cs,O studies.

The sources of thermionic emission (dark current) from the GaAs(O–Cs) type of photocathode have been considered by Bell (1969). The components of this dark current are:

(a) electrons thermally excited by the phonon field in the bulk of the p^+ base, diffusing to the edge of the depletion layer, the "diffusion component";
(b) electrons generated in the depletion layer itself via trapping centers, the "generation component";
(c) electrons generated via surface traps, the "surface state component"; and
(d) electrons excited in the bulk by the ambient infrared radiation at the operating temperature.

The generation components (b) and (c) tend to be dominant and are estimated (on a worst case basis) to be $3–7 \times 10^{-15}$ A cm^{-2}. This may be compared with the usual values of 10^{-12} A cm^{-2} for the S1 (Ag–O–Cs) surface or 10^{-16} A cm^{-2} for the S20 (multialkali) surface.

8.3 Barrier Studies of III–V Photocathodes to Extend the Infrared Response

Since Cs has a work function of about 1.95 eV, the fact that very thin layers of this element on other metals produce work functions of as low as 1.5 eV needs comment. The accepted explanation involves dipole action and is as follows.

The ionization potential of valence electrons in Cs is low (3.9 eV). Cesium atoms on a metallic surface tend therefore to give up valence electrons and adhere as ions. Thus a monolayer of Cs which is partially ionized acts as a sheet of electric dipoles which are polarized in such a direction as to accelerate electrons out of the metal. The effective work function therefore is lowered. The values observed for metals such as Ag, Cu, and W are of the order 1.5–1.6 eV. If the Cs coating becomes much in excess of a monolayer the work function observed begins to approach the value characteristic of pure Cs, as expected.

When Cs layers are applied to a semiconductor the results observed depend on the semiconductor and on the layer thickness. On Si, a vacuum cleaved (111) surface initially p-type becomes intrinsic and subsequently n-type with succeeding fractional monolayer depositions of Cs (Allen and Gobeli, 1966). On cleaved GaAs surfaces (Zn doped 5.5×10^{16} cm^{-3}) a heavy layer of Cs gives a Schottky barrier, 0.63 eV, at the metal–semiconductor interface (Uebbing and Bell, 1967). On heavily doped p–GaAs however, the effect of about a monolayer of Cs is to cause considerable band bending and E_A in Fig. 8.5(c) becomes 1.4 eV. Therefore, as shown in Fig. 8.12(a), the effective electron affinity for the p–GaAs–Cs structure is zero.

If the Cs layer is interacted with O_2, Cs_2O is formed. This is an n-type semiconductor with a band gap of 2.0 eV and an electron affinity of 0.6 eV or some value thereabouts. The donor level in Cs_2O is about 0.5 eV below the conduction band edge. Sonnenberg (1969a,b) therefore has postulated a heterojunction-type band diagram for GaAs–Cs_2O of the form shown in Fig. 8.12(b). It is seen that this represents a negative electron affinity of about -0.5 eV if we neglect the energy spike. Since the total thickness of the Cs_2O layer is less than 40 Å, one may question the validity of representing the Cs_2O by a band diagram (Sommer *et al.*, 1970). Layers thicker than a few tens of Å are not desirable, of course, because of the increased probability of hot electron energy loss by scattering in passing through the Cs_2O. However the heterojunction approach may be a constructive interim concept, even if not valid, if it suggests questions that otherwise would not be asked. For instance, one question raised is whether there is evidence for a barrier to electron flow such as the $\Delta\chi$ barrier of Fig. 8.12(b). The answer appears to be in the affirmative, at least for some structures involving Cs–O layers.

In a study of Cs_2O layers on GaSb, James and Uebbing (1970) conclude that the work function of the layer may be as low as 0.7 eV. Therefore with p–GaSb the long-wavelength threshold for high-efficiency photoemission should be 1.8 μ. However the yield threshold is found to correspond to an interfacial barrier of 1.23 eV (1.0 μ). The threshold was obtained from a Fowler type plot by extrapolating the straight-line portion of a (yield)$^{1/2}$ versus photon energy line to zero yield. The work function was determined

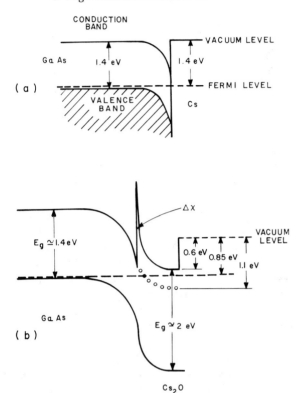

Fig. 8.12. *Energy diagrams postulated for* GaAs(Cs) *and* GaAs(Cs$_2$O).
(a) An effective electron affinity of zero for GaAs(Cs), (b) heterojunction band diagram suggested by Sonnenberg for GaAs–Cs$_2$O. (After Sonnenberg, 1969b.)

by a Kelvin vibrating probe type measurement. Figure 8.13 shows the effect of increasing the Cs$_2$O thickness by successive O–Cs treatments. This shows that for very thin layers the yield threshold and the work function are the same, but that for thicker layers the threshold tends to settle at 1.23 eV.

The effect of this kind of interfacial barrier is also seen from photoemission measurements for GaAs$_{1-x}$Sb$_x$ samples of various band-gap energies. The best quantum yield observed for 1.17 eV photons (corresponding to 1.06 μ Nd-laser wavelength) was from a specimen with a band gap of 1.20 eV. For semiconductors with band gaps less than the effective heterojunction barrier, the photoemission must be by hot electrons going over the barrier or by thermalized electrons tunneling through it. Neither of these processes leads to efficient photoemission. Therefore reducing the material band gap below the interfacial barrier reduces the escape probability by a large amount and the yield decreases in spite of increases in the optical absorption coefficient.

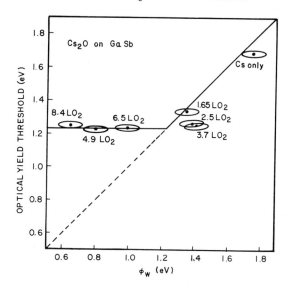

Fig. 8.13. *Comparison of the work function (Kelvin) and the photoyield threshold for* Cs_2O *on a vacuum-cleaved single crystal of* p^{++}*–GaSb.*

The Cs_2O thickness is indicated for each point by the total oxygen exposure in Langmuirs (1 Langmuir = 10^{-6} Torr sec^{-1}). One Langmuir of O_2 exposure would correspond to approximately 7 Å of Cs_2O for a unity O_2 sticking coefficient. (After James and Uebbing, 1970.)

It appears that the work function of Cs–O coated surfaces can be reduced to about 0.6–0.7 eV somewhat independently of the III–V material on which the Cs–O layers are applied. The interfacial barrier height has been measured for a number of semiconductor substrate materials. For heat cleaned, liquid epitaxial grown samples of GaAsSb, InGaAs, and InAsP the thresholds are 1.21, 1.20, and 1.17 eV, respectively. The efficiency near threshold may be expected to depend on the band structure and James *et al.* (1970) find InAsP to be most efficient, in agreement with calculations of the Γ electron escape probability. The results obtained with InAsP–Cs_2O cathodes include: sensitivities of 1200 μA lm^{-1}; 100 μA with a lumen source through a 2540 infrared filter (compared with 24 μA for GaAsSb); quantum yields larger than 10% per incident photon for all photon energies above 1.27 eV and a quantum yield of 2% per incident photon at 1.17 eV (1.06 μ).

Cs–O activation of p–Si of resistivity 0.005–0.01 Ω-cm has been examined by Martinelli (1970). The threshold was about 1.1 eV photon energy for a yield of 10^{-4} electrons per absorbed photon, and for photons of energy above 1.5 eV the yield was greater than 10^{-1}. A condition of effective negative electron affinity was therefore achieved, which is an advance on earlier results for Cs-coated Si. The escape depth for thermal photoelectrons was estimated to be about 5.5 μ. From Auger measurements of (100)Si during treatment by Cs and O_2 it appears that the O_2 penetrates the Cs and goes to the Si interface. The merits of Si negative-affinity photocathodes relative to III–V compound photocathodes need evaluation in specific applications.

Much of the high-yield photocathode work reported here has been concerned with reflection type performance; however thin film, transmission-type systems are under study.

Other investigations of III–V photoemitters include:

InGaAs–Cs–O, Uebbing and Bell (1968), B. F. Williams (1969), Klein (1969);
InP–Cs–O, Bell and Uebbing (1968);
$GaAs_{1-x}P_x(Cs)$, Simon *et al.* (1969);
InAsP–Cs–O, Sonnenberg (1970);
GaAsSb–Cs–O, Antypas and James (1970). The concept of field aided photoemission in a graded band-gap sample was examined in this GaAsSb work.

A drawing of the kind of apparatus that is desirable for emission studies is given in Fig. 8.14. The quantum yield spectrum measured for an $InAs_{0.15}P_{0.85}-Cs_2O$ structure (Sonnenberg, 1970) is shown in Fig. 8.15. The quantum efficiency obtained at 1.06 μ represents more than an order of magnitude improvement in yield over S1 photosurfaces. Yields of 2% have now been observed for InAsP–Cs$_2$O at this wavelength.

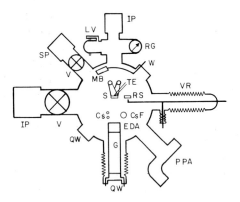

Fig. 8.14. *Ultrahigh vacuum chamber for photoemission experiments.*
S, sample with sample holder for operation between liquid nitrogen and room temperatures; TE, thermocouple; RS, reference sample; Cs, Cs$_2$CrO$_4$ channels; CsF, quartz crucible for evaporation of Cs halides; MB, quartz crystal microbalance; EDA, energy distribution analyzer; G, Cs ion gun; W, window; QW, quartz window; PPA, mass spectrometer; VR, vibrating reed for work function measurements; RG, Redhead gauge; V, valve; LV, leak valve; IP, ion pump; SP, sorption pump. (After Klein, 1969.)

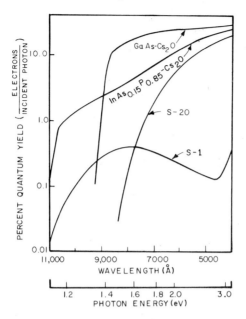

Fig. 8.15. *Spectral response of* InAs$_{0.15}$P$_{0.85}$–Cs$_2$O *compared to* GaAs–Cs$_2$O, S20 *and* S1 *responses.*
(After Sonnenberg, 1970.)

This is a rapidly developing field and further improvements in longwavelength response, possibly to 1.7 μ (~0.7 eV), may be hoped for as more is learned about the factors controlling the barriers.

One approach that may be worth study is to retain the proven GaAs(Cs) or GaAs(Cs–O) negative electron affinity structure and to achieve sensitivity at photon energies below 1.4 eV by the addition of a p–Ge–ZnSe junction. The energy band diagram expected (without external bias applied) would then be as in Fig. 8.16(a). The external bias proposed is as in Fig. 8.16(b) and the resulting band diagram is Fig. 8.16(c). Notice that the electron photoexcited in the Ge has no energy barriers to its movement through the ZnSe and the GaAs conduction bands. The valence band barrier between the ZnSe and the GaAs is substantial even with J_{23} forward biased and therefore there should be no hole injection from the p^+–GaAs into the ZnSe that might represent undesirable current flow. (Undesirable, because it might reduce the electron lifetime in the ZnSe and because it would produce voltage drops and heating.) In principle, if interface-recombination effects are small, this heterostructure may be expected to have efficient emission of electrons with a photoexcitation threshold down to 0.7 eV, the band gap of the Ge. Similar structures of other semiconductors with good lattice and electron

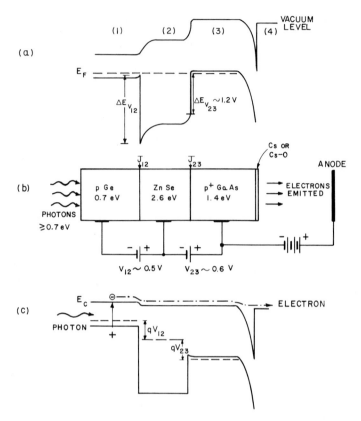

Fig. 8.16. Cs(O)/p^+–GaAs/νZnSe/p–Ge *electron-emitting photocathode: sensitivity to* 0.7 eV (1.75 μ) *expected.*
(a) Energy diagram without bias voltages; (b) bias circuit for operation of the structure; (c) energy diagram with bias voltages applied.

affinity match conditions may be envisaged, and these also should be examined.

8.4 Secondary Emitters for Photomultipliers

The low work-function conditions that have been discussed for photo-emission are also of interest with respect to the emission of electrons excited by exposure of the surface to bombardment by high-energy electrons. The principal concepts of secondary-electron emission have been reviewed by Dekker (1957, 1958).

The secondary-emission yield in a simple model may be written as

$$\delta = \int(-\epsilon^{-1}\,dE/dx)f(x)\,dx \tag{8.6}$$

where E is the energy remaining in a primary electron at a depth x from the surface and ϵ is the energy required to produce a secondary excited electron in the material. The function $f(x)$ is the probability that an excited electron produced at x reaches the surface and is emitted. This probability may be written as

$$f(x) = B_1 B_2 \exp(-x/L) \tag{8.7}$$

where the coefficient B_1 is the fraction of the excited electrons that diffuse towards the surface, B_2 is the probability that escape occurs when an electron reaches the surface, and L is the mean free path (escape depth) for recombination of the excited electrons while diffusing towards the surface. If the energy loss per unit path length is constant for a primary electron with a penetration depth R, then

$$\delta = (B_1 B_2/\epsilon)(E_0 L/R)[1 - \exp(-R/L)] \tag{8.8}$$

where E_0 is the initial energy of the primary electron.

This model results in a yield curve that increases with primary energy up to a certain value of E_0 and then declines somewhat, since the secondary electrons created by very deeply penetrating primaries are unable to reach the surface. The model is in reasonable agreement with experimental results, as shown in Fig. 8.17. For metals the yield tends to be low because the escape length is small ($\sim$30 Å) and only a small fraction of the excited electrons have energies sufficient to cross over the energy barrier into vacuum. The mean free path of an electron 3 eV above the Fermi level may be less than 100 Å and energy loss occurs by both electron–electron and electron–photon collisions.

For a semiconductor such as Si, a mean free path of 60 Å has been inferred before an optical phonon-producing collision results in an energy loss of 0.06 eV (Lee *et al.*, 1964). Therefore about 17 collisions are required for the electron to lose 1 eV of energy and so it travels a distance of about 1000 Å (17×60) in a random-walk fashion in a time of about 10^{-12} sec. Hence from a simple three-dimensional random-walk formula, if N is the number of collisions and λ is the mean free path between collisions, the net distance traveled is

$$l = (N/3)^{1/2}\lambda \tag{8.9}$$

This has the value 140 Å for the example considered above (Simon and Williams, 1968).

For GaP the mean free path is 25 Å and the energy loss per collision is
0.05 (B. F. Williams and Simon, 1967), and so for 1 eV loss the value of l
from Eq. (8.9) is about 65 Å. The application of Cs to p–GaP creates band
bending and a negative electron affinity condition as indicated in Fig. 8.18(a).
If the p doping level is high, the width of the bent-band region is small and
so the electrons on their way to the surface make few collisions. Thus electrons
which have been excited and then thermalized into the conduction band
are able to escape unless they recombine with holes. Such recombination
depends on the electron lifetime or diffusion length. Even for high p-type
doping it is possible to achieve lifetimes of 10^{-10} sec and diffusion lengths of
1 μ or more in III–V semiconductors. Therefore thermalized electrons can
diffuse for times that are two orders of magnitude longer than for conventional secondary emitters and still be emitted if the effective electron affinity
barrier for conduction band electrons is zero or negative. For p–GaP(Cs),
Simon and Williams (1968) have reported secondary-emission yields of
130 and predicted that values of 250 should be obtainable, as shown in
Fig. 8.18(b). This is an order of magnitude better than the values for MgO,
the highest gain secondary emitter previously known.

Typical photomultiplier arrangements are shown in Fig. 8.19. In the
circular electrostatic multiplier tube shown in Fig. 8.19(a), the light falls on
the front face of the photocathode and the electrons emitted are multiplied

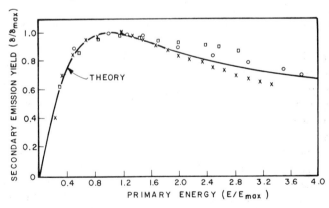

Fig. 8.17. *Theoretical curve of secondary-emission yield as a function of primary energy in normalized*
coordinates.
Also shown are experimental data for Pt, Ge, and MgO. (After Simon and Williams, 1968.)

	δ_{max}	E_{max}
($\times$) MgO	24.0	1200 V
(O) Ge	1.15	400
($\square$) Pt	1.8	700

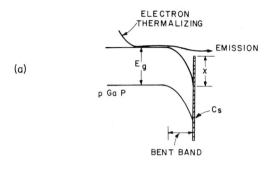

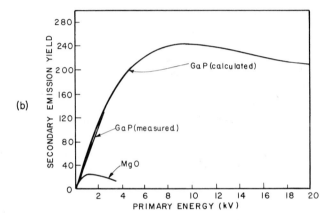

Fig. 8.18. *Secondary emission from p–GaP(Cs).*
(a) Band diagram, showing negative effective electron affinity; (b) yield as a function of primary electron energy. The GaP (calculated) curve is derived from Eq. (8.8) and $R = (1.15/\rho \times 10^{-6} E_0^{1.35}$, where $B_1 B_2 = 5$, $\epsilon = 8.7$ eV, $L = 2000$ Å, and $\rho = 5.35$ g cm^{-3}. (After Simon and Williams, 1968.)

by nine dynode stages. In the conventional widely used 1P21 phototube, the photocathode is of the S4 class and the sensitivity is about 40 μA lm^{-1}. The overall gain for the nine dynode stages is 2×10^6 corresponding to an average gain of about five per stage. Conventional dynode materials are Cs_3Sb, Mg–Ag, or Be–Cu. Figure 8.19(b) shows a partition-type electron multiplier in which the photocathode is transparent and mounted in the end of the tube. It is screened from the secondary-emission dynode electrodes by a partition and aperture that provides convenient separation during activation. Several other electron multiplier structures are known such as Venetian screen, box-type, cross-field, and diode arrangements (Zworykin and Ramberg, 1949; Morton, 1949, 1956; Leverenz and Gaddy, 1970; Fertin *et al.*, 1968).

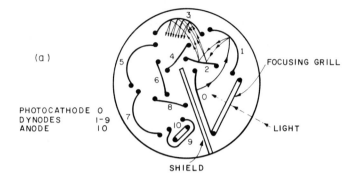

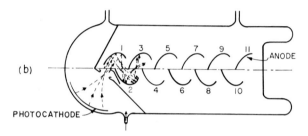

Fig. 8.19. *Typical photomultiplier electrode arrangements.*
(a) Circular configuration of electrodes and a nontransparent photocathode, (b) linear configuration of the dynode section, with a partitioned transparent photocathode.

Since the gain per stage is only of the order of 5 for conventional secondary emitters, the statistical fluctuations in the number of secondary electrons emitted by the first dynode limits the tube performance. To provide discrimination between signals representing the emission of one or of two photoelectrons it is necessary that the first dynode have a gain greater than 15 to 20. Even higher gains are needed to distinguish between n and $(n + 1)$ photoelectrons when n is greater than 1. For a GaP(Cs) first dynode at 600 V, Simon *et al.* (1968) have observed gains of between 20 and 40. With these tubes, Morton *et al.* (1968) have been able to achieve outstanding pulse-height resolution capabilities and have distinguished the emission of one, two, and up to five photoelectrons. The pulse-height resolution, expressed as the fractional full width at half maximum (FWHM), is a function of the dynode gains. If $\bar{N}_e$ photoelectrons leave the photocathode as the result of a series of scintillations, then

$$\text{FWHM}_{(N_e)} = 2.35 \, (\bar{N}_e)^{-1/2} \{1 + \Delta^{-1}[c\delta/(\delta - 1)]\}^{1/2} \qquad (8.10)$$

where Δ is the gain of the first dynode, δ is the gain of the remaining dynodes and c is an empirical coefficient of about 1.6.

The pulse-height spectrum of some very weak light flashes is given in Fig. 8.20 and shows peaks corresponding to one, two, and up to five electrons.

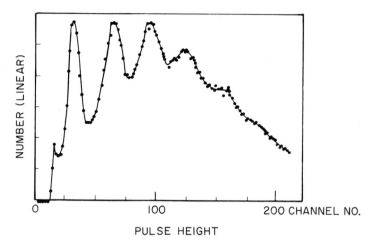

Fig. 8.20. *Pulse-height spectrum showing peaks corresponding to one, two, and up to five electrons.* (After Morton *et al.*, 1968.)

The observed FWHM for a single electron peak is 0.63. Resolution of this order has not been achievable previously and this new technology should have many applications in the detection and measurement of very low light-level scintillations (for example in tritium counting, carbon counting, and Cerenkov-type counters).

8.5 Cold-Cathode Electron Emitters

An efficient room-temperature (or near room-temperature) cathode, capable of emitting electrons for electron beam applications on the application of a voltage to a junction, is a device objective that has been under study for many years with only limited success until recently.

One approach that has been examined is the emission of hot electrons from reverse-biased shallow p–n junctions in Si (Bartelink *et al.*, 1963). The energy band diagram for such a structure is shown in Fig. 8.21 with reverse-bias voltage V_{app} applied. The n-type region through which photoexcited hot electrons must pass is ~1000 Å in thickness. However the mean free path for optical-photon emission in Si is only about 60 Å and for impact ionization about 190 Å. Therefore the efficiency of emission is very low (~10^{-6} electrons per photon).

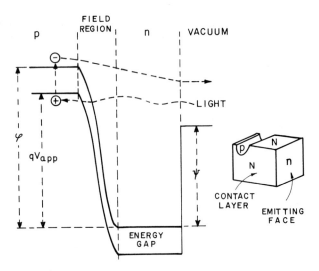

Fig. 8.21. *Energy band diagram and experimental* Si *electron emitter.*
Two modes are possible: (1) carriers created by light can be emitted from the *p*-type
conduction band through the field and the thin *n* region if they retain sufficient energy;
(2) carriers created in the junction at avalanche breakdown can be emitted if they are
energetic enough. (After Bartelink *et al.*, 1963.)

Another approach has been the study of emission from thin film tunnel
structures. One such structure was a thin film sandwich of Al–Al$_2$O$_3$–Pt,
where the Pt was 25 Å thick with a Pd grid evaporated over it (Cohen,
1962 a,b) and a Cs treatment added. For a sandwich current of about 50 mA
the emission current was 15 μA, which corresponds to a transfer ratio of
3×10^{-4} and a current density of 8 mA cm^{-2}. Stable operation was
observed for a period of two months under continuous operation in a sealed-
off vacuum tube.

However with the recent availability of high quantum yield electron
emitters of GaAs and related compound semiconductors and with the
present availability of highly efficient light emitting diodes of GaAs and GaP,
the expectation is that these structures may be coupled or integrated to
provide for the first time a cold-cathode emitter that is relatively efficient.
The concept is that application of current to the light emitting section creates
a photon flux that is absorbed in a region of smaller band gap which emits
the photoexcited electrons. One such structure that has been studied
(Kressel *et al.*, 1970c) is an *n–p* Al$_x$Ga$_{1-x}$As electroluminescent diode with
the *p* side covered with an absorbing *p*-type GaAs (1 μ) layer having a Cs–O
treated negative electron affinity surface. The initial studies produced a
differential efficiency of about 10^{-3} for the ratio of current emitted into

vacuum to the input current at current densities above 0.2 A cm^{-2}. More recently an improvement of about an order of magnitude in this efficiency has been obtained. An efficiency of 10^{-2} for a diode operating at 50 A cm^{-2} would mean a cathode current density of 0.5 A cm^{-2}. Whether stable long-term operation can be achieved under such conditions remains to be seen.

Heterojunctions for the most part have been formed by growing epitaxially one semiconductor material onto a different semiconductor material. Many methods of growth have been used such as vapor epitaxy, evaporation, alloying, and solution growth. The method of fabrication is particularly important as the heterojunction properties which are measured are found to be often a function of the manner in which the junction was fabricated. In this chapter we will discuss the problems of fabrication, the various types of systems applicable to heterojunction formation and their merits and failings, some measurements used for determining the resulting material properties, and the techniques for making ohmic contact to the various semiconductor materials.

9.1 Considerations in Heterojunction Fabrication

The fact that the junction is formed between materials having slightly different properties poses some problems not encountered normally in semiconductor fabrication. Other problems common to semiconductor layers formed by epitaxial growth, whether on parent semiconductor substrates or insulating substrates, are also discussed with reference to heterojunction formation.

9.1.1 *Crystal Structure*

In order to form junctions in which the bulk material properties on either side of the junction determine the resulting electrical properties, the two materials must have a similar crystal structure and be closely matched in lattice spacing. A list of semiconductors showing their lattice structure and spacing is included in Table 1.2. The mechanical misfit between two similar lattices can be described in terms of the density of edge dislocations required to accommodate the lattice mismatch. The edge-type dislocations must be very close together, and there must be dislocations for each of several different Burger's vectors in order to produce a lattice fit. Krause and Teague (1967) have used scanning x-ray microscopy to reveal a cross grid of dislocation lines at the interface between Ge deposited on (001) GaAs. They attribute these misfit dislocations to a partial accommodation of the lattice mismatch between the materials. The dislocation lines occur in the [110] and [1$\bar{1}$0] directions and have Burger's vector components along [1$\bar{1}$0] and [110] respectively. The largest percentage of the dislocations formed are most likely to be present in the material, either overgrowth or substrate, which is more plastic at the growth temperature. For growths of Ge on Si at 400°C the Ge only had significant dislocations whereas for GaP grown on GaAs, Oldham (1965) found that both materials showed a high number of dislocations near the interface. The "dangling bonds" which result will be involved in producing interface charge. Order of magnitude estimates of the charge density extrapolated from low densities of dislocations in homogeneous materials give values of $\sim 10^{14}$ for Ge–Si where the lattice mismatch is 4% and $\sim 2 \times 10^{12}$ for Ge–GaAs, where the lattice mismatch is 0.08%. The edge dislocations may also act as very active recombination centers for holes and electrons crossing the junction.

On the basis of interface conductance measurements on Ge–GaAs heterojunctions, Esaki *et al.* (1964a,b) conclude that the trapped electron density per square centimeter at the interface is less than 5×10^{10} cm^{-2}. Donnelly and Milnes (1967) from capacitance studies of *n–p* Ge–Si junctions concluded that the effective trapped electron density at the interface was about 10^{12} cm^{-2} (see Table 2.1). The dangling bond density expected from the 4% lattice mismatch, however, is 6×10^{13} cm^{-2} (Table 4.1). Therefore the trapped electron density at the interface may be less than the dangling bond density expected from simple geometrical considerations.

Oldham and Milnes (1964) in studies on *n–n* Ge–Si heterojunctions (see Chapter 4) found the interface states sufficient to affect greatly the resulting barrier and electrical properties. It may be concluded that lattice mismatches of the order of 1% or greater normally cause sufficient interface states to bend the bands at the interface and, hence, largely determine the device characteristics. (If the lattice mismatch exceeds 8–10%, epitaxy is difficult to obtain.) The exact effect of the misfit is dependent on the thickness of the materials involved. For a very thin layer on a thick substrate, no dislocations may be involved but rather the thin layer will elastically deform in order to fit better. The elastic deformation energy can, for the thin structure, be less than the energy needed to produce a dislocation. Since we do not know the interface dislocation energies with much precision, it is difficult to state the thickness of the layer needed to produce dislocations. However, in Ge–Si for example, only four or five atomic layers are probably needed if they have bulk properties. A detailed analysis of the energy associated with dislocations at the interface is contained in papers by van der

Merwe (1963, 1964). The effects of dislocations on the electrical properties of germanium are discussed by Matare (1959).

9.1.2 Thermal Expansion of the Lattice

A difference in the thermal expansion coefficient usually exists between the two different materials which form the heterojunction. Thus, on cooling from the usually high temperatures of fabrication to room temperature, the lattice is strained and dislocations may be produced. Riben *et al.* (1966) report severe strain in Ge–Si junctions which resulted in cracking of the Ge over-layer. Hovel and Milnes (1969) have observed similar cracking in thick ZnSe layers grown on Ge. Experiments on the plastic deformation of Ge layers grown on GaAs are presented by Light *et al.* (1968). They conclude that no interface dislocations are produced for their films but the layers are plastically deformed. The driving force for the strain is the difference in thermal expansion between the Ge and GaAs.

Table 1.2 lists the (average) linear thermal expansion coefficients for many semiconductors. The data must be used with care, however, as the coefficients vary as a function of temperature. The linear thermal expansion coefficients have been measured by several authors as a function of temperature. Some results are shown in Fig. 9.1.

It is apparent from experiments in our laboratory that the effects of the differences in thermal expansion can be minimized by (a) growing the epitaxial layer at as low a temperature as possible consistent with good material properties, (b) growing thin layers, and (c) cooling the layer slowly from the growth temperature.

9.1.3 Cross-Doping Effects

The growth system and the growth conditions must be carefully chosen to minimize cross-diffusion effects of the heterojunction elements and the dopants and autodoping of the overgrowth and possibly the substrate. A heterojunction pair such as GaAs–Ge is particularly susceptible to this problem as Ge is a dopant in GaAs, and Ga and As are dopants in Ge. As discussed by Vieland and Seidel (1962), Ge is an amphoteric dopant in GaAs with $N_D > N_A$. In n-type GaAs the incorporation of Ge into the lattice does not alter the conductivity type but greatly reduces the mobility of carriers. In Ge, As diffuses more rapidly than Ga so conceivably one could end up with one or two homojunctions in the germanium. Incorporation of Ge in GaAs and Ga and As in Ge has been observed by Rediker *et al.* (1964) in alloyed Ge–GaAs heterojunctions. Under some growth conditions we have observed n-type As

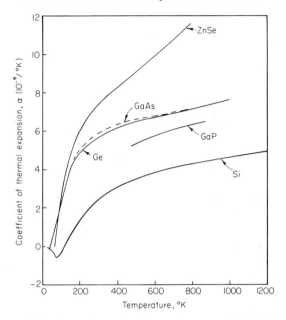

Fig. 9.1. *Linear coefficient of thermal expansion* (α) *versus temperature for* Ge, Si, GaAs, GaP, *and* ZnSe.
Ge, Si after Kirby (1963); GaAs after Feder and Light (1968) and Bernstein and Beals (1961); GaP after Bernstein and Beals (1961); ZnSe after Singh and Dayal (1967) and Novikova (1961b).

doped layers in p-type Ge adjacent to the GaAs interface. This has occurred for junctions formed by growing Ge on GaAs using GeH_4 at temperatures in excess of 600°C and for GaAs grown on Ge via HCl transport above 500°C. Analysis of these layers has shown the As concentrations to be approximately 10^{19} cm^{-3} (Ladd and Feucht 1970a), which agree with observations by Schulze (1966) for GaAs grown on Ge using an iodine transport system.

Since diffusion is a rapidly increasing function of the growth temperature the effects can be minimized by using low growth temperatures, 400–600°C, and conditions which do not cause etching of the substrate. There is a limit imposed, however, by the surface mobility of the depositing atoms which, if too low, will not yield good single-crystal growths, and by the stability and formation rate of a surface oxide.

9.1.4 *Substrate Preparation*

Preparation of the substrate is particularly important for good epitaxial growth and for good junction properties. This is critical for heterojunction formation since the substrate-growth interface coincides with the electrical junction. In order to reduce the cross-doping effects the junction fabrication is done at as low a temperature as possibly commensurate with good growth morphology. *In situ* etching of the substrate for cleaning just prior to growth is desirable but it is often difficult or impossible to arrange, particularly for a system using a very low growth temperature.

An example of the sensitivity of the electrical characteristics of hetero-junctions to surface preparation is shown in Fig. 9.2. In this work Riben *et al.* (1966) found, for Ge–GaAs junctions fabricated at 400°C by the epitaxial deposition of Ge using an iodine system, that cleaving the substrate *in situ* produced significantly better junctions than chemically or iodine vapor etched substrates. The chemically prepared GaAs substrates were etched in $H_2O : H_2O_2 : H_2SO_4$ (1 : 1 : 10 ratio) immediately before insertion in the growth tube. An improved polishing technique for Ge and GaAs has been reported by Reisman and Rohr (1964) using NaOCl. This, together with improvements in the iodine transport process, has resulted in mirror-smooth Ge layers as reported by Berkenblit *et al.* (1968).

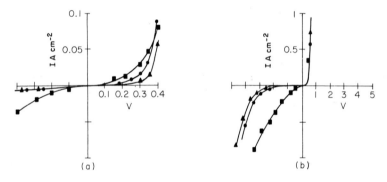

Fig. 9.2 *V–I characteristics of vapor-grown n–p* Ge–GaAs *junctions for* (▲) *cleaved in situ while the growth is in process,* (●) *cleaved in air, and* (■) *chemically etched* GaAs *substrates.* (a) Low voltage region, (b) high voltage region.

A typical procedure used in our laboratory for preparing Ge substrate seeds for the subsequent growth of a GaAs or ZnSe layer is as follows: Ge substrates are cut and mechanically lapped down to a 1 μ finish, then mounted on quartz disks and chemically polished to scratch-free mirror smoothness using a solution of NaOCl and Breck shampoo (50 ml NaOCl, 100 ml H_2O, and 5 ml shampoo) on a rotating disk of Pellon cloth. The shampoo serves to prolong the life of the cloth and results in scratch-free surfaces. The samples are then cleansed in boiling trichlorethylene, acetone, and methanol and finally etched in modified Superoxol ($1HF : 1H_2O_2 : 4H_2O$) : ($20H_2SO_4$) in a rapidly rotating beaker in order to remove several microns from the Ge surface. This last etch removes shallow surface damage at the cost of a slight rounding at the edges.

For the fabrication of heterojunction transistors, GaAs or ZnSe is grown on Ge seeds, which have a thin diffused p-layer to form the base. The preparation of the Ge seeds then requires additional processing in order to obtain the thin p-layer, which serves as the transistor base region, on the surface of an n-type substrate. The diffusion must provide a 1–2 μ p-layer with a high surface concentration in the order of 10^{18}–10^{19} cm^{-3} without causing any significant surface deterioration of the Ge surface. The following process developed by Jadus (1967) in our laboratory has met the above conditions.

The Ge substrate is prepared as above and placed with the surface to be diffused face up on a quartz platform. A box is then constructed around the sides and top of the seed using pieces of Ge slightly thicker than the seed for the sides of the box and a slab of Ge for the top. The single crystal p–Ge that forms the box has uniform p-type Ga doping at twice the density that is desired for the surface of the seed. The assembled quartz platform, seed, and doped box are then placed in a clean quartz tube. This quartz tube is provided with an adequately large connection to a vacuum system. After a system vacuum of 4×10^{-8} Torr is indicated, the samples are heated by a furnace around the tube. The vacuum system pumps continuously while the platform, seed, and box assembly are raised to the diffusion temperature and held there for the diffusion time. After each diffusion, the Ga depleted material on the surface of the Ge box is etched away in order to ready it for use as a source in subsequent runs. Ga doped layers, 1–2 μ thick, with surface concentrations ranging from 2×10^{17} to 2×10^{19} cm^{-3} may be diffused at a temperature of 850°C with no significant substrate surface degradation using this technique.

Substrates of GaAs are polished satisfactorily by mechanical lapping to 0.05 μ, then heated in $2H_2O : 1NaOCl$ for one hour at 60 to 70°C. This removes about 0.5 mils of material and produces mirror smooth surfaces. ZnSe substrates are prepared by the same mechanical lapping followed by a 10–15 min etch in a rapidly rotating beaker of bromine methanol solution. The amount of material removed by this etch is determined by the bromine concentration, and can be varied over a wide range. The resulting surfaces are mirror smooth to the eye for a low bromine concentration but show some defect structure under high magnification. According to the studies of Yep and Archer (1967), every chemical treatment of the ZnSe surface leaves a residual film on the surface. For a final treatment with bromine methanol this film is approximately 135 Å thick while a final etch in HCl leaves a 343 Å film. They report that the thickness of this remnant film can be markedly reduced by etching the surface for 30 sec in $K_2Cr_2O_7 + H_2SO_4$ at

80°C, followed by a rinse in warm KCN for 5 min and a rinse in hot de-ionized water for 5 min. This results in a film thickness of approximately 14 Å (compared to a film thickness of 9 Å for a surface cleaved in air and measured in a N_2 ambient).

9.1.5 Construction of the Growth System

The construction and contamination-free properties of the system for fabricating the heterojunction are as important as the preparation of the substrates. Most growth systems used have been either vapor-growth methods using a halogen for a transport agent or metal-solution growth systems. In halogen transport systems, in order to eliminate sources of contamination from reaction with the halogen, all parts of the system coming into contact with it should be constructed of pyrex, quartz, and inert materials such as Teflon or Kel-F. Stainless steel should not be exposed to the halogen.

The connections in the system should be fused glass connections where possible and Teflon or Kel-F fittings for the others. No sealing compounds such as vacuum grease, etc., should be used on any of the connections as these materials are readily attacked by a halogen plus H_2 ambient. However, Teflon or Kel-F fittings are permeable to He to such a degree that this constitutes a gross leak as measured by a He leak detector. The available information about Teflon indicates that the rate of permeation of each component of air is substantially the same as that for He. Thus it is advisable to provide each critical fitting with a polyethylene jacket with dry N_2 flowing through it to exclude O_2 from the system. The system components should also be arranged so that it is convenient to leak-check the system periodically with a portable He leak detector. Use of the above techniques helps in the elimination of numerous small hillocks and bumps in the surface of epitaxial layers. The heterojunctions then obtained are more reproducible in their electrical characteristics.

Hillocks may also be caused by crystallites detaching from the walls of the growth tube and dropping onto a seed if it is horizontal. Mounting of the seed vertically or in a clamp horizontally so that growth is on the underside is usually desirable.

9.2 Growth Systems Using Iodide Disproportionation

Some of the earliest work on the vapor growth of Ge by the iodine transport method was done by Marinace (1960a,b) in both open- and closed-tube systems. In both systems the reversible disproportionation reaction

$$2GeI_2 \rightleftharpoons Ge + GeI_4 \qquad (9.1)$$

is used to transport Ge. For a fixed amount of iodine the equilibrium shifts to the right as the temperature is reduced.

In the closed-tube system the germanium is transported from a high temperature source region to a lower temperature by allowing the reaction products in each temperature region to circulate to the other region in a cyclic manner. The tube is loaded with the seed and source Ge and an amount of iodine determined by the desired vapor pressure. It is then evacuated and sealed and positioned in the furnace as shown in Fig. 9.3.

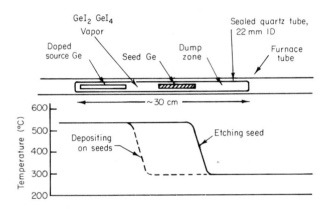

Fig. 9.3. *Closed-tube apparatus for the deposition of* Ge.
[After Marinace (1960a), *IBM J. Res. Develop.* **4**, 248.]

Initially the temperature profile is adjusted so that both the source and seed material at 500–550°C etch away, and the Ge being removed is deposited in the "dump zone" at ~300°C. After about 1 to 2 h, the temperature profile is changed so that the seed region drops quickly to 300–400°C. Deposition on the substrates is then carried out for 4 to 30 h at temperatures in the range of 300–400°C. The rate of transport depends on the two temperatures—the circulation rate and the iodine pressure. For Ge seed and source temperatures of 400 and 550°C, respectively, the growth rate varies from 5–10 μ day^{-1} for an I$_2$ density of 0.03 mg cm^{-3}, to 200 μ day^{-1} for 4.5 mg cm^{-3}.

Marinace's work showed that n-type dopants such as P, Sb, and As and p-type dopants B and Ga could readily be transported in the system. Indium did not transport very well and Al stopped the growth process. An advantage of the closed-tube system is that the purity of the deposit does not depend very heavily on the purity of the iodine. A major disadvantage of this type of system is pointed out in the work of Kasano and Iida (1967) on the preparation of GaAs–Ge and InAs–GaAs heterojunctions. They found an intermediate layer of 10–20 μ thickness between substrate and growth. Their results suggest that this layer is caused by species which are initially vaporized from the substrate and then precipitated in the initial growth in the closed tube. Arizumı and Nishinaga (1966b) conclude, on thermodynamical arguments, for GaAs–Ge junctions prepared by the closed tube process that a great deal of As will be incorporated in the grown Ge layer.

Marinace (1960b) in addition to closed-tube studies also used an open-tube system for iodine transport of Ge, to grow abrupt Ge–GaAs heterojunctions. A similar open-tube system, used by Riben et al. (1966), is shown in Fig. 9.4 with the corresponding temperature profile. Ge–GaAs and Ge–Si abrupt junctions were grown in this system at 400°C and the electrical

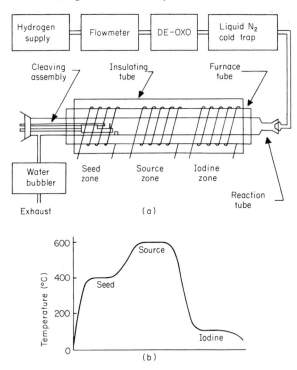

Fig. 9.4. *Apparatus used for the vapor growth of germanium by the germanium di-iodide disproportionation reaction.*

(a) Furnace, cleaver, and gas supply; (b) temperature profile of the furnace during the growth process.

characteristics evaluated. In order to grow Ge on Si at this low temperature, the Si was cleaved *in situ* during steady state Ge deposition using the cleaving apparatus shown in Fig. 9.5. L-bar samples similar to those of Gobeli and Allen (1960) were fabricated and cleaved in the growth tube in the seed zone as shown in Fig. 9.4. This technique enabled Riben to obtain an oxide-free silicon surface at a temperature of 400°C.

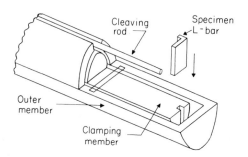

Fig. 9.5. *Detail of the all-quartz cleaving apparatus used to obtain cleaved surfaces during the growth.*

With this system the grown Ge films had n-type carrier concentrations greater than 10^{16} cm^{-3}. Using a multizone furnace, Riben *et al.* (1965) were able to reduce this level to 10^{15} cm^{-3}. From their results they concluded that the source of impurities was the iodine and that the n-type level observed, which was 0.2 eV below the conduction band, was most likely oxygen. As discussed earlier in Section 9.1 the Ge layers grown at 400°C on Si developed cracks upon cooling to room temperature due to the difference in thermal expansion coefficient for the two materials.

A more detailed study of the growth parameters and the thermodynamical considerations for the type of system is contained in a series of papers by Arizumi and co-workers (Arizumi and Akasaki, 1963, 1964; Arizumi and Nishinaga, 1966a). Their system was different from that in Fig. 9.4 in that a dopant zone was added between the I_2 and Ge source zone. They found that Sb was quite successful as an n-type dopant and BI_3 as a p-type dopant for heavily doped layers.

More recently Reisman and co-workers (Reisman and Alyanakyan, 1964; Reisman and Rohr, 1964; Reisman and Berkenblit, 1965, 1966) and Berkenblit *et al.* (1968) studied the kinetics of the GeI_2 system. Their work resulted in mirror smooth growths of Ge on GaAs with undoped carrier concentrations of approximately 10^{14} cm^{-3}. In their system the I_2 is converted to HI before entering the growth tube and higher flow velocities and higher HI concentrations are used. A diagram of a similar system, used in our laboratory, is shown in Fig. 9.6. A mixture of H_2 and He is passed through the iodine tower (65–75°C) and then through a platinum wool column (350–450°C) where it is converted to HI. The HI is then converted in the baffle Ge source region (600°C) to GeI_2 where it travels to the cooler seed region (350°C) and Ge is deposited out. The Ge epitaxial layer can be doped n-type by adding AsH_3 and p-type using BI_3. The resultant p-type growths on (110) GaAs had no intermediate n-type layer and were quite smooth.

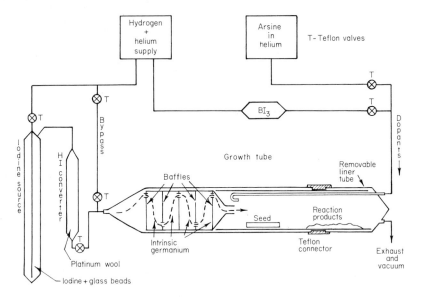

Fig. 9.6. *Schematic diagram of germanium iodide growth system.*

Zeidenbergs and Anderson (1967) have used a disproportionation reaction involving Te to grow n–silicon on GaP. Deposition was accomplished in a closed tube system at a GaP seed temperature of 850°C. Si was transported as $SiTe_2$ from the source to the seed zone. Because of the large difference in thermal expansion coefficients between the two materials the GaP often cracked upon cooling to room temperature. Te is an n-type dopant in Si and therefore the Si layers were heavily n-type.

9.3 HCl Transport Systems

Several types of vapor transport systems using HCl as a transport agent have been described for growing epitaxial layers. These may be grouped into three categories, (1) close-spaced HCl systems, (2) systems using HCl directly in which the source and substrate are separated by an appreciable distance, and (3) systems in which HCl is produced from a chloride such as AsCl$_3$, GeCl$_4$, etc.

The close-spaced system first described by Nicoll (1963) and Robinson (1963) used water vapor as the transport agent. These were adapted by Jadus *et al.* (1967) and Hovel and Milnes (1969) for growing GaAs and ZnSe using HCl as the transport agent. The close-spaced system has the advantage of high efficiency transport which enables growth at relatively low temperatures. A schematic diagram of this type of system for growing GaAs on Ge is shown in Fig. 9.7. The substrate and source are separated by approximately 250 μ by quartz spacers. The source is maintained at a higher temperature than the seed, typically 720 and 600°C, respectively, by focusing the light of a sun gun on each of two SiO$_2$ coated Si blocks. Chlorine, by the dissociation of HCl, acts as the transport agent. Epitaxial growth proceeds via the following reactions as discussed by Fergusson and Gabor (1964):

$$GaCl + Cl_2 \rightleftharpoons GaCl_3 \tag{9.2}$$

$$2GaCl + \tfrac{1}{2}As_4 \rightleftharpoons 2GaAs + Cl_2 \tag{9.3}$$

High purity HCl, Pd diffused H$_2$ and a leak-tight quartz, Teflon, and Kel-F system were used to obtain good growths.

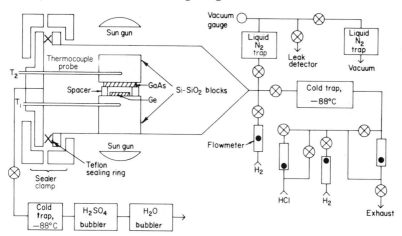

Fig. 9.7. *Schematic diagram of a close-spaced epitaxial growth system.*
(After Ladd and Feucht, 1970a.)

The morphology of the grown GaAs layer on Ge depends to a large extent on the growth conditions. For conditions of rapid growth (source 650°C, substrate 550°C, and HCl concentration 0.7%) the GaAs has a structural surface and is of moderately high resistivity. The high resistance is a result of a very low mobility and poor crystal structure. For other conditions (source 720°C, seed 600°C, and HCl concentration 0.1%), the GaAs has a very smooth surface and is of much lower resistivity. These conditions, however, result in very heavy autodoping of the GaAs near the interface with Ge and a high As content (greater than 5×10^{19} cm^{-3}) at the Ge interface. For the conditions of rapid growth this As content is less than 7×10^{18} cm^{-3}. From our measurements the doping by As of the Ge substrates is caused both by diffusion from the GaAs layer and by incorporation from the vapor phase during the initiation of the growth reaction. Vapor phase incorporation is accentuated by substrate orientations other than near the (111) plane and by high HCl concentrations. Low growth rates at the beginning of the growth process are also thought to contribute to high arsenic concentrations. Diffusion of As from the GaAs layer is shown to be less of a problem than incorporation from the vapor.

The quality of the grown layer is quite dependent on the orientation of the substrate. A comparison of the surfaces of GaAs grown at 650°C on Ge substrates oriented on or near a (111) plane is shown in Fig. 9.8.

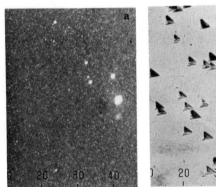

Fig. 9.8. *Surfaces of epitaxial* GaAs *grown on* Ge *at* 650°C.
(a) Substrate (111); (b) several degrees off (111) toward (100); (c) 8° off (111) toward (100). Major division (e.g., 20–30) equals 200 μ.

Growths on the (111) plane at all substrate temperatures showed a fine, matte finish, and the layers typically had poor mobilities. Growths on (100) substrates at 600–650°C showed good structure and those on a (311) substrate at 600°C resulted in mirror smooth layers. Growths on 8° off (111) substrates exhibited a matte finish up to 500°C changing to a very smooth background film with pyramidal defects at 550°C. Hall measurements on thick GaAs layers on 8°–(111) substrates had mobilities greater than 3500

$cm^2 \, V^{-1} \, sec^{-1}$ and carrier concentrations ranging from 1 to $5 \times 10^{16} \, cm^{-3}$, presumably due to residual Ge.

The ZnSe close-spaced system shown in Fig. 9.9 is similar to the GaAs system except for the reactions and operating conditions and the incorporation of a Zn zone. In this system the source and seed temperatures are 620–760°C and 500–660°C, respectively, and the HCl concentration is approximately 0.01%.

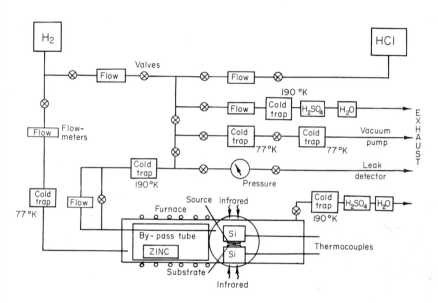

Fig. 9.9. *Schematic diagram of a close-spaced* ZnSe *growth system.*
(After Hovel and Milnes, 1969.)

Single-crystal layers of ZnSe have been grown on (111) oriented substrates of ZnSe, GaAs (As face), and Ge using both pure H_2 and H_2–HCl atmospheres. Strong dependence of the growth rate upon the substrate material was observed; the rates were typically 10 to 40 times higher on Ge than on either of the other two materials. This can probably be attributed to a chemical reaction occurring between the Ge and the reaction products from the ZnSe etching.

Figure 9.10a shows the surface of a 15 μ ZnSe layer grown under "fast" conditions, i.e., high growth rates and high HCl concentrations. A large density of irregular stacking faults is apparent, which is a result of severe etching of the Ge substrate by the simultaneous action of HCl and H_2Se (one of the ZnSe–HCl reaction products) during the initial stages of the ZnSe growth. The situation is greatly improved by reducing the HCl concentrations and growth rate by an order of magnitude (Fig. 9.10b). Under these "slow" growth conditions the concentration of etching gases is drastically reduced and the surface of the Ge substrate is unaffected.

Growths of ZnSe on GaAs substrates had a cobblestone appearance while those of ZnSe on ZnSe were mirror smooth. A summary of the growth conditions for various parameters of the system are given in Table 9.1. After the ZnSe–Ge and ZnSe–GaAs layers have been grown, cracks may develop upon cooling due to the difference in thermal expansion coefficients of the two materials. These have been eliminated by growing thin layers ($<4 \mu$) and cooling the junctions slowly ($\sim 1°C$ min^{-1}).

TABLE 9.1

Growth of ZnSe upon ZnSe, Ge, and GaAs[a]

Growth conditions	Source temp.	Substrate temp.	Growth rate
ZnSe upon ZnSe	760°C	660°C	8 μ h^{-1}
(0.2% HCl)	620°C	580°C	0
ZnSe upon Ge	760°C	620°C	160 μ h^{-1}
(0.2% HCl)	620°C	570°C	8 μ h^{-1}
ZnSe upon ZnSe	760°C	660°C	0.05 μ h^{-1}
(pure H$_2$)			
ZnSe upon Ge	760°C	660°C	0.3 μ h^{-1}
(pure H$_2$)			
ZnSe upon Ge	680°C	570°C	1 μ h^{-1}
(0.02% HCl)			
ZnSe upon GaAs (B)	750°C	640°C	0.5 μ h^{-1}
(0.02% HCl)	700°C	590°C	0
ZnSe upon GaAs (B)	700°C	580°C	1.0 μ h^{-1} on GaAs
(0.02% HCl with			3.0 μ h^{-1} on Ge
Ge seed also present)			

[a] All substrates were (111) oriented. The total flow rate was 200 cm^3 min^{-1} (reaction tube 43-mm diameter). Source to substrate spacing was 0.012 in.

Difficulties are encountered in doping II–VI compounds such as ZnSe because of the self-compensation which occurs. For each intentionally added *n*-type impurity a corresponding Zn vacancy may be formed which, being acceptor-like, compensates the donor. It is therefore necessary to provide Zn atoms to the crystal ambient in order to control the Zn vacancy density and allow the donors to become electrically active. This is accomplished in this system by placing a Zn-filled quartz boat near the input end of the tube (see Fig. 9.9). The Zn is kept at room temperature during the growth period and is therefore inactive during the growth but is heated to temperatures of 700–800°C after the growth has finished, establishing a Zn partial pressure of

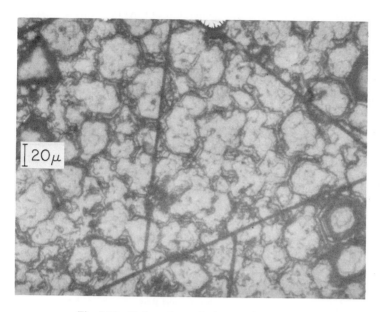

Fig. 9.10a. ZnSe *surface under fast growth conditions.*
Source temperature 640°C; substrate temperature 570°C; 15 μ h^{-1}; 0.2% HCl.

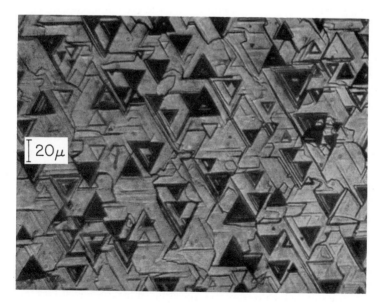

Fig. 9.10b. ZnSe *surface under slow growth conditions.*
Source temperature 700°C; substrate temperature 580°C; 2 μ h^{-1}; 0.02% HCl.

10–150 Torr in the vicinity of the grown layer. Zinc may then diffuse into the layer from the vapor while the Ge or GaAs substrate is partially shielded by its contact with the SiO_2-covered Si block. Doping of the ZnSe layer is accomplished by using a ZnSe source crystal which is heavily doped. Gallium has been found to work better than In or Al, but the lowest resistivity layers grown to date have been of the order of 10^3 Ω cm after the Zn treatment step.

Presumably Ge is incorporated in the grown layers. The measured mobility is 50–100 cm^2 V^{-1} sec^{-1} compared with 300–400 cm^2 V^{-1} sec^{-1} for the mobility of bulk ZnSe. When n–ZnSe is grown on n–Ge substrates, the zinc treatment step seems to result in very little incorporation of Zn in the Ge near the interface. Usually, however, the n–ZnSe layers have been grown on p–Ge substrates or on n–Ge collector substrates already provided with a heavily doped p-layer as a diffused base region. In growth on n–GaAs substrates the Zn treatment step causes a thin p-type diffused layer to develop in the GaAs at the ZnSe interface. Since the thickness and doping levels are suitable this p-layer has been used as the base for ZnSe–GaAs transistor action. This is discussed by Sleger (1971) and Sleger et al. (1970).

Baczewski (1965) used HCl to transport ZnSe onto GaAs in an open tube system where the ZnSe source was separated from the seed. For these growths NH_4I was added to the system to remove any free Cl and thus to prevent the etching of the GaAs substrate even with high concentrations of HCl.

Several other workers have used HCl for the growth of epitaxial layers in systems where the source or sources and seed are separated by large distances. They differ in their method of reacting the HCl with the source material and in their source of HCl. The system of Fig. 9.11 was used by Amick (1963) to grow GaAs on Ge and metallic substrates. In this system As is vaporized at 400–450°C and carried into the reaction zone by H_2. HCl reacts with Ga at 700–800°C to form $GaCl_3$ and GaCl vapor which is also carried into the reaction zone. In the reaction zone at 600–700°C the GaCl and As_4 form GaAs at the surface of the substrate. In this type of system the GaCl and As_4 partial pressures can be independently varied. Amick concludes that one of the most critical steps in the growth of GaAs on foreign substrates is the preparation of the substrate surface. In order to obtain a mirror smooth surface, etching of the seed with HCl vapor at 800°C is provided. The character of the surface of the GaAs deposits depends on the growth temperature. For temperatures below 700°C the surface is grainy while around 750°C the surface is smooth.

Investigation of Amick's growths indicated that the strain energy due to mismatch in lattice constant and thermal expansion coefficient was accommodated in a highly localized region near the Ge–GaAs interface. Ge and Si in concentrations of 100 ppm and 50 ppm, respectively, were found in the growth along with a few other impurities at levels of 3 ppm. The Ge was a

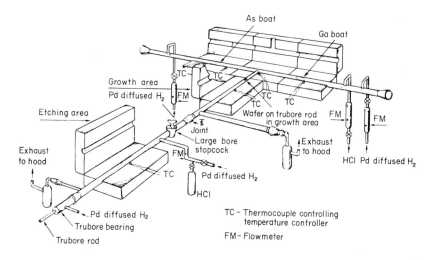

Fig. 9.11. *Sketch of* GaAs *deposition apparatus with furnaces open.*
(After Amick, 1963.)

result of autodoping from the substrate while the Si was probably a result of a reaction between liquid Ga and the quartz boat.

Ing and Minden (1962) and Weinstein *et al.* (1964) have used similar open-tube systems but with the Ga and As or P source in series to grow GaAs and GaP. Weinstein reports the growth temperature for the growth of good epitaxial layers of GaP is much more critical than that for GaAs. Conrad *et al.* (1967a) used a similar system to that of Amick to grow $Ga_xIn_{1-x}As$ on (100) semi-insulating GaAs. The proportion of Ga to In was controlled by regulating the HCl flow over the two sources independently. They found that when these alloys were deposited on very thin GaAs substrates (less than 250 μ thick) the samples were plastically distorted. The effect was more pronounced as the In content of the alloy was increased and was attributed to the mismatch of the lattice constants.

GaAs or GaP may also be grown using arsine or phosphine in conjunction with HCl and Ga. The diagram of such a system is shown in Fig. 9.12. Tietjen and Amick (1966) report the growth of $GaAs_{1-x}P_x$ on Ge and GaAs substrates for a seed temperature of 725–750°C. Typical flow rates for their system were 15 cm³ min⁻¹ total for arsine and phosphine, 5 cm³ min⁻¹ of HCl and 2.5 liters min⁻¹ of hydrogen. Doping was accomplished by introducing H_2Se for *n*-type material and Zn for *p*-type doping. Growths with donor concentrations from 5×10^{16} to 2×10^{19} electrons cm⁻³ or acceptor concentrations from 10^{18} to 10^{19} holes cm⁻³ have been obtained. Using similar conditions in our laboratory we have grown GaAs epitaxially on

Ge and GaAs substrates at temperatures between 620 and 660°C. Layers with no internal doping added have been n-type with a carrier concentration of approximately 5×10^{16} cm^{-3} and a room temperature mobility of 5200 cm^2 V^{-1} sec^{-1}. The impurities in the grown layer come most likely from the HCl or arsine. Recently Burmeister and Regehr (1969) have used a similar system to grow GaAs$_{1-x}$P$_x$ on Ge for $0.3 < x < 0.4$. They found that the best growth occurred on (311) oriented Ge substrates and that there was substantial incorporation of Ge in the GaAs$_{1-x}$P$_x$ layer.

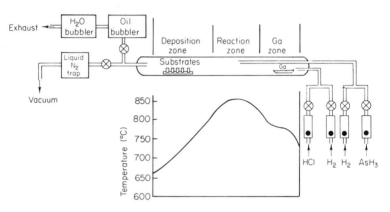

Fig. 9.12. *Schematic diagram of* GaAs *vapor growth system showing seed placement and temperature profile.*

An alternate scheme to using HCl for the growth of epitaxial layers of GaAs and GaP is to react the trichloride of As or P with Ga. A typical system for this type of growth is shown in Fig. 9.13 after E. W. Williams and Blacknall (1967). In the high temperature region the trichloride reacts with Ga to form GaCl and free As or P. In the cooler region these combine to form GaAs and GaCl$_3$. The quality of the grown layer has been studied in a series of papers by Bobb *et al.* (1966a–c), and Holloway *et al.* (1965,1966) and Holloway and Bobb (1967). They find that under particular conditions epitaxial layers may be grown with X-ray rocking curves close to that for bulk material. They point out the importance of saturating the Ga with As prior to growth in order to get better layers. Effer (1965) also finds that GaAs grown under conditions of excess arsenic produces better material. The quality of the grown layer is a direct function of the purity of the AsCl$_3$. AsCl$_3$ which is further purified by gradient zone freezing or distillation has produced high quality material as reported by Effer (1965) and Conrad *et al.* (1967b). Maruyama *et al.* (1969) have grown high purity n-type GaAs using commercially obtained very pure starting materials. Electron

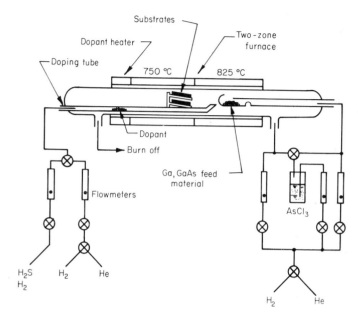

Fig. 9.13. *Typical* Gaas *growth system using* AsCl₃ *and gallium.*
(After E. W. Williams and Blacknall, 1967.)

mobilities of 9180 cm² V⁻¹ sec⁻¹ at room temperatures and 164,000 cm²
V⁻¹ sec⁻¹ at 77°K were obtained.

Oldham (1965) and Kamath and Bowman (1967) have used PCl₃ and Ga
as starting materials for growing GaP. They used GaAs as seed material
with seed and source temperatures of 800–840°C and 950–1000°C, respectively. Oldham reports that the GaP layers show considerable strain throughout and that they contain 1–5% As when grown on GaAs seeds. Kamath and
Bowman find that their undoped layers are *p*-type with a carrier concentration of ∼10¹⁴ cm⁻³ and an As contamination of only a few hundred ppm. By
adding small amounts of water vapor they have been able to vary the
resistivity of the *p*-layers from 10² to 10¹⁰ Ω cm. Taylor *et al.* (1968) have
grown GaP on the 111A and 111B faces of GaP and GaAs using PCl₃. They
found that the GaP grown on the GaP has a room temperature mobility of
187 cm² V⁻¹ sec⁻¹ and is of better quality than that grown on the GaAs
substrates. The material grown on GaAs substrates shows some As incorporation and a gradient in the impurity concentrations. Kesperis *et al.* (1964) used
Ga and PCl₃ to grow GaP on Si. SiO₂ on the substrate surface prevented the
formation of a uniform layer and the difference in thermal expansion coefficient between GaP and Si caused cracking in some layers.

The thermal decomposition of Ge or Si tetrachloride has been used by Theurerer (1961) and Cave and Czorny (1963) for the epitaxial growth of Ge and Si layers. A schematic diagram of such an epitaxial deposition system is shown in Fig. 9.14. The vapor phase reaction for the reduction of Si or Ge is given by:

$$XCl_4 + 2H_2 \rightleftharpoons 4HCl + X(s) \qquad (9.4)$$

where X represents Ge or Si. The thermodynamics of this reaction has been studied in a series of papers by Schafer (1953) and Schafer *et al.* (1956). Oda (1962) and Miller and Grieco (1962) have used a similar system for growing Si–Ge alloys up to 30 atomic percent Ge on Si substrates. Oda was able to dope the Si–Ge crystals *n*-type using PCl_3 and *p*-type using BBr_3.

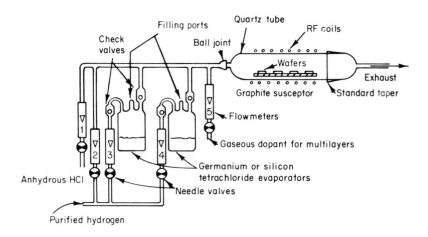

Fig. 9.14. *A typical* Ge *or* Si *tetrachloride epitaxial deposition system.*
(After Cave and Czorny, 1963.)

A discussion of the theory, techniques and methods of evaluation of the epitaxial growth of Si is included in the books by Burger and Donovan (1967), and Kane and Larrabee (1970). A recent discussion of the status of Si epitaxy by Runyan (1969) includes a discussion and references to many systems for obtaining Si epitaxy. In addition to the above work, Holonyak *et al.* (1962) have used either halogens or halides to transport compounds of GaAs, GaSb, and InSb on various substrates in a closed tube. In a recent review article, Tietjen *et al.* (1970) describe the use of HCl for the vapor growth of a broad spectrum of III–V compound semiconductors. Yim *et al.* (1970) describe an extension of this work to II–VI and III–V quaternary alloys and Dismukes *et al.* (1970) to semiconducting mononitrides.

9.4 Growth from Hydrogen Compounds and Water Vapor

9.4.1 *Growth by the Pyrolysis of Germane and Silane*

The pyrolysis of GeH_4 and SiH_4 offers the advantage, compared to the other methods discussed to this point, that there is no possibility of vapor back-etching due to the presence of halogens or halides. The reaction which occurs for the deposition of Ge or Si by this method is

$$XH_4(g) \xrightarrow{\text{Heat}} X(s) + 2H_2(g) \tag{9.5}$$

A typical system for the pyrolytic decomposition of GeH_4 or SiH_4 is shown in Fig. 9.15.

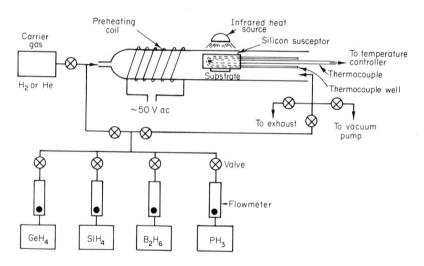

Fig. 9.15. *Growth system for the pyrolytic decomposition of* GeH_4 *and* SiH_4.

Roth *et al.* (1963) obtained mirror smooth growths on Ge substrates over the temperature range 700–900°C using germane which they synthesized. Below 700°C crystalline defects began to appear. Undoped growths were 1 Ω cm *p*-type. Doped layers of *n*- and *p*-type were readily obtained using arsine or diborane, respectively, as dopant sources.

The growth of silicon on silicon substrates by the decomposition of silane has been reported by Bhola and Mayer (1963). They grew 5–30 μ epitaxial Si layers with a high degree of crystal perfection on (111) seeds. For a hydrogen–silane mixture of 0.2% by volume they observed growth rates of

0.8 μ min^{-1} at 1050°C and 1.2 μ min^{-1} at 1140°C. Below a substrate temperature of 1000°C the growth tended to become polycrystalline. Undoped layers were n-type with a resistivity of 25–35 Ω cm. Layers n- and p-doped ranging from 5 × 10^{14} cm^{-3} to 5 × 10^{19} cm^{-3} were readily obtained using phosphine and diborane as dopant sources. Good surface preparation was essential to obtain good growth quality and a low carrier concentration.

In some recent work Richman and Arlett (1969a,b) and Richman *et al.* (1970) have achieved the epitaxial growth of Si on Si at temperatures as low as 800°C using He as the carrier gas instead of H$_2$. The resulting layers were quite smooth and the growth rate was as high as 0.5 μ min^{-1}. They conclude that the presence of excess H$_2$ inhibits the ability to obtain single crystal growth at low temperatures. Joyce and Bradley (1963) indicate that there is a kinetic limitation of removing H$_2$ from the growth surface which causes the low growth rate at low temperatures.

Papazian and Reisman (1968) have grown Ge on semi-insulating GaAs by the pyrolysis of germane for seed temperatures between 600 and 850°C. Good epitaxy was obtained between 650 and 750°C with typical H$_2$ gas stream velocities of 150–720 cm min^{-1} and GeH$_4$ concentrations of 0.02 to 0.14% by volume. Below 650 and above 750°C the growth became structured rather than smooth and above 800°C the Ge alloyed with the GaAs. Preparation of the GaAs substrates was critical. NaOCl polished wafers were treated in a 95H$_2$SO$_4$: 5H$_2$O solution immediately prior to insertion in the growth tube to remove any residue from the NaOCl polish. The purity of the H$_2$ and GeH$_4$ was particularly important for smooth growths and the GeH$_4$ needed to be pumped out when changing tanks to prevent the formation of GeO$_2$. While growths of epitaxial layers on high resistivity Ge substrates always were approximately 7 Ω cm p-type they exhibited both p- and n-type regions when grown on the GaAs. For growths below 750°C a profile of the Ge showed a p-layer for about 0.1 μ from the interface and n-type for the remainder of the layer. This may be attributed to the out-diffusion of Ga and As, plus some As incorporation from the vapor phase due to the dissociation of GaAs. At 700°C the diffusion constants of Ga and As in Ge are $D_{As} \simeq$ 2.5 × 10^{-12} cm^2 sec^{-1} and $D_{Ga} \simeq$ 5 × 10^{-15} cm^2 sec^{-1}.

In our laboratory we have been growing Ge$_x$Si$_{1-x}$ material on Ge substrates by the simultaneous pyrolysis of GeH$_4$ and SiH$_4$. Using a GeH$_4$ concentration of 0.04% and a SiH$_4$ concentration of 0.01% an alloy containing 7% Si was grown on a Ge substrate at 700°C. The growth rate was 11 μ min^{-1} and the resulting material was of good crystalline quality. Growths on substrates oriented 8° off (111) toward the (100) were smoother than those on the (111) face and showed only a low density of slightly tilted triangles. Recent growths in which the H$_2$ carrier gas has been replaced by He have resulted in improved growths at lower temperatures.

9.4.2 Epitaxial Growth Using H_2O Vapor

Water vapor has been reported by several authors to be an effective transport agent in a close-spaced growth system. Nicoll (1963) and Robinson (1963) found that GaAs, GaP, Ge, InP, and InAs could be transported in a flow of H_2. They attributed this to small quantities of water vapor in the H_2 supply or absorbed in the system. Gottlieb and Corboy (1963) investigated the growth of GaAs onto Ge substrates for various concentrations of H_2O flowing through the system. A schematic diagram of the type of system used is shown in Fig. 9.16. The concentration of water vapor was controlled by passing pure hydrogen over ice maintained at a subzero temperature. The substrate was heated to a temperature T_C (800°C) by a quartz iodine lamp and the source to a temperature T_H (900°C) by a molybdenum heater strip.

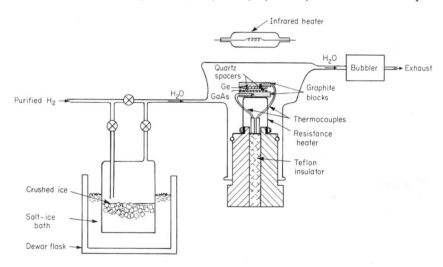

Fig. 9.16. *Close-spaced water vapor growth system.*
(After Gottlieb and Corboy, 1963; and Robinson, 1963.)

The variation of growth rate with vapor pressure of the water vapor is shown in Fig. 9.17. The transport of GaAs is assumed to proceed in the following manner:

$$2GaAs(s) + H_2O(v) \rightleftarrows Ga_2O(v) + H_2(v) + As_2(v)$$

The reaction proceeds to the right at the higher source temperature and to the left at the lower seed temperature. As in all growth systems the quality of the growth depends upon the surface preparation. The efficiency of mass transport was found to be a function of the H_2 flow rate but ranged from 50–95%.

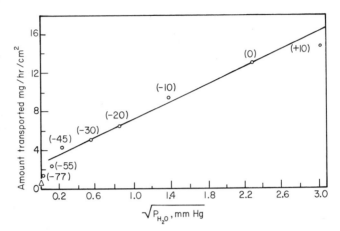

Fig. 9.17. *Amount of* GaAs *transported as a function of water-vapor pressure.*
△ = no water vapor added to system. ( ) = temperatures corresponding to the indicated
vapor pressures (°C); T_H = 800°C, T_C = 700°C. (After Gottlieb and Corboy, 1963.)

The grown layers, which were predominantly *n*-type, frequently were doped
with Ge from the substrate. In addition Robinson (1963) reports that in his
growths of GaAs on Ge he saw a thin *p*-region due to Ga and a thicker
n-region due to As in the Ge. Recently Luther (1969) has used this type of
system to grow zinc doped crystals of GaP on GaAs and GaP substrates.

9.5 Growth of Heterojunctions by Vacuum Evaporation and Sputtering

While most of the effort during the past few years has been expended in the
growth of semiconductor layers from the vapor phase there have also been
studies of the growth of epitaxial films by vacuum evaporation and sputtering.
This work has been concerned with both single element semiconductors
such as Ge and Si and compound semiconductors such as GaAs, CdS, etc.
Much of this work has been concerned with evaporating the semiconductor
material onto an amorphous substrate such as quartz. We are primarily
interested, however, in the growth of one semiconductor material onto
another with a similar lattice structure and reasonably close lattice spacing.

The evaporation of Ge and Si is rather straight forward in that they are
placed in a vacuum system and heated in some suitable manner until their
vapor pressure is sufficiently high that atoms are transported to a heated
substrate. The temperature of the source material may be raised by resistance
heating, rf heating or electron bombardment. In the first two the crucible

which holds the source material is also heated and can be a source of impurities. For this reason heating the source by electron beam bombardment is preferred. The quality of the grown layer is a function of the vacuum obtained, the substrate temperature, the perfection of the substrate surface, and the deposition rate. It has been found by many investigators that regardless of whether *n*- or *p*-type Ge is used as the source material the deposited film is usually *p*-type. This has been attributed by Kurov *et al.* (1957) to the formation of vacancy acceptor levels.

9.5.1 *Growth of* Ge *by Evaporation*

Sakai and Takahashi (1963) evaporated Ge from a tungsten basket at a rate of 10 Å sec^{-1} in a vacuum of 2×10^{-5} Torr onto (111) Ge substrates held between room temperature and 800°C. They found the best quality films were obtained at 450–550°C. The films were heavily doped *p*-type (10^{18}–10^{19} cm^{-3}) even for 40 Ω cm source material but could be made *n*-type by simultaneously evaporating P or As to overcompensate the acceptor centers. The resulting donor concentration was 10^{18}–10^{20} cm^{-3} and the electron mobility was typically 800 cm^2 V^{-1} sec^{-1} at 300°K. When the deposition rate was increased to 100 Å sec^{-1} the crystalline quality was degraded.

Davey (1962), by annealing chemically etched Ge substrates in a vacuum between 500 and 575°C, was able to deposit good Ge films at a 300°C substrate temperature. The vacuum used was 2–8 $\times$ 10^{-6} Torr with a deposition rate of less than 300 Å sec^{-1}. The 30 Ω cm source Ge was heated by electron bombardment and the Ge substrates were (111) or (100) oriented. Butorina and Tolomasov (1966) experimentally established that the quality of Ge grown at 500°C could be greatly improved by the introduction of iodine vapor during the growth process. The iodine serves to clean the substrate *in situ* and also brings the vacuum crystallization conditions closer to equilibrium.

Ryu and Takahashi (1965) deposited Ge films on GaAs substrates in a vacuum of 2×10^{-5} Torr. The films were generally *p*-type unless intentionally doped with P and had the best structure for substrate temperatures of 800°C. They conclude that at an 800°C substrate temperature the GaAs substrate is cleaned by a slow evaporation of itself. Using these conditions they were able to obtain *n–p*, *p–p*, *p–n*, and *n–n* Ge–GaAs heterodiodes. Davey (1966) obtained successful epitaxy of Ge on GaAs at temperatures as low as 300°C but found that a substrate temperature between 475–500°C was best for minimizing twinning. The Ge was heated by electron bombardment in an outgassed and vacuum-loaded graphite boat and evaporated at a rate up to 40,000 Å min^{-1}. The resulting films were *p*-type with a carrier concentration of 3 to 11×10^{17} cm^{-3} and a mobility near 500 cm^2 V^{-1} sec^{-1}. J. D. Williams and Terry (1967) found in their work on vacuum deposited Ge films that the film structure is a strong function of the substrate temperature and the deposition rate, which could account for the wide range of conditions reported by others for achieving epitaxial films. Vasil'ev and Tikhonova (1967) considered the change of substrate temperature during the growth cycle and found that there may be a marked change in substrate temperature. This may have major effects on the defect distribution and on the variation of properties within the film.

Haq (1965) successfully deposited epitaxial n- and p-type Ge films on Ge substrates using asymmetric ac sputtering. The electrode geometry used is shown in Fig. 9.18 together with 60 Hz asymmetric voltages which were applied to the source, substrate, and shields during sputtering. Initially a vacuum of 7×10^{-6} Torr was established while sputtering occurred at an argon pressure of 45×10^{-3} Torr. Prior to deposition the substrate was cleaned by 1000 eV ions for 5 min. For properly adjusted sputtering parameters n- and p-type films of approximately the same carrier concentration

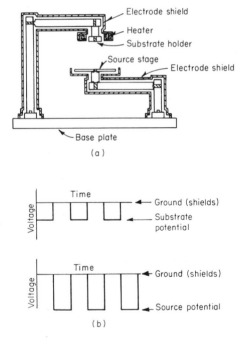

Fig. 9.18. (*a*) *Electrode geometry used for deposition of germanium film by sputtering.* (*b*) *Characteristics of voltages appearing on the source, substrate, and the shields during sputtering.* (After Haq, 1965.)

as the source could be deposited at substrate temperatures as low as 350°C. Krikorian and Sneed (1966) have characterized the structural properties of Ge films deposited onto Ge and CaF_2 substrates as a function of sputtering and evaporation parameters. They found that neither the particular deposition conditions which control the growth rate (voltage and current in sputtering, source temperature, and source–substrate distance in evaporation) nor the deposition techniques influence the transition from amorphous to polycrystalline to monocrystalline films as long as the background pressure and growth rates are identical.

9.5.2 *Growth of* Si *by Evaporation*

The quality of evaporated Si growth layers has improved as impurities and oxygen in the vacuum system have been eliminated and the pressure at which evaporation takes place has been reduced. With these improvements the substrate temperature at which epitaxial growth is obtained has been correspondingly lowered. Some of the first work in epitaxial growth of Si on Si in a vacuum was done by Unvala (1962, 1963, 1964). He obtained single-crystal Si layers on (111) Si substrates in a vacuum of 10^{-5}–10^{-6} Torr and at a maximum deposition rate of 3 μ min^{-1}. The critical substrate temperature for good single-crystal layers was in the range of 1100–1250°C. He found that the faster the deposition rate the higher the substrate temperature required for good quality films. A schematic diagram of his system is shown in Fig. 9.19. The substrate is shown heated by a tungsten filament, which unfortunately may contaminate the deposited film with W. Replacing the heater with a tantalum strip or an electron beam source eliminates this problem.

Hale (1963) obtained similar results for a starting vacuum of 5×10^{-7} Torr. Using doped Si sources he was able to obtain both p- and n-type doping. Boron transported with an efficiency of about 10% and P transported without any apparent loss. Newman (1964) points out that if a glass system is baked out to obtain a better vacuum the substrate surface is converted to p-type as a result of B transfer from the walls.

Nannichi (1963) using vacuum sublimation of Si was able to obtain good growths on (111) Si at temperatures as low as 830°C. The pressure of the system was reduced to 10^{-8}–10^{-9} Torr and the source and substrate were first heated to 1250°C for 1 h to bake out any absorbed gases and to obtain a clean surface. At 830°C it required 50 h, however, to obtain a 2-μ thick layer. Handelman and Povilonis (1964) used the sublimation technique to grow Si epitaxially on Si substrates at a temperature of 1100°C. The reaction chamber contained only Si and quartz and was evacuated to 10^{-8}–10^{-9} Torr. A Si pedestal acted both as the support for the substrate and as the doped

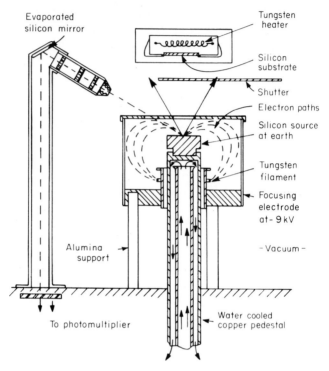

Fig. 9.19. *Equipment used for growing evaporated silicon films using electron bombardment heating of the source.*
(After Unvala, 1963.)

Si source. Both n-type (P doped) and p-type (B doped) films were grown at a rate of ~0.3 μ min^{-1} with a dopant transfer of approximately 100%. The system was also used to grow Si on Ge and Ge on Si, but with a much lower growth rate.

Using Si as the filament and a vacuum pressure as low as 10^{-10}–10^{-12} Torr, Widmer (1964) obtained single-crystal silicon films on (111) Si at a substrate temperature of 550°C. The minimum substrate temperature for epitaxial growth was a function of the manner in which growth was initiated. In one case the substrate was heated to above 1000°C for several minutes and then quenched to the growth temperature before the growth was initiated. Using this technique excellent single crystal growth was obtained at a substrate temperature of 800°C but defects appeared for a substrate temperature of 550°C. If a thin layer was evaporated onto the substrate above 1000°C prior to quenching it to the deposition temperature, good single-crystal layers were obtained at 550°C.

By using O_2-free Si source material Weisberg (1967) was able to obtain epitaxial Si films at similarly low temperatures in a vacuum of 3×10^{-8} Torr. The "Lopex" grade Si source was sublimed at a rate of $1-2$ μ h^{-1} and the substrate was cleaned *in situ* by heating it above 1300°C for 5 min using electron bombardment. On (111) substrates single-crystal growth was obtained as low as 450°C and on (100) Si as low as 380°C. Weisberg concludes that once oxygen is eliminated the effects of the vacuum pressure are greatly reduced.

Two other factors which hinder the orientation of thin films particularly at low temperatures are: (1) the formation of nuclei at different positions on the substrate which may be misoriented from each other and (2) the lack of sufficient surface migration so that the film can reach an ordered state with respect to the deposition rate. In order to overcome these factors Clark and Alibozek (1968) have deposited Si on Si through a moving quartz mask as shown in Fig. 9.20. The fixed Si mask has a "V" shaped opening to encourage growth from a single nucleation site. Silicon was sputtered onto the substrate using the moving mask at a substrate temperature of 200°C. Single-crystal films were obtained on Si substrates at 200°C and highly oriented films on oxidized substrates at the same temperature. Braunstein *et al.* (1968) have combined the vapor-liquid-solid growth mechanism with the moving-mask technique to grow single-crystal films of Si by vacuum evaporation on fused quartz substrates held at 800–900°C.

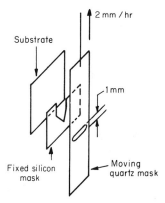

Fig. 9.20. *Schematic diagram of moving mask arrangement.*
(After Clark and Alibozek, 1968.)

9.5.3 *Vacuum Deposition of Compound Semiconductors*

Evaporation of compound semiconductors is usually quite different from that of Ge and Si because of the disparity of the vapor pressures of their components. Direct heating of the compound results in an evaporation of the more volatile component prior to that of the less volatile one, resulting in a

film which is nonstoichiometric. To overcome this problem, flash evaporation, multisource evaporation, and sputtering have been investigated.

9.5.3.1 *Flash Evaporation of Compound Semiconductors.* In this technique the
material to be evaporated is fed into the system in the form of very fine
grains and the temperature of the evaporator upon which the grains drop is
set as high as necessary to evaporate the least volatile component. Because of
the low thermal capacity of each small grain, it evaporates rapidly and nearly
completely so that when the grains are continuously fed into the evaporator,
the vapor produced will be approximately in the same proportion as that in
the source material. If the substrate temperature is also kept low enough to
prevent the reevaporation of the more volatile component, then the condensed film will be of the same composition as that of the source material.
Furthermore, if the substrate is a suitable oriented crystal then epitaxial
growth can take place.

A typical experimental arrangement used by Holloway *et al.* (1966) for the
flash evaporation of InSb is shown in Fig. 9.21. Granules of InSb were
fed continuously from a hopper onto a Ta or W boat that was held at
1600°C. The vacuum pressure was maintained around 5×10^{-7} Torr. The

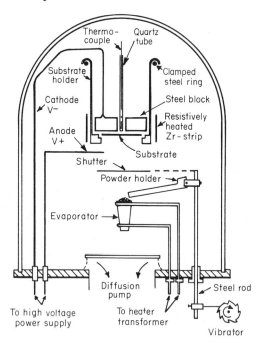

Fig. 9.21. *Apparatus for vacuum deposition by flash evaporation.*
(After Holloway *et al.*, 1966.)

InSb substrates were bombarded before evaporation to remove approximately 0.6 μ from the surface and subsequently annealed at 450°C to remove any damage. Growth took place at a rate of 10 μ h^{-1} at a substrate temperature of 440–460°C. Above 460°C two phases resulted. Measurements of X-ray rocking curves indicated the epitaxial layers had a crystal perfection similar to that found in bulk InSb.

Richards *et al.* (1963) used the same type of system for their study of the evaporation of several III–V compounds primarily on Ge substrates. A summary of their results is given in Table 9.2. The evaporations took place at

TABLE 9.2

List of Some Important Parameters in Epitaxy of III–V Compounds by Flash Evaporation

Compound	Lattice para-meters (Å)	Mismatch with Ge (%)	$T1$ (°C)[a]	$T2$ (°C)[b]	$T3$ (°C)[c]	$T4$ (°C)[d]	Evaporator temperature (°C)
GaP	5.450	− 3.7	400	500	540	560	1500
GaAs	5.653	0	400	400	475	535	1300
				− 475	− 525	− 550	− 1800
GaSb	6.095	+ 7.7	400	450	500	550	1650
InP	5.869	+ 3.7	100	200	300[e]	300	1400
				− 250	—	− 400	− 1650
InAs	6.058	+ 7.1	350	450	500	500	1500
InSb	6.479	+ 14.5	200	250	300	450	1650
				− 300	− 400		
AlSb	6.135	+ 8.4	625	700	—	—	1400
			− 650				− 1600

[a] $T1$—onset of oriented growth. [b] $T2$—epitaxy with twinning. [c] $T3$—untwinned epitaxy.

[d] $T4$—onset of reevaporation of group V element.

[e] Reevaporation of P also began at this temperature.

a vacuum pressure of 10^{-5} Torr. As can be seen the crystalline nature of the deposited film depended strongly on the temperature of the substrate but variations in the deposition rate between 10–100 Å sec^{-1} had no effect on the quality of the film. Using the same type of apparatus Müller (1964) did a detailed electron diffraction study of GaAs films deposited on (111), (110), and (100) Ge substrates. He found that for substrate temperatures above 300°C the films showed extensive orientation. Above 600°C films on the (111) and (100) surface were free of twins and epitaxial in nature. Below 600°C he found stacking faults on (111) planes and the appearance of a metastable hexagonal phase.

The growth of highly oriented layers of Cr doped semi-insulating GaAs on Ge and GaAs substrates has been reported by Light *et al.* (1968). The flash evaporation technique was used with Cr doped GaAs source material and varying amounts of additional As. They obtained the best results for substrate temperatures between 525 and 575°C and a 1 : 1 volume ratio of As to GaAs for the source material. From electrical measurements the films were similar to bulk Cr doped GaAs.

Ludeke and Paul (1967) using flash evaporation techniques have grown high quality epitaxial films of CdTe, HgTe, and CdSe on cleaved BaF_2 substrates. Epitaxial growth of CdTe was obtained for substrate temperatures of 70–350°C with the best films being grown at 200°C followed by an anneal at 350°C. The maximum temperature for growth was limited to 350°C by evaporation from the substrate. Epitaxial films of CdSe were grown with the wurtzite structure below 250°C and the sphalerite structure above this temperature. Cubic layers of CdS have been grown on InSb and CdS substrates by Holloway and Wilkes (1968). The cubic CdS films were grown on the (100), (110), and both (111) InSb faces at substrate temperatures between 350 and 375°C. Growths were also obtained on the (001)S and (001)Cd faces of CdS at 300°C.

9.5.3.2 *Sputtering of Compound Semiconductors.* In sputtering, an electrical discharge is passed between electrodes at a low gas pressure and the cathode electrode is slowly disintegrated under the bombardment of the ionized gas molecules. The disintegrated material leaves the cathode surface either as free atoms or in chemical combination with residual gas molecules. Some of the liberated atoms are condensed on surfaces surrounding the cathode. Using this technique, Moulton (1962) deposited InSb and GaSb onto substrates of fused silica and aluminosilicate glass. The depositions were carried out in an Ar back-pressure of $50–150 \times 10^{-3}$ Torr and a sputtering potential of 2–4 kV. The films were of the same conductivity type as the source and showed increasing Hall mobilities for substrate temperatures increasing from 50–250°C for InSb and 110–350°C for GaSb

Francombe (1963) has successfully sputtered epitaxial films of InSb, PbTe, Bi_2Te_3, and GaAs. The GaAs films were grown to a thickness of 1–12 μ on Ge substrates at 525°C. Molnar *et al.* (1964) made a more detailed investigation of the growth of sputtered films of GaAs onto silica and Ge substrates. GaAs films were epitaxially deposited on (100), (110) and (111) Ge substrates in the temperature range of 450–600°C at Ar pressures of 2.5 to 12×10^{-2} Torr. The best films were obtained on Ge substrates heated to 560–580°C. Above 580°C there was a breakdown of the GaAs crystal structure which was attributed to the loss of As.

9.5.3.3 *Thermal Evaporation.* The most common method for the thermal evaporation of compound semiconductors is the two-source three-temperature technique. In this technique each element of the compound is heated separately, as is the substrate on which the film is to be grown. In general the vapor pressure of one of the components is quite high compared to the other so that the temperatures of the two sources must be controlled independently so the vapor densities of both components are in the stoichiometric ratio. A discussion of the criteria for evaporation of good compound films is given by Gunther (1967). A typical experimental arrangement is shown in Fig. 9.22. This type of system has recently been used by Goodman (1969) to evaporate photoluminescent films of ZnSe on sapphire (1$\bar{1}$02) substrates. The films were grown onto the substrates at 670°C and were cubic in nature.

A different type of system arrangement shown in Fig. 9.23 was used by Davey and Pankey (1964, 1968) in their evaporation of GaAs films. Their earlier work was concerned with the structural and optical properties of GaAs films deposited on amorphous substrates. Steinberg and Scruggs (1966)

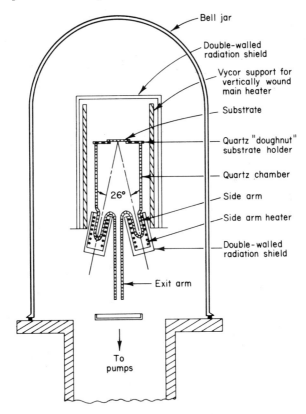

Fig. 9.22. *Cross-section schematic view of a multiple-arm evaporation chamber.* (After Goodman, 1969.)

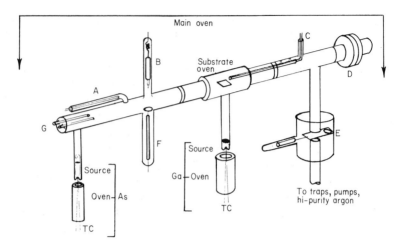

Fig. 9.23. *A GaAs film evaporation system.*
(A) Pirani gauge; (B) electrical contact to diode structure; (C) thermocouple; (D) metal flanges and viton gaskets as an entrance port for loading system; (E) particulate valve; (F) circular Ta plate, positive electrode in diode structure; (G) quartz rods which extend the length of the envelope and which guide the substrate carrier. (After Davey and Pankey, 1968.)

have similarly grown epitaxial films of GaAs on rock salt which has a lattice constant close to that of GaAs. Davey and Pankey (1968) in their recent work have grown GaAs epitaxially on Ge and GaAs substrates. The substrates and evaporated films were compared using reflection electron diffraction techniques. For a vacuum of 10^{-7} Torr and substrate temperature of 425–450°C, single-crystal films with little or no twinning were obtained on the low order faces of Ge, on the nonpolar faces of GaAs and on the (111)B GaAs face. Below 400°C the films were polycrystalline and above 450°C films on the (111) Ge and GaAs faces were hcp. The existence of the hcp structure was attributed to excess Ga or to special conditions of the substrate.

Cadmium sulfide has been evaporated by several investigators in their study of heterojunction diodes and transistors. Muller and Zuleeg (1964) and Dutton and Muller (1968) evaporated CdS on CdTe, Al_2O_3, and SiO_x for the investigation of the resulting diode properties. They used the technique described by Zuleeg and Senkovits (1963) where CdS was evaporated from a single source. As long as the CdS source zone was maintained below 750°C the deposited films retained their stoichiometry. The vacuum pressure was approximately 10^{-5} Torr and the substrate temperature around 150°C. After deposition the 2–3 μ films were baked at 300°C in a H_2S atmosphere for 10–15 min. This increased their resistivity by a factor of a hundred to 10^4 Ω cm and changed their color from orange-yellow to a lemon-yellow. Page (1965) has grown thin polycrystalline films of CdS on Si to form a space-charge limited transistor. These were grown in a vacuum to 10^{-7} Torr and a seed temperature of 200°C at a rate of 25 Å sec^{-1}. The resulting electron mobility was 6 cm² V^{-1} sec^{-1} compared to a bulk value of 200 cm² V^{-1} sec^{-1}.

Aven and Garwacki (1963) used the furnace arrangement of Fig. 9.24 to grow epitaxial films of CdS on ZnTe to form heterojunction diodes. In the crystal etch position in a H$_2$ atmosphere (600 Torr) the ZnTe evaporated at a slow rate cleaning the substrate surface. The CdS source temperature was then raised to 840°C and when a thin film began depositing on the walls of the tube the ZnTe seed was moved into the film deposition position. At a 600°C seed temperature the rate of growth was 1.5 μ min^{-1}. The quality of the resulting film depended greatly on the orientation of the substrate. Only the (111) Zn surface produced single-crystal CdS deposits. This was to be expected as only this orientation gives a good lattice match and the (111) Te surface showed deep furrows as a result of the thermal etching.

9.6 Solution Growth of Heterojunctions

9.6.1 *Discussion of Basic Principles of Solution Growth*

Solution growth processes were first successfully applied to semiconductors by Nelson (1963) for the fabrication of Ge–Ge tunnel diodes. Recently they

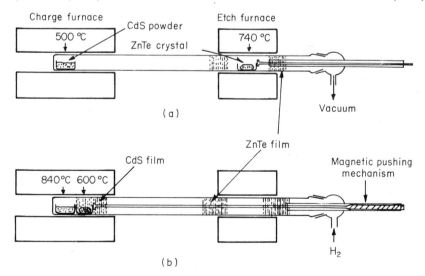

Fig. 9.24. *Experimental arrangement for growing epitaxial* CdS–ZnTe *junctions.*
(a) Crystal etch position. (b) Film deposition position. (After Aven and Garwacki, 1963.)

have become very important for the growth of high quality materials for injection light sources and lasers, Gunn devices, and ohmic contacts.

The physical basis for this process is that solute will precipitate from a saturated solution upon cooling. The binary phase diagram of a typical combination such as In–Ge is shown in Fig. 9.25. If an In–Ge mixture,

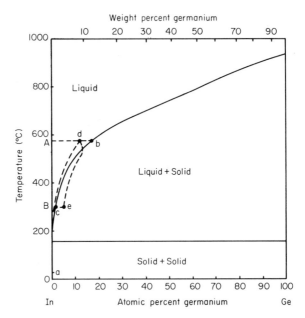

Fig. 9.25. In–Ge *binary phase diagram.*
(After Hansen, 1958.)

containing approximately 15% Ge by weight, is slowly heated to temperature *A*, a saturated liquid solution at point *b* on the solid–liquid curve will result, with the excess of Ge floating on top of the solution. If the solution is now slowly cooled to temperature *B*, Ge solute will precipitate from the solution, usually in the form of a single-crystal epitaxial growth if an appropriate seed is made available.

Since the temperature cycles used are usually too fast for the solution to remain in quasi-equilibrium, the composition of the solution will not always fall on the solid–liquid curve shown in Fig. 9.25. On heating, the solution probably will not be saturated and the composition will follow the dashed curve a–d instead of a–b. On cooling, a supercooled solution is possible and the composition may follow the dashed curve d–e instead of b–c.

Figure 9.26 shows a diagram of a solution growth system. It consists of a quartz furnace tube with a surrounding furnace. The temperature of the

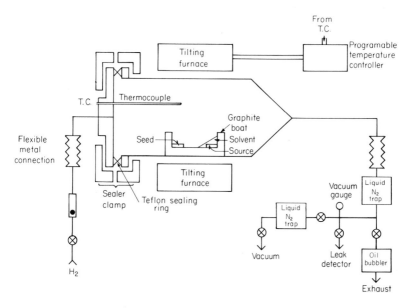

Fig. 9.26. *Schematic diagram of a system used to grow* Ge *by the solution growth technique.*

furnace is closely controlled by a thermocouple controller. The entire furnace is free to rotate to either side of the horizontal as indicated. The seed crystal is fixed by an appropriate clamping arrangement at one end of a graphite boat, which is locked in position at the center of a constant temperature zone by quartz rods. The solution consisting of the solvent, the semiconductor to be deposited, and a small amount of impurity, if desired, is free to run from one end of the boat to the other depending on the tilt of the furnace. The growths are carried out in hydrogen obtained from a palladium diffuser to prevent oxidation.

A typical growth cycle is shown in Fig. 9.27. With the furnace tilted so that the solution is at the end of the graphite boat opposite the seed, power is applied to the furnace. As the temperature increases, part of the excess semiconductor goes into solution. At a suitable maximum temperature, the power is turned off. As the furnace cools, the solution becomes saturated and semiconductor solute begins to precipitate. At an empirically determined temperature the furnace is tipped so the solution covers the seed crystal. During this growth period, semiconductor atoms precipitating from the solution grow epitaxially on the exposed surface of the seed crystal. In order to achieve good epitaxial growth it has been found necessary to actually wet (dissolve a few atomic layers of) the seed crystal. At the end of the growth period, the furnace is tipped back causing the molten metal solution to roll

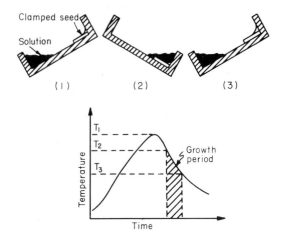

Fig. 9.27. *Typical solution growth cycle.*
(1) Heating period. Power on: $T < T_1$; power off: $T_1 > T > T_2$. (2) Growth period.
$T_2 > T > T_3$. (3) Cooling period. $T < T_3$.

off the seed to the other end of the boat. In actual practice some of the solution may cling to the seed, so at a slightly lower temperature it may be desirable to move the boat to the end of the furnace and wipe any remaining solution from the seed. When the temperature approaches room temperature, the seed is removed and cleaned in a chemical reagent, usually hot concentrated hydrochloric acid, to remove any remnants of solidified solution.

A different boat scheme, which has been used by Hayashi *et al.* (1970b) to grow up to four different epitaxial layers, is shown in Fig. 9.28. In this configuration, when solution 1 reaches the desired growth temperature the sliding solution holder is moved to bring the seed in contact with the solute. After the desired growth period, the solution holder is moved on so that the seed is under the graphite portion of the holder between solutions 1 and 2. Any excess solute left on the seed is wiped off by the bottom of the sliding solution holder. The above may be repeated then for solutions 2, 3, and 4 to grow a total of four layers. A small groove may be machined in the seed holder on either side of the seed to scrape off any surface film from the solute. Hayashi *et al.* (1970a) have used this method to grow GaAs and $Al_xGa_{1-x}As$ layers on GaAs seeds (see Section 5.7.3). An alternate configuration would hold the solution holder stationary and slide the seed holder to put the seed in contact with the various solutes.

While solution growth systems work well for the growth of a layer of the same material as the seed, the growth of a layer of a different material than the substrate, as in the fabrication of heterojunctions, involves an additional consideration. The conditions of growth usually should be such that the

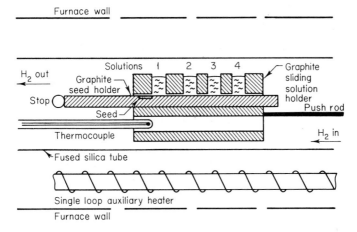

Fig. 9.28. *Solution growth system for the epitaxial growth of four layers using a sliding solution holder.* (After Hayashi *et al.*, 1970b.)

substrate material is not dissolved to any appreciable extent by the solution. This is necessary to prevent the grown layer from being either a eutectic structure or a regrown homojunction at the interface.

Kurata and Hirai (1968) discuss the conditions necessary to achieve abrupt heterojunctions. Consider for example the growth of Ge on Si from an Ag solution. As shown in Fig. 9.29, the Ge–Si alloy system exhibits complete solid solubility, and the lowest solidus temperature in this system is equal to the melting point of Ge. Therefore, to prevent the dissolving of the base Si, it is essential that the eutectic temperatures of the Si–metallic solvent and Ge–metallic solvent systems are lower than the melting point of Si and that the alloying temperature is still lower than the eutectic line of the Si–metallic

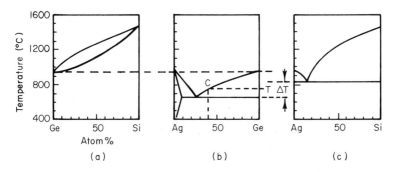

Fig. 9.29. *Phase diagrams involved in the growth of* Ge *on* Si *from an* Ag–Ge *solution.* (After Kurata and Hirai, 1968.)

solvent system. Assuming that the metallic solvent is silver and that the Ge–Ag system is heated to a temperature T, according to Fig. 9.29(b), the composition of the initial Ge–Ag solution will be C. The conditions for not dissolving the Si substrate are satisfied as long as T lies within the temperature interval ΔT as shown in Fig. 9.29(b) and (c).

Another very important consideration for successful solution growth is the temperature gradient which exists across the solute. For example, if the solute is poured over the seed and a small temperature gradient exists such that the top of the solute is cooler than the seed as the system is cooled, the growth from solution will be primarily as a crust on the surface of solute. This is undesirable since there may be no growth on the seed or harmful etching of the seed before growth begins. In an open tube system, as described earlier, this type of gradient may exist due to the cooling nature of the flowing gas and the close proximity of the seed holder to the furnace windings. It is important that the substrate is 0.5–1°C cooler than the solute surface to ensure growth occurs on the substrate. This may be accomplished in a horizontal system by a furnace with split windings where the upper winding is maintained a few degrees hotter than the lower. It also may be achieved in a vertical arrangement of the solution-growth system as described later.

9.6.2 Epitaxial Growth of Ge on GaAs and Si

Using a carbon boat and the tipping procedure described in Fig. 9.27, In–Ge–Ga solutions were used to grow p-type Ge and Sn–Ge–As solutions were used to grow n-type Ge on GaAs. Because of the need to actually wet the GaAs, solutions inevitably become contaminated with Ga and As, which are both shallow impurities in Ge. This contamination has a large effect on the resistivity of the germanium overgrowths. The effect is minimized by starting the growth at as low a temperature as possible and by having the initial growth rate quite large.

Using an In–Ge solution, consistent reliable results could only be obtained on the (111) Ga plane. Growths were initiated at 400°C and the solute poured off at 300°C, resulting in 25 μ overgrowths. With Sn–Ge solutions, consistent results were obtained on the (111) Ga and (111) As planes of GaAs. Growths were also obtained on the (110) plane, but the growth structure was usually very poor. Best results, as far as metallurgical quality and device characteristics were concerned, were usually obtained on the (111) As face. Growth occurred between 500 and 380°C and was approximately 50 μ thick. The grown layers were single crystal in nature and the junctions were abrupt and flat for either solution. As far as surface quality, dislocation density, and stacking faults were concerned, however, germanium overgrowths obtained using In–Ge solutions were far superior to those obtained using Sn–Ge solutions.

As noted previously, the cross-doping effect of Ga and As from the GaAs has a large effect on the resistivity of the Ge overgrowths. If no additional impurities were added to the solution, either In or Sn, the Ge overgrowths were found to be heavily *n*-type. In order to achieve *p*-type material, Ga has to be added to the solution. All this indicates that the As from the GaAs dominates the impurity concentration of the Ge overgrowth and any *p*-type material obtained is highly compensated and often has a graded impurity profile.

In an alloying-type solution growth process, Kurata and Hirai (1968) used Ag or Al as solvents for Ge. They placed the substrate material on a quartz plate and the appropriate Ge alloy on top of the substrate. This was heated to a temperature slightly higher than the melting point of the alloy and then cooled to the desired temperature with argon flowing through the furnace. The junction formed from the Al–Ge alloy appeared as an interface line and the transition from GaAs to Ge was abrupt with no intermediate layers. However, for the Ag–Ge alloy melt, an intermediate phase layer between the seed GaAs and the crystallized Ge was formed. From this result, it is supposed that a eutectic line of the Ag–GaAs alloy system may exist at a lower temperature than expected, and so the necessary conditions discussed earlier are not satisfied.

The growth of Ge on Si by the solution growth method is complicated by the stable oxide on Si which inhibits wetting by most solvents. Donnelly and Milnes (1966b) used an *in situ* cleaving technique to achieve an oxide-free Si surface to overcome this problem. Silicon L-bars with their major axis in the $\langle 111 \rangle$ direction were used as seeds and placed in the graphite boat as shown in Fig. 9.30. The L-bar was clamped at one end and a quartz rod was used to cleave the bar while it was surrounded by solution. Growth was initiated when the crystal was cleaved at 600°C and continued to 400°C when the solution was removed by tilting the boat. Back-scattered Laue X rays indicated good single-crystal Ge overgrowths in the $\langle 111 \rangle$ direction. However, slight elongation and splitting of some of the Laue spots were noted due to thermal strain and cracking, respectively, which occurred because of the large thermal mismatch between Ge and Si. The type and resistivity of the Ge overgrowths were controlled by adding small amounts of Ga and As to the solution for *p*-type and *n*-type Ge, respectively. Since there was no cross-doping effect as with Ge on GaAs, uniform doping levels as low as 10^{17} cm^{-3} were obtained.

Kurata and Hirai (1968) have grown Ge on Si using Ag as the metallic solvent. The Si substrate was (111) oriented and the Ag–Ge alloy contained 40% Ge (atomic percent). Growth took place between 770 and 650°C in an

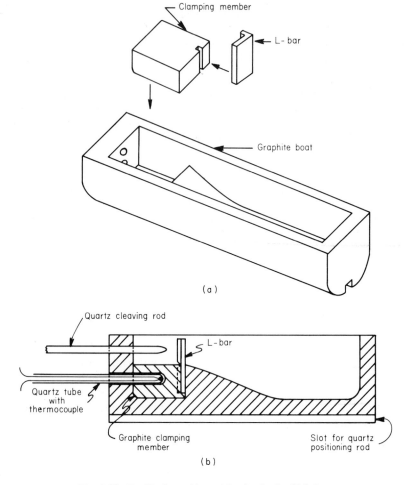

Fig. 9.30. *Graphite boat with provision for cleaving* Si *L-bars.*

Ar–20% H_2 ambient to promote wetting. For a cooling rate of $100°C$ h^{-1} the growth rate was 0.4 mm h^{-1}. The epitaxial Ge overgrowth was found to be heavily strained and cracked because of the expansion coefficient mismatch of Ge on Si.

9.6.3 *Solution Growth of III–V Compounds*

The solution growth technique has been particularly important in the growth of III–V compounds for optical and microwave device applications.

Its advantages are (1) that very high purity material can be grown presumably due to the gettering action of the metallic solvent, (2) the impurity concentration of the grown layer may be controlled, and (3) mixed III–V crystals can readily be grown. For optical diodes of GaAs and GaP this type of system has produced high efficiency light emitting devices.

Kang and Greene (1967) used a modified tilt tube technique to grow high quality GaAs on Cr doped semi-insulating GaAs substrates from a Ga solution. A graphite boat was used which was vacuum baked at 1400°C prior to use. They saturated the Ga solution with GaAs at 850°C and after tilting cooled at the rate of 200°C h⁻¹ to 400°C. Single-crystal layers 50–100 μ thick were obtained with doping densities $(N_D - N_A)$ as low as 1.65×10^{15} cm⁻³ and room temperature mobilities as high as 9300 cm² V⁻¹ sec⁻¹. No photoemission was observed due to deep-lying defect states for the epitaxial layers whereas two such peaks were seen for the original source material. This indicates a significant reduction in the related imperfections during the growth process. Harris and Snyder (1969) using a similar system have grown homogeneously tin doped GaAs using high purity Ga (99.99999% pure) and Sn (99.99999%). The solution was cooled from 860 to 500°C at rates between 6 and 90°C h⁻¹. Epitaxial layers were grown on (111), (110), and (100) GaAs faces. There were large differences in the cooling rates necessary to avoid surface instabilities and gallium inclusions in the epitaxial material. The (111) crystals were the least sensitive to growth instabilities while the (110) crystals required a much slower cooling rate, less than 8°C h⁻¹, to prevent instabilities. The "as grown" surfaces of epitaxial layers grown on (100) substrates at a cooling rate of 8°C h⁻¹, were mirrorlike and could be used for devices without further surface preparation. The carrier concentrations were reproducible between 3×10^{14} and 5×10^{16} cm⁻³. For $N_D - N_A$ equal to 1.3×10^{15} cm⁻³ the room temperature mobility was 7200 cm² V⁻¹ sec⁻¹.

The solubility of GaAs and GaP in various metallic solvents is given in Fig. 9.31 from the work by Rubenstein (1962, 1966a,b). A solution-growth system has been used by Lorenz and Pilkuhn (1966) to grow GaP onto GaP substrates using a Ga solution. Tellurium was used as the dopant for *n*-type growths and Zn for *p*-type. The Ga–GaP solute was heated to 1140°C, poured over the substrate, and cooled to 700°C in 40 min. Best growths were obtained on well oriented (111)B faces. They concluded that Ga inclusions were probable at steps in misoriented crystals. Trumbore *et al.* (1967) using the same technique obtained *n*-type Te doped GaP layers with carrier concentrations of $0.4–1.3 \times 10^{18}$ cm⁻³ on *p*-type GaP substrates, and *p*-type Zn doped layers with $1–4.5 \times 10^{17}$ cm⁻³ carriers on *n*-type substrates. The injection luminescence efficiency for these devices was higher than previously reported ($1–5 \times 10^{-3}$).

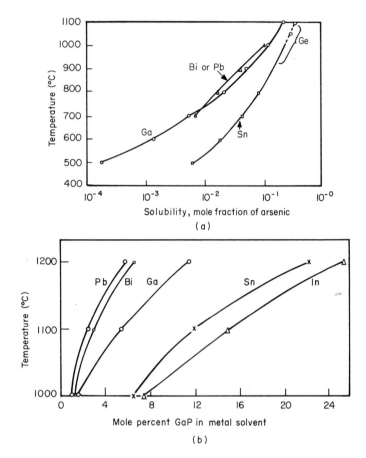

Fig. 9.31. (a) *Solubilities of* GaAs *as a function of temperature.* (b) *Solubilities of* GaP *as a function of temperature.*
[(a) After Rubenstein, 1966a; (b) After Rubenstein, 1962.]

Saul (1968) has studied the defect structure of GaP crystals grown by liquid and vapor phase epitaxy. He found GaP grown by liquid phase epitaxy had no dislocations introduced at the substrate-growth interface for diodes fabricated on dislocation free substrates. This was in contrast to those grown from the vapor phase. Recently, GaP solution-grown diodes have been reported (Saul, 1969a,b) with injection luminescence efficiencies of 6% (external) at 300°K, achieved by giving special attention to doping profiles and layer perfection.

Woodall *et al.* (1969) have used the vertical solution-growth system of Fig. 9.32 to grow $Ga_{1-x}Al_xAs$ on GaAs substrates. A typical time–temperature curve for their growths is also shown. The system was all quartz except for the substrate holder which was graphite and the Al_2O_3 crucible. This was necessary since Al reacts with quartz at temperatures as low as 750°C causing

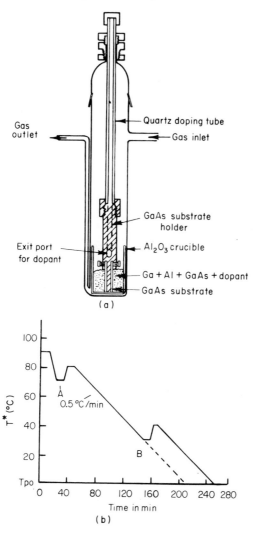

Fig. 9.32. $Ga_{1-x}Al_xAs$ *epitaxy from* $Ga+Al$ *solution.*
(a) Liquid phase epitaxy apparatus used for the $Ga_{1-x}Al_xAs$ system. (b) Cooling and heating schedule for $Ga_{1-x}Al_xAs$ epitaxy. (A) substrate is dipped into melt; (B) dopant is inserted into melt or cooling is continued along dashed line; (T^*) is temperature above pull out temperature, T_{po}; $(T_{po}) = 875°C$ for Al/Ga mole ratio $= 0.0013$–0.0104; $(T_{po}) = 910°C$ for Al/Ga mole ratio $= 0.013$–0.026. (After Woodall *et al.*, 1969.)

Si contamination. In this system the GaAlAs solute was heated to a given temperature and the substrate dipped into it at time A in Fig. 9.32(b). The temperature of the solute was then raised for a few minutes to dissolve any growth which occurred during the dipping and to etch-clean the seed. The temperature was then allowed to fall with a cooling rate of $0.5°C$ min^{-1} and growth occurred. If a $p–n$ junction was to be formed, dopant of the other type was added at B. The temperature was again increased as before, followed by cooling and growth. Epitaxial growths were obtained on (111) and (100) faces. The growth rate on the latter was 0.9μ min^{-1} between 955 and 850°C. The mole fraction of Al in the grown layer varied between 0.5 and 0.7 depending on the starting solution and growth temperature and decreased as one traveled away from the junction.

$Al_xGa_{1-x}P$ diodes have been grown on GaP substrates by Kressel and Ladany (1968) using the horizontal boat technique. From a Ga solution n-type material was grown by doping with Al or Te. The p-region was formed by out-diffusion of Zn from the GaP substrate. The resulting epitaxial $Al_xGa_{1-x}P$ layer contained 25% Al but the Al concentration was found to decrease away from the substrate.

A review of additional work on the growth of various compounds from metallic solutions is given by Luzhnaya (1968).

9.6.4 Solution Growth of the II–VI Compounds

The work on the solution growth of II–VI compounds has been mainly connected with the growth of the material without the aid of a substrate. Rubenstein (1966b, 1968) and Rubenstein and Ryan (1967) have grown ZnS, ZnSe, ZnTe, CdS, CdSe, and CdTe from Bi, Sn, Zn, and Cd solutions. The solubilities of these compounds in the various solutions as a function of temperature are given in Fig. 9.33. Crystals of the compounds were grown from solution by sealing the material and solvent in an ampoule with a background pressure of 10^{-5} Torr. The system was raised in temperature and a saturated solution formed with excess solid floating on top. With a temperature difference between the top solid and the bottom of the ampoule, crystals of the compound grew on the cooler bottom while the solid at the top supplied material to the solute. The temperature difference was in the range of 10–100°C. Hársy et al. (1967) have grown ZnS single crystals from Ga and In solutions. The ZnS and solvent was sealed in an evacuated ampoule. After reaching a temperature of 1180–1200°C they slowly cooled the solute to 4–500°C at the rate of 15–20°C h^{-1}. During this period small crystals grew on the walls of the evacuated tube.

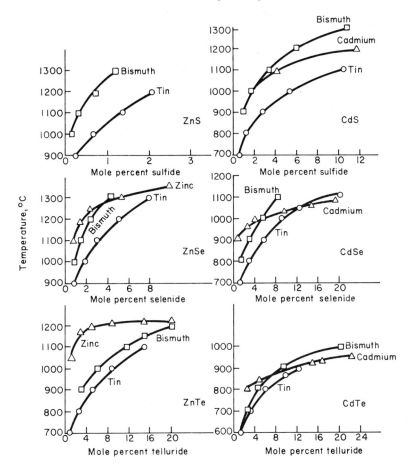

Fig. 9.33. *Liquidus solubilities of six II–VI compounds in* Bi, Sn, *and their parent metals.*
(After Rubenstein, 1968.)

Solution-growth epitaxy studies of II–VI compounds are in progress, and papers may be expected on these topics in the next few years.

9.6.5 *Traveling Solvent and VLS Methods of Solution Growth*

The traveling solvent and vapor–liquid–solid (VLS) techniques are two methods of application of solution growth. The traveling solvent method (TSM) was used by Weinstein *et al.* (1964) to prepare abrupt GaP–GaAs heterojunctions. A schematic representation of the method used is shown in

Fig. 9.34. Because of the increased solubility of GaAs in Ga at higher temperatures the GaAs tends to dissolve at the upper liquid–solid interface and deposit out at the lower interface. Using this technique with a ΔT of 170°C and a lower interface temperature of 650°C, the GaAs grew epitaxially on GaP. The transition between the materials was abrupt since the rate of solution of GaP in Ga at 650°C is much less than that of GaAs at 820°C. By doping the liquid Ga with Ge, p-type GaAs was obtained.

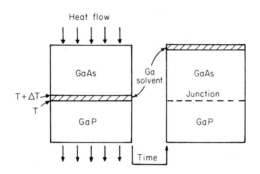

Fig. 9.34. *Schematic representation of traveling solvent method for growth of* GaP–GaAs *heterojunction.*
(After Weinstein *et al.*, 1964.)

In the VLS system first used by Wagner and Ellis (1964) to grow silicon, the crystal grew from a solution present as a layer or droplet consisting of the solvent Au and crystal component. Crystal growth began when the Au solution became supersaturated with the silicon material. The crystal material was transported to the solution by provision of a suitable vapor phase component. Gallium has been used as the solvent to grow crystals of GaAs and GaP by this technique. Ellis *et al.* (1968) have used controlled VLS growth to grow small GaP crystals of high perfection on GaP substrates near 1000°C.

9.7 Alloying of Heterojunctions

Single-crystal junctions between different semiconductors can be produced by alloying techniques. The alloying techniques used may be divided into three categories: interface alloying, alloying from a metallic solution, and alloying by completely melting one of the materials. Dale (1966) in his review paper on alloyed heterojunctions makes a comparison of these various alloying techniques as shown in Table 9.3.

TABLE 9.3

Comparison of Alloying Techniques for the Fabrication of Semiconductor Heterojunctions

Process	Advantages	Disadvantages
Interface alloy	Simple equipment and technology Short process time Wide range of heterojunction pairs possible	High temperatures required Size limited (few mm^2) Generally highly strained (dependent on components) Crystal alignment may be difficult
Solution growth (from alloy system)	Simple equipment and technology Low alloying temperatures Wide range of heterojunction pairs possible including solid solutions	Complex metallurgy Long process time High risk of contamination by fast diffusing impurities Size limited (few mm^2)
Solution growth (traveling growth solution)	Simple equipment and technology Simple metallurgy Relatively low process temperatures Lowly stressed layers Large area heterojunctions possible	Long process time High risk of cross diffusion Limited range of compatible materials
Vapor transport -interface alloy	Large area heterojunctions possible Relatively short process time	Very high temperatures required Technology and equipment more complex than by other processes Limited range of suitable materials High risk of contamination by impurities At present highly strained layers

9.7.1 *Interface Alloying*

In interface alloying the heterojunction is formed by placing two semi-conductor materials in contact with each other and heating the sandwich so that a temperature gradient exists across the interface. A typical system which uses a carbon heater strip is shown in Fig. 9.35. In this arrangement the higher melting-point material is placed adjacent to the heater strip so it is held at the higher temperature. The temperature of the strip is raised until 5–20 μ thickness of the lower-melting-point semiconductor adjacent to the interface becomes liquid. In order to assure wetting of the two materials they

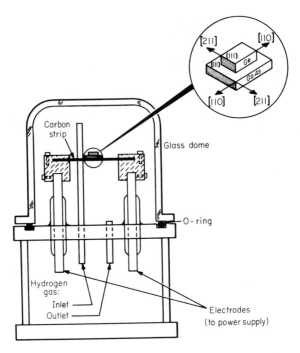

Fig. 9.35. *Schematic representation of the hot stage used for the fabrication of heterojunctions.* Insert shows the sample placement and alignment. (After Mroczkowski *et al.*, 1965.)

should be properly cleaned and the alloying done in a hydrogen ambient. As soon as wetting is observed and before the complete wafer of the lower melting-point material can liquefy the temperature of the heater is reduced. This causes the melted region to recrystallize forming the heterojunction, Rediker *et al.* (1964, 1965) and Hinkley *et al.* (1965) have grown GaAs–Ge. GaAs–GaSb, GaAs–InSb, and InAs–GaSb heterojunctions using this technique. In all cases the regrown material was found by Kossel line patterns to be single crystal in nature even though the lattice mismatch between GaAs and InSb is 14% and that of GaAs–GaSb is 8%. Rediker *et al.* (1964) report that there is a rotation of the (111) faces of GaAs and GaSb of 20° with respect to each other to accommodate for the 8% mismatch.

Rediker *et al.* (1964) used an electron beam microprobe to determine the chemical composition across the interface of their GaAs–Ge and GaAs–GaSb junctions. The results of a typical electron beam microprobe analysis for Ge alloyed to GaAs to form a heterojunction is shown in Fig. 9.36. The interface region is seen to be approximately $4\,\mu$ wide and the change in Ge, Ga, and As content is not monotonic. This can be explained by assuming that during the alloying cycle a small amount of GaAs is dissolved by the Ge and that on refreezing the segregation coefficient for GaAs is less than unity. Since Ga and As are dopants in Ge, and Ge is a dopant in GaAs, a complex interface structure is possible. Mroczkowski *et al.* (1965) in a study of their GaAs–Ge heterojunctions found using an electron probe that the boundary layer

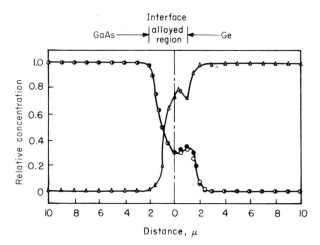

Fig. 9.36. *A typical electron beam microprobe analysis of an alloyed* GaAs–Ge *heterojunction.* △: Ge; ○: Ga; ●: As. The 100% lines correspond to the intensity of the GeK$_\alpha$ line from the bulk Ge and the intensities of the As and GaK$_\alpha$ lines from the bulk GaAs. (After Rediker *et al.*, 1964.)

between the GaAs and Ge consisted of a four-layer structure. On the other hand an investigation of the GaAs–GaSb junction by Rediker *et al.* (1964) showed that the transition from As to Sb atoms was abrupt, without structure, and occurred within 2–3 μ.

It is clear from the above that any interpretation of alloyed heterojunction properties must be made with caution and only after a thorough investigation of the interface structure.

9.7.2 *Alloying by Complete Melting of One of the Materials*

A similar method to that described above for obtaining a heterojunction is the method in which one of the materials is raised to a high enough temperature so it is molten throughout. The molten semiconductor may be obtained by heating a solid pellet above the melting point or depositing it from the vapour phase on a substrate whose temperature is sufficiently high.

Shirafuji and Nakayama (1964) obtained rectifying junctions between GaAs and Ge by melting a Ge pellet onto an etched GaAs surface. In order to obtain a smooth interface upon cooling they found the cooling rate needed to be less than 1–3°C min⁻¹. An analysis of these alloyed diodes showed them to have a *p–i–n* structure indicating a complex interface.

Germanium–silicon heterojunctions were fabricated by Thompson and Reenstra (1967) and Shewchun and Wei (1964) by alloying Ge into Si. In the Thompson and Reenstra system the alloying fixture consisted of two heater strips separated by a quartz spacer. The Si substrate was placed on the lower one and the Ge on top of the Si. The temperature cycle shown in Fig. 9.37 was used with the heating and cooling rates indicated. Above 900°C

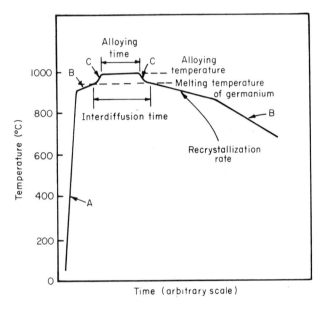

Fig. 9.37. *Typical fabrication temperature cycle for alloying* Ge *into* Si.
A—10°C sec⁻¹, B—1°C sec⁻¹, C—2°C sec⁻¹. (After Thompson and Reenstra, 1967.)

a temperature difference of 20°C was maintained between the heaters. The time at the alloying temperature was varied as well as the cooling rate during recrystallization. Measurements of the resulting *n–n* junctions showed that a *p*-layer was present in a Ge–Si alloy transition region. The width of the *p*-layer, which was between 3 and 7 μ increased with increasing inter-diffusion times and with decreased recrystallization rates. They conclude the *p*-layer results from energy levels caused by dislocations in the alloy region due to the 4% lattice mismatch.

In their fabrication process, Shewchun and Wei place a small chip of Ge on a Si substrate which sits on a strip heater. Their alloy cycle consists of an abrupt heating to a temperature between 950°C and 1200°C for a period of

3–15 sec followed by an abrupt cooling. The short alloying times were employed in order to form abrupt junctions with minimum cross diffusion. Measurements of the resulting junction showed it to be 10 μ wide and to consist of an abrupt transition from Si to Ge through a 1 μ Ge–Si regrowth layer. From electrical measurements they conclude that large dislocation densities exist at the interface which give rise to tunneling currents via defect states.

Germanium–silicon heterojunctions have been fabricated by Brownson (1964, 1965) by depositing Ge from $GeCl_4$ on a Si substrate held at a temperature considerably above the melting point of Ge. The initial growth was with the Si substrate held at 1100°C in order to have an oxide-free Si surface. After 60 sec at 1100°C the temperature was reduced to 970°C for 180 sec. The germanium halide was then turned off and regrowth on the Si surface took place from the liquid Ge–Si alloy as the system cooled. The regrown $Ge_{0.75}Si_{0.25}$ alloy was single crystal but under considerable strain. The lattice and thermal mismatch between the alloy and substrate led to cracking of thick layers and only low resistivity growths were possible. The reverse characteristics of these junctions were quite poor, presumably due to a poor interface region.

A method of producing heterojunctions between compound semiconductors by alloying and a substitution reaction has been described by Kodera *et al.* (1966). In this system an alloying material A was placed on a wafer of a compound semiconductor BC and the whole assembly was heated to a predetermined temperature so as to form a molten mixture of A and BC. If the formation energy of a compound semiconductor AC were smaller than that of BC at that temperature, the compound AC would be formed by the substitution reaction and would be crystallized on the substrate BC during the cooling process, thus producing a heterojunction between two compound semiconductors AC and BC. The necessary conditions for obtaining heterojunctions by this process are: (1) the formation energy of AC should be smaller than that of BC at the alloying temperature, and (2) the alloying material A and the substrate material BC should form a molten alloy below the melting point of BC. Such conditions are satisfied by the combinations, Al–GaP, Al–GaAs, Al–GaSb, and Cd–HgSe, resulting in heterojunctions AlP–GaP, AlAs–GaAs, AlSb–GaSb, and CdSe–HgSe, respectively. A three-layer structure can be formed by using a compound AD in place of A. Upon cooling, layers of AC and BD can crystallize successively on the substrate BC.

Both two- and three-layer structures have been obtained experimentally by alloying Al and AlSb, respectively, to GaAs. In the former the Al was alloyed at 900°C in flowing H_2 while in the latter the alloying took place between AlSb and GaAs in a sealed quartz ampule at 1150°C. These resulted in AlAs–GaAs and GaSb–AlAs–GaAs heterojunctions.

9.7.3 *Alloying from a Metallic Solution*

In this technique low melting point alloys of a metal and the semiconductor to be grown are heated in contact with a semiconductor substrate. Alloys with melting points below 600°C are normally used in order to avoid diffusion effects between the substrate and grown material. The process is very similar in nature to that of solution growth (see Section 9.6) but the alloy, a pellet, or sphere, is much smaller than the substrate.

Diedrich and Jotten (1961) used various alloys of In, Sn, Bi, Al, Au, and Si in order to form a wide-gap Si emitter on Ge. In many cases lattice disturbances, cracks, and poor recrystallization resulted in very poor junction characteristics. This was particularly true of Au–Si, Al–Si, Zn–Si, and Au–Ge–Si alloys. Bismuth alloys were not particularly useful because of a very small penetration depth and Sn alloys resulted in poor electrical properties. The best transistors were made when In was one constituent of the alloy; In–Si, In–Ge–Si, and In–Au–Si (Ga) alloys all yielded good emitter junctions and transistors. No measurements were reported which determined the composition of the recrystallized region, so it is not clear what type of junction existed.

More recently Dale and Josh (1965) have used Bi based alloys containing various amounts of GaSb, InAs, and Mn to grow, by alloying, epitaxial layers of varying composition on n–GaAs substrates. In their work, they first premelted the alloy pellets onto the GaAs and then alloyed them in evacuated silica tubes at 500–600°C for 30–120 min. This was followed by slow cooling for 1–8 h in order to prevent cracking of the grown layers. A summary of their work is presented in Table 9.4. Metallographic and microprobe analysis of the regrown layers from Bi–GaSb and Bi–InAs alloys showed that the regrown layers were solid solutions of $GaSb_xAs_{1-x}$ or $Ga_xIn_{1-x}As$ on GaAs and were abrupt single-crystal p–n heterojunctions. Similar studies of the junction formed from the alloys containing Mn showed two distinct regrown layers. The first adjacent to the GaAs was presumed to be a solution of $Ga_xIn_{1-x}As$ ($\sim\frac{1}{2}\ \mu$ thick) while the second was found to be Mn_2As. This presumably gave a metal–semiconductor junction in series with a semiconductor heterojunction.

9.8 Measurement Techniques

9.8.1 *Techniques for Detecting a Hidden Homojunction*

In the fabrication of heterojunctions one must be concerned whether or not a spurious homojunction is present near the metallurgical interface between the

TABLE 9.4

Alloy Systems for the Solution Growth of Solid Solution–GaAs Heterojunctions[a]

Alloy composition (part by weight)	Minimum prewetting[b] temperature (°C)	Solution-grown layer	Approx. composition of solution-grown layer	Remarks
1. 90 Bi, 10 GaSb	350	$GaSb_xAs_{1-x}$	Arsenic content dependent on alloying temperature, x generally 0.9 when alloyed at 500 to 600°C	Good diode characteristics —photosensitive
2. 80 Bi, 20 GaSb	350	$GaSb_xAs_{1-x}$		
3. 90 Bi, 10 InAs	350	$Ga_xIn_{1-x}As$	x generally of the order of 0.5 for an alloying temp. of 550 to 600°C	Poor diode characteristics. Alloy 4 more constant than 3
4. 80 Bi, 20 InAs	350	$Ga_xIn_{1-x}As$		
5. 90 Bi, 5 InAs, 5 Mn	400	$Ga_xIn_{1-x}As–Mn_2As$	x generally of the order of 0.5 for an alloying temp. of 550 to 600°C	Two-layer structure interface layer thought to be $Ga_xIn_{1-x}As$. 6. Good diode characteristics photosensitive
6. 85 Bi, 10 InAs, 5 Mn	400	$Ga_xIn_{1-x}As–Mn_2As$		
7. 75 Bi, 15 InAs, 10 Mn	450	$Ga_xIn_{1-x}As–Mn_2As$	x of the order of 0.05	Interface layer $\frac{1}{4}\mu$ thick Mn_2As up to 50 μ. Good diode characteristics —photosensitive

[a] After Dale and Josh (1965).
[b] The minimum prewetting temperature is the minimum temperature at which the alloy will "wet" the surface of a GaAs substrate.

two semiconductors comprising the heterojunction. In the n–p GaAs–Ge case, for example, such a spurious homojunction may occur in the Ge because of As present in the epitaxial reactor. In the fabrication of n–p–n GaAs–Ge–Ge transistors the p-type base doping was made very high and the operation of the epitaxial reactor was adjusted to make unlikely the formation of a spurious homojunction.

Several experiments were performed to determine whether or not an unwanted homojunction existed in the Ge near the GaAs surface. The principal experiment consisted of lapping a sample at a small angle (100 to 1), contacting the n–Ge collector or bulk region with a fixed W probe, and contacting the n–Ge collector, p–Ge base, or n–GaAs emitter regions with a movable W probe. The sample was illuminated with a focused microscope

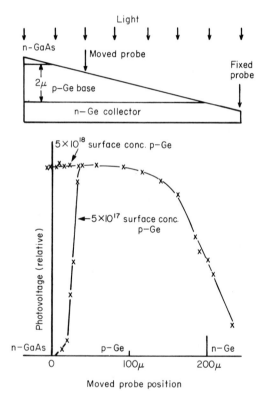

Fig. 9.38. *Photovoltage measured between a fixed collector probe and a moved probe versus position of the moved probe across the base region.*

The probe position shown on the horizontal axis corresponds directly with its position on the angle lap shown above. A horizontal movement of 10 μ corresponds to a vertical penetration of 0.1 μ in the base. (After Jadus and Feucht, 1969.)

lamp, and the photovoltage between the two probes was recorded as a function of the moved probe position. Figure 9.38 shows the sample-probe geometry along with two typical results. The vertical axis of the plot is the measured photovoltage. The horizontal axis is the position of the moved probe which corresponds directly with the probe position along the lapped sample above the curves. The curve labeled 5×10^{18} cm^{-3} was the typical result for devices in which the surface concentration of acceptors in the Ge base region was 5×10^{18} cm^{-3} or greater (up to 5×10^{19} cm^{-3}). For this case, the photovoltage rose as the probe moved through the collector depletion region. The photovoltage then stayed nearly constant up to and just beyond the Ge–GaAs interface. For the curve corresponding to a Ge base-region acceptor-surface concentration of 5×10^{17} cm^{-3}, there was a marked decrease in photovoltage as the Ge–GaAs interface was approached. This decrease is due to the spurious homojunction found when the p–Ge doping was too low. It was also observed that even 7×10^{18} cm^{-3} uniformly doped p–Ge seeds were insufficient to prevent a spurious junction unless the reactor was properly operated.

The absence of any spurious homojunction and attendant n–Ge surface layer was also indicated by measurements of the potential difference between the interior of the p–Ge base and the surface of the Ge adjacent to the GaAs when current was flowing. In order to gain access to the surface of the Ge, the GaAs was etched off a portion of the Ge using a methanol–1% Br solution. This etch easily removes the GaAs exposing the Ge surface but does not etch the Ge. Solutions which etch one semiconductor material rapidly without significantly attacking another are useful in the investigation of hetero-junctions. A list of etches and their effects on semiconductors that we have investigated in our heterojunction studies are shown in Table 9.5.

9.8.2 *Electron Microprobe Studies*

An electron microprobe may be used as an analyzer to determine the profile across a junction or, through Kossel line patterns, to determine the crystalline perfection of a layer. As an analyzer the electron beam is focused to a 1 μ-size spot on the surface. It is operated at a voltage such that characteristic X rays are generated at the spot where the beam strikes. The emitted X rays are analyzed with respect to wave length and intensity and compared with a known standard. From these measurements the percentages of various elements within the volume which interacts with the electron beam can be determined. Rediker *et al.* (1964) used this technique to determine the concentrations of Ge, Ga, and As across an alloyed Ge–GaAs junction (see

TABLE 9.5

Etches and Their Effects on Semiconductors[a]

	Etch									
Semi-conductor	HCl	H$_2$SO$_4$	HNO$_3$	White-etch[b]	Methanol bromine[c]	NaOCl[d]	NaOH[e]	H$_2$O$_2$	Chromic acid	Other etches
Ge	N	N	S	V	N	S	M	S	N	
Si	N	N	N	V	N	N	M[f]	N	N	
SiC	N	N	N	N	N	N	N	N	N	
GaAs	S, N	S	S, N	V	S, M	S	S	N	N	
GaN	N	N	N	N	N	N	S[f]	N	N	
GaP	N	N	N	S	S	S[g]	N	N	N	
GaSb	N[g]	N	V	V	M	S	N	N	S	M[h]
InAs	M[i]	N	M	M	M	N	N	N	S	M[j]
InP	V, M	N	N	N	M	N	S	N	N	
InSb	N[k], S[g]	N	V	V	M	N	N	N	N	M[l]
CdS	M	N	V	S	S	N	N	N	S	
CdSe	S[i,m]	N	V	V	M	N	N	N	S[m]	M[n]
ZnS	S	N	N	N	S	N	N	N	S[g]	
ZnSe	S[g]	N	M[m]	M	V	S[k], M[g]	M[g]	V	S[g]	
ZnTe	S[i]	N	M[m]	V	M	N	V	N	S	

[a] All solutions are at room temperature and concentrated unless otherwise noted. V—etches material vigorously; M—etches material moderately; S—etches material slightly; N—does not perceivably etch material.

[b] 1 HF : 4 HNO$_3$.

[c] Approximately 5% Bromine.

[d] 30% NaOCl.

[e] Approximately 20% NaOH and warm 40–50°C unless noted otherwise.

[f] 50% NaOH and hot 90–100°C.

[g] Hot 90–100°C.

[h] Etch: 1 HF : 2 HNO$_3$: 1 CH$_3$COOH.

[i] 20 to 100°C.

[j] Etch: 1 HF : 5HNO$_3$: 1 CH$_3$COOH.

[k] 20°C.

[l] Etch: 2 HF : 1 HNO$_3$: 1 CH$_3$COOH.

[m] Tends to form surface layer which inhibits etching.

[n] Etch: (12 K$_2$Cr$_2$O$_7$ sat. sol. : 4 H$_2$SO$_4$) + (3 HCl) at room temp. Rinse: 1 Na$_2$S$_2$O$_4$: 1 NaOH : 3 H$_2$O at 85°C.

Fig. 9.36). Similarly Ewing and Smith (1968) used this type of analysis to determine the composition of Ga(AsP) epitaxial layers grown on GaAs.

In the determination of crystal structure using Kossel lines the pattern of diffracted X rays is used for the analysis. The X rays which are generated at the spot where the beam strikes diverge in all directions as from a point source. Wherever the Laue conditions are satisfied, diffraction effects occur forming a set of cones. The resultant conics form the complete self-diffraction pattern for the crystal and are called Kossel lines. These patterns contain information on the crystal structure and unit cell dimensions of the material under study. For a detailed description of the mechanism of Kossel diffraction see Lonsdale (1947) and Hanneman *et al.* (1962).

In their study of the metallurgical properties of Ge–GaAs heterojunctions, Mroczkowski *et al.* (1965) used Kossel patterns. They obtained patterns at various points on the junction and superposed the resulting patterns to find that the junction region had the identical orientation of the substrate. They conclude from their work that approximately 10 μ of material is required in order to obtain an unambiguous pattern. Heise (1962) has reported the precision determination of lattice constants using a Kossel line technique.

9.8.3 *X-Ray Studies*

A common method for investigating the crystallographic orientation and single crystal nature of epitaxial layers is to use the back-reflection Laue method. A continuous beam of X-ray radiation, usually from a tungsten target, is allowed to fall on the fixed crystal under investigation. X rays are diffracted whose wave length satisfies Bragg's law for the particular values of d, the separation of parallel planes, and θ, the Bragg angle. The diffracted beams corresponding to different wavelengths appear as a characteristic pattern on film placed between the X-ray source and the crystal. The most usual information obtained from this is the orientation of the sample under consideration and whether it is single crystal. Riben *et al.* (1966) used this technique to verify the single-crystal nature of epitaxial layers of Ge on GaAs and Si. Studies of Ge/GaAs structures indicated no distortion of the Ge film or GaAs substrate. However, Laue photographs of Ge/Si structures show considerable distortion in the form of elongated Laue spots with striations as previously reported by Ruth *et al.* (1960). The latter attributed the distortion to the lattice mismatch between Ge and Si; however, such distortion in the Laue photographs has not been found to be present in other heterojunction pairs with equally large lattice mismatches (InP–GaAs and InAs–GaAs), nor in heavily dislocated Ge layers grown on any Ge or GaAs substrates. In addition, it has been observed that, in Laue photographs taken at 350°C, some of the distortion disappears, suggesting that the greater shrinkage of the Ge overgrowth during cooling from the growth temperature due to differences in thermal expansion coefficients is responsible for the distortion. The striations in the spots are considered to be reflections from the various slightly misoriented facets separated by cracks. This has been further confirmed by an experiment in which the X-ray beam diameter was increased and the striations became more pronounced. For thin epitaxial films one must be careful the layer is sufficiently thick to insure that one is looking at diffraction from the grown layer rather than from the substrate. The use of a Cu target, instead of W, produces longer wavelength radiation which has a smaller penetration depth so it is often used for thin layers.

X-ray reflection topography can be used to characterize nondestructively a crystal revealing imperfection in the crystalline structure. Barrett and Massalski (1966) describe the use of the reflection Berg–Barrett geometry, as shown in Fig. 9.39 for making topographs. In this system the K_a radiation is reflected from the crystal under study onto a fine grain photographic film. The intensity of the reflected radiation is a function of the crystalline structure so that twin regions, dislocations, etc., can be seen on the film.

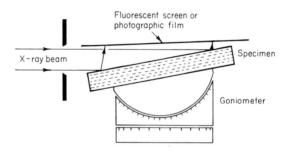

Fig. 9.39. *A crystal (or polycrystalline specimen) mounted on a goniometer in position for Berg–Barrett photographs.*
(From "Structure of metals," by C. S. Barrett and T. B. Massalski. Copyright 1966, McGraw-Hill, New York. Used with permission of McGraw-Hill Book Company.)

Krause and Teague (1967) and Meieran (1967) used this technique to observe dislocations at the GaAs–Ge interface which they regard as misfit locations resulting from the lattice mismatch. Holloway and Bobb (1968) and Light *et al.* (1968) have observed plastic deformation in Ge layers grown on GaAs and slip on the (111) planes. They concluded that the cause for the stress was the differential coefficient of thermal expansion between Ge and GaAs. Light *et al.* found that slip occurred only along two (111) planes for samples deposited at 350°C but heated to 500°C. For Ge deposited at 600°C they observed slip along all four (111) planes when the specimen was cooled to room temperature. Bobb *et al.* (1966a–c) and Holloway *et al.* (1966) have reported, in a series of papers, the use of X-ray rocking curves for determining the crystal perfection of epitaxial deposits. A double-crystal diffractometer, shown schematically in Fig. 9.40, was used in their work. CuK$_a$ radiation was diffracted first by a low-dislocation density reference crystal R and the α_2 component eliminated by a slit. The second crystal W was the epitaxial specimen. This was mounted on an eccentric goniometer whose arc was positioned so that rotation of the specimen did not alter the position of intersection of the beam with the surface. The specimen was rotated about an axis lying in the reflecting planes to vary the angle *g*. The intensity of the detected beam reached a maximum at one angle and decreased rapidly on either side of it. The width of the rocking

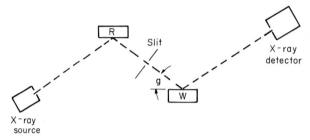

Fig. 9.40. *Schematic diagram of double-crystal X-ray rocking curve system.*

curve was a measure of the perfection of the grown layer. For the best layers of GaAs grown on GaAs and Ge substrates they found that the rocking curves were similar to those for good bulk crystals. The perfection of the layers grown were found to be quite dependent on the surface preparation used, both for growth of GaAs on GaAs and Ge and InSb on InSb. The width of the curve was also broadened for crystals containing high dislocation densities.

9.8.4 *Electron Diffraction*

Electron diffraction has been used by Müller (1964) and Holloway *et al.* (1965) to study the structure of films of GaAs grown on Ge. In this system a beam of electrons is accelerated through a high potential and impinges on the sample under study. The pattern of diffracted electrons is then recorded photographically on a suitably placed film. This is carried out in a high vacuum so the electrons are not deflected by gaseous atoms. The depth of penetration of the electron is a function of the accelerating voltage which varies from 50 V to 100 kV and of the material properties. For low voltages (50–1000 V) we have low energy electron diffraction (LEED) which is capable of looking at single atomic layers. For higher voltages the diffraction takes place up to depths of several hundred angstroms. Since only material near the surface is involved, the cleanliness and preparation of the surface is of particular importance. Contamination of grease, impurities, oxides, etc., can greatly alter the diffraction pattern obtained.

Müller (1964) investigated films of GaAs deposited on Ge by flash evaporation while Holloway *et al.* (1965) used HCl to grow GaAs on Ge. Electron diffraction patterns of the layers showed stacking faults and twinning for low growth temperatures. Müller found twin free deposits above 600°C on (111) and (100) surfaces while Holloway *et al.* found the temperature for twin-free deposits was a function of orientation in the following order (001) < (110) < (111).

The observation of Auger electrons emitted from a surface due to an impinging electron beam provides a sensitive method for investigating the surface of a material. Harris (1968a) and Chang (1971) describe the technique and general principles for using this as a probe to study contaminants on a surface. Since the energy of the bombarding electrons is less than 1–2 keV, the mean free path of the electrons is quite small and the results are very sensitive to the surface properties. Chang (1970) has investigated Si surfaces and detected O_2 and C in concentrations as low as 5×10^{12} atoms cm^{-3} and 2×10^{13} atoms cm^{-3}, respectively.

N. J. Taylor (1969) has used Auger electron measurements to study the photoyield of GaAs–O–Cs photocathodes. He found that there was a linear relationship between the decrease in photoyield and an increase in C contamination.

9.9 Ohmic Contacts to Semiconductors

In the investigation of the properties of semiconductor devices, the transition from the semiconductor to the metal connections is extremely important. Usually it is desired that this contact be ohmic. In principle, this means a contact which is noninjecting and which has a linear $I–V$ relationship in both directions. In practice, a contact is considered ohmic if the voltage drop across it in either direction is much smaller than that across the device and hence does not perturb significantly the device characteristics. Actual linearity of the contact $I–V$ relationship is therefore not important if the voltage is small.

There are three major approaches to achieving an ohmic contact: (1) by choosing a metal with the proper relative Fermi level so that the barrier is small for thermally excited currents, (2) by heavily doping the semiconductor near the junction so that the current can be carried by quantum mechanical tunneling through the barrier, and (3) by introducing numerous recombination centers in the interface region on the semiconductor side of the junction.

Heterojunction studies tend to involve contact problems, since many structures involve semiconductors for which the contact technology problems have not been as extensively examined as for Si and Ge. Various methods for fabricating contacts such as alloying, electroplating, and evaporation and the problems associated with these will be discussed. Various materials reported as giving ohmic contacts to particular semiconductors are listed to provide convenient access to the literature.

9.9.1. *Basic Principles for Ohmic Contacts*

In principle, when a metal is brought into intimate contact with a semiconductor the type of electrical barrier which results is a function of the difference in the work function ϕ_m of the metal and the work function ϕ_s of the semiconductor.

The types of energy band diagrams that are expected for metal contacts on n- and p-type semiconductors are depicted in Figs. 6.1 and 6.2. As shown by Fig. 6.1, for a metal contact to be ohmic to an n-type semiconductor, ϕ_m should be less than ϕ_s and for $\phi_m > \phi_s$ the barrier should increase linearly according to the work function difference. Conversely, as shown in Fig. 6.2, for a metal contact to be ohmic to a p-type semiconductor, ϕ_m should be greater than ϕ_s.

In practice, however, for the covalent semiconductors such as Ge, Si and GaAs the resulting barrier energies do not follow this simple model. The barrier tends to remain nearly the same for all metals because of surface states which have not been taken into account. For the more ionic semiconductors, such as ZnS, the barrier energy is a function of the work function (or electronegativity) of the metal and the semiconductor. For an example of the variation of barrier heights for the two types of semiconductors as a function of the electronegativity of the metal see Fig. 6.7.

For most semiconductors, because of interface states, ohmic contacts are not obtained by the proper choice of work functions. The technology generally used is therefore the provision of a very heavily doped semiconductor region between the metal and bulk semiconductor. The heavily doped region is of the same type as the semiconductor region to be contacted thus forming an n^+–n or p^+–p ohmic structure. Between the metal and the heavily doped semiconductor skin a barrier exists, but since the skin region is heavily doped, the depletion width of the barrier will be sufficiently thin that quantum mechanical tunneling can take place. Thus the barrier becomes essentially transparent and the contact requires very little voltage drop for current flow. The heavily doped region is usually obtained by alloying or diffusing a suitable impurity or forming a compound. While this method works quite well for Ge, Si, and GaAs, it is difficult to obtain the required heavy doping for many wide-band-gap semiconductors because of their tendency to compensate for the introduced foreign atoms by the formation of native defects of the opposite type.

A third principle which may be used to form an ohmic contact is to introduce a sufficient number of recombination centers at the metal–semiconductors interface so that thermodynamic equilibrium is maintained throughout the interface. This recombination or disturbed layer may result from damaging or straining the surface of the semiconductor by sand blasting or lapping or by introducing impurities which produce effective recombination centers. Plating of almost any metal to such a damaged surface will produce a suitable contact while the alloying process itself may cause the straining or introduce the recombination levels.

9.9.2 Fabrication Methods for Obtaining Ohmic Contacts

The methods for achieving ohmic contacts to semiconductors may be divided into five broad classifications: alloying, liquid regrowth, plating, evaporation, and bonding. Within each category there are many variations since each investigator has his own particular recipe which he follows to obtain a good contact.

The alloying process for contacts is similar to that described for forming heterojunctions in Section 9.7. A small metal sphere or a thin metal layer is heated to a temperature such that a small portion of the semiconductor is dissolved. The system is cooled and the semiconductor regrows, incorporating some of the metal and any doping impurities originally in the metal. Usually in this method the metal or impurity in the metal forms a heavily doped layer of the same type as the bulk semiconductor to be contacted (n^+ on n, p^+ on p). Many of the wide-gap semiconductors, however, present contact problems because they cannot be doped heavily enough.

Particularly important in making contact is the wetting of the semiconductor by the metal. This is discussed in detail by Chung (1962) for Al and Si, and by Zettlemoyer (1969) in general. In order to promote wetting and a good contact the metal and semiconductor must be clean. A flux is often used during alloying to remove any remaining surface film and enhance wetting (Schwartz and Sarace, 1966). The thickness of the surface film has been measured for Si, GaAs, and ZnSe by Yep and Archer (1967) and Archer et al. (1968) for various etchants used in the final surface preparation. They observed thicknesses in the range 9–340 Å. Attention must also be paid to the difference in thermal expansion coefficient between the semiconductor and contact material so that residual stresses are not developed in the semiconductor upon cooling from the alloy temperature to room temperature. This requires the materials to be closely matched in thermal expansion coefficients or for the contact material to have a yield sufficient to absorb the difference in expansion.

Liquid regrowth, which is discussed in Section 9.6, is used to grow a heavily doped layer on a similarly doped substrate (n^+ on n or p^+ on p) of the same material. This has been used recently for ohmic contacts to GaAs for Gunn effect devices (Harris *et al.*, 1969).

Contacts to semiconductors can also be made by several different plating techniques such as electroplating, displacement plating, or electroless plating. A review of the plating of metals onto semiconductors is given by Hillegas and Schnable (1963). A clean relatively oxide-free surface is desirable prior to plating. The plated metal then either serves as the solder for an external lead or is contacted by another metal, by soldering or bonding. In electroplating, a metallic salt is reduced to the desired metal at the cathode when current passes between the electrodes. Since semiconductors have a higher resistivity than metals, care must be taken to obtain a uniform current distribution and hence a uniform plating. Electroplating is discussed in detail by Blum and Hogaboom (1949) and Lowenheim (1963). In displacement plating, the semiconductor to be plated is oxidized, removing an amount of semiconductor equal to the thickness of the plating. The plating stops when the semiconductor is covered, resulting in a very thin deposit. This may be increased in thickness by subsequent electroplating. The other type of plating which is widely used for contacting to semiconductors is electroless plating. In electroless plating a metallic salt is reduced to the metal at the semiconductor surface by a chemical reducing agent which is present in the plating solution. Nickel, gold, and platinum are most widely used for semiconductor contacts. A detailed description of electroless Ni plating on Si is given by Sullivan and Eigler (1957). As they show, the adherence of the Ni film is much better on a lapped surface than on a chemically polished surface.

Another contacting technique which is widely used is vapor plating or evaporation of a suitable metal. This technique is particularly useful for integrated circuits where complex contact patterns are formed by evaporating the metal followed by photoresist definition of the contact pattern. General techniques for vacuum plating are discussed in detail by Holland (1960) and Powell *et al.* (1966). Vacuum plating as applied to the growth of heterojunctions is discussed in Section 9.5. The metal may be evaporated by a heated filament, an electron-beam gun, or by sputtering. The heated boat or filament technique is the simplest but suffers from the fact that the heated container may contaminate the evaporated metal by reaction or outgassing of impurities. A shutter mechanism is often used to block the initial vapors which may be contaminated with the more volatile impurities. If multiple boats or filaments are used, two or more metals may be evaporated simultaneously with independent control of each.

The electron-beam gun melts the evaporant alone, which also serves as its own crucible. This eliminates contamination from the crucible and gives purer deposited metal films. In the sputtering system, positive gas ions bombard the source (cathode), emitting metal atoms. These ejected atoms traverse the vacuum chamber and are deposited on the substrate. This type of system has the advantage that the polarity of the system may be reversed so that sputtering may occur from the substrate to a remote anode thereby cleaning the substrate surface. This back sputtering is particularly useful for removing any thin residual oxide or other layer on the substrate surface while it is in the vacuum system and can be followed by immediate deposition of the metal contact.

In making ohmic contacts by vacuum and chemical plating, often the deposited metal is heated so that it fuses with the semiconductor. For some of the materials with low melting points such as In, this may cause balling of the film and spotty contacts. Often this problem can be alleviated by depositing on top a second metal with a higher melting point, such as Ni, before the semiconductor is heated.

Bonding may be used to form a contact between a thin wire and a metallic layer or the semiconductor directly. Thermocompression and ultrasonic bonding are the two types of bonding processes normally used. In thermo-compression bonding, heat and pressure are used to attach very fine wires, usually of Au or Al, to a contact pad or to the semiconductor itself. In ultra-sonic bonding, a combination of pressure and ultrasonic (60 kHz) vibration is used. The ultrasonic vibration causes a scrubbing action which breaks up any thin surface film, thus causing intimate contact between the two materials to be bonded. The mechanical force and scrubbing action cause a molecular mingling of the two materials in contact and thus forms a bond. Again, fine Au or Al wires are used, but in this case no heat is required so any previous bonding is not affected. A general discussion of the types of bonding is given by Fogiel (1968) and a more detailed description of ultra-sonic bonding by Peterson *et al.* (1962).

Low resistance contacts to semiconductors are most easily made if the portion to be contacted is very heavily doped (n^+ or p^+). Then a deposited metal may make good contact to a clean surface without any further heat treatment. In order to obtain a heavily doped surface in the contact region, diffusion is often used. More recently for materials for which heavy doping by diffusion is impossible, ion implantation has been proposed.

9.9.3 *Tables of Various Metals Reported as Ohmic Contacts to Semiconductors*

TABLE 9.6

Metals Reported as Ohmic Contacts to Ge

Metal	n- or p-type Ge	References
Ag	p	Borneman et al. (1965), Matsuura et al. (1962), Turner (1959)
Ag–Al–NiCr–Sb	n	Michel et al. (1969)
Ag–Al–Sb	n	Michel et al. (1969)
Ag–Au–Mo–NiCr–Sb	n	Michel et al. (1969)
Ag–Pb	n	Looney (1958)
Al	n, p	Christiensen (1959), Heck and Looney (1965), Matsuura et al. (1962), Pikhtin et al. (1970)
Al–Au–P	n	Schmidt and Wernick (1966)
Al–In	p	Biondi (1958)
Al–Pd	p	Schmidt (1962a,b)
As	p	Borneman et al. (1955)
As–Pb	n	Biondi (1958)
As–Sn	n	Fang and Howard (1964)
Au	n, p	O. L. Anderson et al. (1957), (1958), Christiensen (1959), Matsuura et al. (1962), Mehl et al. (1963)
	n	Heck and Looney (1965)
	p	Turner (1959), Bridgers et al.
Bi	n	Matsuura et al. (1962)
	p	Borneman et al. (1955), Turner (1959)
Cd	p	Borneman et al. (1955), Turner (1959), Pikhtin et al. (1970)
Co	p	Turner (1959)
Cr	p	Borneman et al. (1955), Turner (1959)
Cu	p	Borneman et al. (1955), Bridgers et al. (1958), Matsuura et al. (1962), Turner (1959)
Ga	p	Matsuura et al. (1962)
Ga–In	p	Biondi (1958)
In	p	Hunter (1956, 1962), Matsuura et al. (1962), Turner (1959), Pikhtin et al. (1970) Biondi (1958), Borneman et al. (1955), Dale and Turner (1963)

TABLE 9.6 (cont.)

Metal	n- or p-type Ge	References
Ni	p	Borneman et al. (1955), Turner (1959)
Ni–Sn		Mayer (1959)
Pb	p	Borneman et al. (1955), Turner (1959)
Pb–Rh–Sn	n, p	Lin (1957)
Pb–Sb	n	Biondi (1958)
Pb–Sb–Sn	n	Bridgers et al. (1958), Hunter (1956, 1962)
Pb–Sn	p	Bridgers et al. (1958), Pikhtin et al. (1970), Reichenbaum (1959)
Pt	p	Borneman et al. (1955), Turner (1959)
Rh	p	Borneman et al. (1955), Turner (1959)
Sb	n, p	Borneman et al. (1955), Bridgers et al. (1958) Matsuura et al. (1962), Pikhtin et al. (1970), Turner (1959)
Sb–Sn	n	Dale and Turner (1963)
Sn	n, p	Borneman et al. (1955), Dale and Turner (1963), Turner (1959), Pikhtin et al. (1970)
Te	n, p	Turner (1959), Pikhtin et al. (1970)
W	p	Borneman et al. (1955)
Zn	p	Borneman et al. (1955), Turner (1959), Pikhtin et al. (1970)

TABLE 9.7
Metals Reported as Ohmic Contacts to Si

Metal	n- or p-type Si	References
Ag	n, p	Matsuura et al. (1962)
Ag–Pb	n, p	Looney (1958)
Ag–Ti	n, p	Lepselter (1960), Springgate (1969)
Al	n, p	Cunningham and Harper (1967), Flores (1964), Heck and Looney (1965), Hooper et al. (1965), Kirvalidze and Zhukov (1960), Matlow and Ralph (1959), Matsuura et al. (1962), Meyer (1969), Sello (1969), Pikhtin et al. (1970), Lane (1970)

TABLE 9.7 (cont.)

Metal	*n*- or *p*-type Si	References
Al–Au	*n, p*	Cunningham (1965)
Al–Au–P	*n*	Schmidt and Wernick (1966)
Al–Au–Mo–PtSi	*n, p*	Sello (1969)
Al–Cr–Cu–Si–Sn	*n, p*	Totta and Sopher (1969)
Al–Pb	*p*	Sullivan and Eigler (1956)
Al–Pd	*n, p*	Schmidt (1962 a,b)
Al–Si–W	*n, p*	May (1969)
Au	*n, p* *p*	Heck and Looney (1965), Matsuura, *et al.* (1962) Mehl *et al.* (1963), Turner (1959), Wurst and Borneman (1957)
Au–B	*p*	Hunter (1956, 1962)
Au–Co–Si–Pt	*n, p*	Shinoda (1969)
Au–Cu	*n, p*	Matlow and Ralph (1961)
Au–Mo	*n, p*	Cunningham (1965), Cunningham and Harper (1967)
Au–Sb	*n*	Flores (1964)
Au–Sb–Si	*n*	Hemment (1966)
Au–Sn	*n*	Bender and Bernstein (1965)
Au–Sn(Al, As, n, Sb)	*n, p*	Jones (1962)
Bi	*n*	Matsuura *et al.* (1962)
Cd	*p*	Pikhtin *et al.* (1970)
Cd–Sb	*n*	Bender (1962)
Chromel	*n, p*	Hooper *et al.* (1965)
Cu	*n, p*	Hooper *et al.* (1965), Kirvalidze and Zhukov (1960), Turner (1959), Matsuura *et al.* (1962)
Fe	*n, p*	Kirvalidze and Zhukov (1960), Wurst and Borneman (1957)
Ga	*p*	Matsuura *et al.* (1962)
Ga–Sn	*n*	Harman and Higier (1962)
Ge	*p*	Gorodetsky *et al.* (1959)

TABLE 9.7 (cont.)

Metal	n- or p-type Si	References
Ge–Sb	n	Gorodetsky *et al.* (1959)
Graphite	p	Harman and Higier (1962)
In	p	Matsuura *et al.* (1962), Pikhtin *et al.* (1970)
	n	Harman and Higier (1962), Wurst and Borneman (1957)
In–Pb	p	Wolsky (1959)
K	n	Wurst and Borneman (1957)
Li	n	Wurst and Borneman (1957)
Mg	n	Wurst and Borneman (1957)
Mo	n, p	Hooper *et al.* (1965)
Na	n	Wurst and Borneman (1957)
Ni	n, p	Hooper *et al.* (1965), Iwasa *et al.* (1968), Kirvalidze and Zhukov (1960), Matlow and Ralph (1961), Sullivan and Eigler (1957), Teramoto *et al.* (1968)
Pb	p	Sullivan and Eigler (1956), Wurst and Borneman (1957), Pikhtin *et al.* (1970)
Pb–Sb–Sn	n	Hunter (1956, 1962)
Pb–Sn	n, p	Reichenbaum (1959), Sullivan and Eigler (1956), Pikhtin *et al.* (1970)
Pt	n, p	Cohen (1965), Wurst and Borneman (1957)
Pt_5Si_2	n, p	Lepselter and Andrews (1969)
Pt_5Si_2–Pt–Au	n, p	Shinoda (1969)
Pt_5Si_2–Pt–Ti	n, p	Lepselter (1966), Sello (1969)
Rh	p	Green (1965), Wurst and Borneman (1957)
Rh–Si	n, p	Lepselter and Andrews (1969)
Sb	n	Matsuura *et al.* (1962), Pikhtin *et al.* (1970)
Sn	n, p	Kirvalidze and Zhukov (1960), Turner (1959), Wurst and Borneman (1957), Pikhtin *et al.* (1970)

TABLE 9.7 (cont.)

Metal	*n*- or *p*-type Si	References
Te	*n*	Pikhtin *et al.* (1970)
V	*n, p*	Hooper *et al.* (1965)
Zn	*n*	Turner (1959), Wurst and Borneman (1957)
	p	Pikhtin *et al.* (1970)
Zn–Si$_2$	*n, p*	Lepselter and Andrews (1969)

TABLE 9.8

Metals Reported as Ohmic Contacts to the III–V Compound Semiconductors

Compound	Metal	*n*- or *p*-type semiconductor	References
AlAs	Au–Sb	*n*	Whitaker (1965)
	In	*n*	Whitaker (1965)
Al$_x$Ga$_{1-x}$As	Au–Sn	*n*	Aven and Swank (1969)
	Au–Zn	*p*	Aven and Swank (1969)
AlP	In	*n*	Richman (1968)
GaAs	Ag	*n, p*	Cunnell *et al.* (1960), Gol'dberg *et al.* (1967b), Klohn and Wandinger (1969), Knight and Paola (1969), Konaplya *et al.* (1965), Matino and Tokunaga (1969)
	Ag–Al	*n*	Ramachandran and Santosuosso (1966)
	Ag–Au–Ge	*n*	Matino and Tokunaga (1969)
	Ag–Bi	*p*	Dale and Josh (1964)
	Ag–Ge–In	*n*	Cox and Hasty (1969), Cox and Strack (1967), Matino and Tokunaga (1969)
	Ag–Ge–Sn	*n*	Matino and Tokunaga (1969)
	Ag–In	*n, p*	Matino and Tokunaga (1969)
	Ag–In–Sn	*n*	Cox and Strack (1967)
	Ag–In–Zn	*p*	Cox and Strack (1967)

TABLE 9.8 (cont.)

Compound	Metal	n- or p-type semiconductor	References
GaAs	Ag–Mn	p	Nuese and Gannon (1968), Wustenhagen (1964)
	Ag–Ni	n, p	Schmidt (1966)
	Ag–Pb	p	Libov *et al.* (1965)
	Ag–Sn	n	Jones *et al.* (1961), Knight and Paola (1969)' Matino and Tokunaga (1969), Lavine (1963)
	Ag–Te	n	Wustenhagen (1964)
	Ag–Zn	p	Libov *et al.* (1965)
	Al	p	Loebner (1963)
	As–Sn	n	Lavine (1963)
	Au	n	Casey (1962), Cunnell *et al.* (1960), Konaplya *et al.* (1965), Matino and Tokunaga (1969), Nathan *et al.* (1962)
	Au–Ge	n	Bernstein (1962), Cox and Hasty (1969), Knight and Paola (1969), Salow and Grobe (1968)
	Au–Ge–Ni	n	Harris *et al.* (1969), Braslau *et al.* (1967)
	Au–Ge–Si	n	Braslau *et al.* (1967)
	Au–Ge–Sn	n	Salow and Grobe (1968)
	Au–In	n	Cox and Hasty (1969), Hakki and Knight (1966), Knight and Paola (1969), Paola (1970)
	Au–Mo–Sn	n	Broom (1963)
	Au–Ni	n, p	Gol'dberg *et al.* (1967b), Logan *et al.* (1962)
	Au–Ni–Sn	n	Logan *et al.* (1962), Gol'dberg *et al.* (1967b)
	Au–Ni–Zn	p	Gol'dberg *et al.* (1967b)
	Au–Sb	n	Lowen and Rediker (1960)
	Au–Si	n	Bernstein (1962)
	Au–Sn	n	Bernstein (1962), Casey (1962), Cox and Hasty (1969), Jones *et al.* (1961), Knight and Paola (1969), Lavine (1963)

TABLE 9.8 (cont.)

Compound	Metal	n- or p-type semiconductor	References
GaAs	Au–Te	n	Pikhtin *et al.* (1970)
	Au–Zn	p	Holonyak and Lesk (1960), Klohn and Wandinger (1969), Lavine (1963)
	Cd	p	Jadus *et al.* (1967)
	Cd–In	p	Fang and Howard (1964)
	Cr	p	Matino and Tokunaga (1969)
	Cu	n, p	Bernstein (1962), Konaplya *et al.* (1965)
	GaAs–Au–Ge–Ni	n	Harris (1969)
	GaAs–Au–In	n	Lawley *et al.* (1967)
	In	n, p	Casey (1962), Cunnell *et al.* (1960), Gol'd-berg *et al.* (1967a), Gunn (1964), Klohn and Wandinger (1969), Konaplya *et al.* (1965), Lavine (1963), Libov *et al.* (1965), Lowen and Rediker (1960), Lucovsky and Cholet (1960), Nathan *et al.* (1962), Pikhtin *et al.* (1970)
	In–Ni	n	Hakki and Knight (1966)
	In–Pb	p	Lavine (1963)
	In–Pb–Zn	p	Lavine (1963)
	In–Te	n	Fang and Howard (1964), Spitzer and Mead (1963)
	In–Zn	p	Cunnell *et al.* (1960), Holonyak and Lesk (1960), Jadus *et al.* (1967), Libov *et al.* (1965), Minden (1963), Spitzer and Mead (1963)
	Mo	n, p	Konaplya *et al.* (1965), Matino and Tokunaga (1969)
	Ni	n	Casey (1962), Konaplya *et al.* (1965)
		p	Klohn and Wandinger (1969)
	Ni–Sn	n	Furukawa and Ishibashi (1967c), Hakki and Knight (1966), Sharpless (1958)
	Pb	n, p	Bernstein (1962), Libov *et al.* (1965), Schwartz and Sarace (1966)
	Pb–Sn	n	Bernstein (1962), Minden (1963), Reichen-baum (1959)

TABLE 9.8 (cont.)

Compound	Metal	n- or p-type semiconductor	References
GaAs	Pb–Zn	p	Holonyak and Lesk (1960), Minden (1963)
	Pd	n	Konaplya et al. (1965)
	Pt	n	Konaplya et al. (1965)
	Sb–Sn		Dale and Turner (1963)
	Sn	n	Allen (1959), Bernstein (1962), Cunnell et al. (1960), Dale and Turner (1963), Furukawa and Ishibashi (1967c), Gol'dberg et al. (1967b), Gunn (1964), Ing et al. (1968), Jadus et al. (1967), Konaplya et al. (1965), Libov et al. (1965), Salow and Grobe (1968), Schwartz and Sarace (1966), Pikhtin et al. (1970)
	Sn (Ge, Te, S)	n	Holonyak and Lesk (1960)
	Sn–Zn	p	Schwartz and Sarace (1966)
	Te	n	Pikhtin et al. (1970)
	Ti	p	Matino and Tokunaga (1969)
	W	n	Konaplya et al. (1965)
	Zn	p	Gol'dberg et al. (1967b), Jadus et al. (1967), Schwartz and Sarace (1966), Pikhtin et al. (1970)
$GaAs_xP_{1-x}$	Au–Sn	n	Aven and Swank (1969)
	Au–Zn	p	Aven and Swank (1969)
GaP	Ag–Te	n	Goldstein and Perlman (1966)
	Ag–Zn	p	Goldstein and Perlman (1966)
	Au–Be	p	Gershenzon et al. (1966)
	Au–Ga–In	p	Cohen and Bedard (1968)
	Au–In	p	Plaskett et al. (1967)
	Au–Ni–Zn	p	Allen and Henderson (1968)
	Au–Si	n	Bergh and Strain (1969)
	Au–Sn	n	Aven and Swank (1969), Lorenz and Pilkuhn (1966), Plaskett et al. (1967), Shih et al. (1968)

TABLE 9.8 (cont.)

Compound	Metal	*n*- or *p*-type semiconductor	References
GaP	Au–Te	*n*	Pikhtin *et al.* (1970)
	Au–Zn	*p*	Aven and Swank (1969), Bergh and Strain (1969), Lorenz and Pilkuhn (1966), Shih *et al.* (1968)
	Ga	*p*	Kamath and Bowman (1967)
	In	*n*	Goldstein and Perlman (1966), Ignatkina *et al.* (1965), Pikhtin *et al.* (1970)
	In–Sn	*n*	Purohit (1967)
	In–Zn	*p*	Ignatkina *et al.* (1965)
	Ni–Sn	*n*	Gershenzon *et al.* (1966)
	Sn	*n*	Allen and Henderson (1968), Hara and Skasako (1968), Ignatkina *et al.* (1965)
	Te	*n*	Pikhtin *et al.* (1970)
	Zn	*p*	Pikhtin *et al.* (1970)
GaSb	Ag–Ni	*n, p*	Burns (1968)
	Au–Ge	*p*	Bernstein (1962)
	Au–Si	*p*	Bernstein (1962)
	Au–Sn	*p*	Bernstein (1962)
	Au–Te	*n*	Pikhtin *et al.* (1970)
	Cu	*p*	Bernstein (1962)
	In	*p*	Dale and Turner (1963), Pikhtin *et al.* (1970)
	Ni	*n*	Daw and Mitra (1965)
	Sb–Sn	*n*	Dale and Turner (1963)
	Sn	*n, p*	Bernstein (1962), Burns (1968), Dale and Turner (1963), Pikhtin *et al.* (1970)
	Te	*n*	Pikhtin *et al.* (1970)
	Zn	*p*	Pikhtin *et al.* (1970)
InAs	Au	*p*	Takahashi *et al.* (1969)
	Au–Ge	*n*	Bernstein (1962)
	Au–Sn	*n*	Bernstein (1962)

TABLE 9.8 (cont.)

Compound	Metal	n- or p-type semiconductor	References
InAs	Cu	p	Bernstein (1962)
	Ni	n	Daw and Mitra (1965)
	Sn	n	Bernstein (1962)
InP	Au–Ge	n	Bernstein (1962)
	Au–Sn	n	Bernstein (1962)
	In	n	Gunn (1964)
	Ni	n	Daw and Mitra (1965)
	Sn	n	Bernstein (1962), Gunn (1964)
InSb	Au–Ge	p	Bernstein (1962)
	Au–Sn	n	Bernstein (1962)
	Ga–In	n	Bott et al. (1967), Clawson and Wieder (1967)
	Ni	n	Daw and Mitra (1965)
	Sn	n	Bernstein (1962)

TABLE 9.9

Metals Reported as Ohmic Contacts to II–VI Compound Semiconductors

Compound	Metal	n- or p-type semiconductor	References
CdS	Ag	n	Sihvonen and Boyd (1958)
	Al	n	Boer and Lubitz (1962), Butler and Muscheid (1954, 1955), Pizzarello (1967), Sihvonen and Boyd (1958)
	Al–Ti	n	Boer and Hall (1966)
	Al–Ti–Pt	n	Boer and Hall (1966), Sihvonen and Boyd (1958)
	Au	n	Boer and Lubitz (1962), Butler and Muscheid (1954, 1955), Sihvonen and Boyd (1958)
	Au–In	n	Boer and Hall (1966)

TABLE 9.9 (cont.)

Compound	Metal	n- or p-type semiconductor	References
CdS	Be–Cu	n	Sihvonen and Boyd (1958)
	Bi	n	Sihvonen and Boyd (1958)
	Cu	n	Sihvonen and Boyd (1958)
	Ga	n	Boer and Lubitz (1962), R. W. Smith (1955a), Walker and Lambert (1957)
	Ga–In	n	Goodman (1964a), Johnson and Darsey (1968), Okada and Matino (1964)
	In	n	Boer and Hall (1966), Boer and Lubitz (1962), Clark and Wook (1965), Jaklevic *et al.* (1963), Johnson and Darsey (1968), Pizzarello (1967), Scholten (1966), Serdyuk and Bube (1967), Sihvonen and Boyd (1958), R. W. Smith (1955a), Spitzer and Mead (1963), Walker and Lambert (1957), Fainer *et al.* (1970)
	$In_2O_3(Sn)$	n	Sihvonen and Boyd (1960)
	Pt	n	Sihvonen and Boyd (1958)
	Ta	n	Sihvonen and Boyd (1958)
CdSe	Au	n	Nakai *et al.* (1964)
	In	n	Tubota (1963)
CdTe	Au	p	DeNobel (1958)
	Au–Indalloy 13	p	Crowder and Hammer (1966)
	In	n	DeNobel (1958)
	In–Ni	p	Ludwig (1967)
	Indalloy 13	p	Crowder and Hammer (1966)
	Pt	p	DeNobel (1958)
	Rh	p	DeNobel (1958)
$Cd_xHg_{1-x}Te$	Au–W		Volzhenskii and Pashkovskii (1966)
	Cu		Volzhenskii and Pashkovskii (1966)

TABLE 9.9 (cont.)

Compound	Metal	*n*- or *p*-type semiconductor	References
HgTe	Au–W		Volzhenskii and Pashkovskii (1966)
	Cu		Volzhenskii and Pashkovskii (1966)
ZnS	Ga–In	*n*	Blount *et al.* (1966)
	Hg–In	*n*	Aven and Mead (1965)
	In	*n*	Alfrey and Cooke (1957), Aven and Cusano (1964)
	In_2O_3–Sn		Williams (1966)
	SnO_2	*n*	Goldberg and Nickerson (1963)
ZnSe	Au	*p*	Nojima and Ibuki (1966)
	Graphite	*n*	R. T. J. Smith (1969)
	In	*n*	Aven and Cusano (1964), Aven and Segall (1963), Nojima and Ibuki (1966), Calow *et al.* (1968), Sagar *et al.* (1968a)
	In–Ga	*n*	Mironov *et al.* (1969), Goodman (1969)
	InHg	*n*	Aven and Kreiger (1970)
	In–Hg–Ni	*n*	Ludwig and Aven (1967)
	In–Ni	*n*	Ludwig (1967), Ludwig and Aven (1967)
	Pt	*n*	Goodman (1969)
ZnTe	Ag	*p*	Takahashi *et al.* (1969)
	Ag–Sn	*p*	Watanabe *et al.* (1964)
	Au	*p*	Aven and Segall (1963), Miksic *et al.* (1964a)
	Au–Li	*p*	Aven and Garwacki (1967a)
	Au–Indalloy 13	*p*	Crowder and Hammer (1966)
	Cu–Sn	*p*	Watanabe *et al.* (1964)
	Indalloy 13	*p*	Crowder and Hammer (1966)
	In–Sn	*p*	Miksic *et al.* (1964a)
	Te	*p*	Kroger and DeNobel (1958)
$ZnSe_xTe_{1-x}$	Au	*p*	Aven (1965)
	Au–Li	*p*	Aven and Garwacki (1967a)
	Hg–In	*n*	Aven (1965)

TABLE 9.10

Metals Reported as Ohmic Contacts to Other Compound Semiconductors

Compound	Metal	*n*- or *p*-type semiconductor	References
PbTe	Hg–In–Th	*n, p*	King (1961)
SiC	Al–Si	*p*	Shier (1970), Pikhtin *et al.* (1970)
	Cu–Ti	*p*	Shier (1970)
	Ni		Raybold (1960)
	Si(Al)	*p*	Hall (1958)
	W	*n*	Pikhtin *et al.* (1970)

Bibliography

1. ABDULLAEV, G. B., and TALIBI, M. A. (1970). Investigations of different effects in selenium $p-n$ heterojunctions. *Abstr. Int. Conf. Phys. Chem. Semicond. Heterojunctions Layer Structures, Budapest, 1970,* p. 3. Hung. Acad. Sciences, Budapest, Hungary.
2. ABDULLAEV, G. B., GAJIEV, N. D., and TALIBI, M. A. (1970). Investigations of physical properties of metal–dielectric–semiconductor structures based on selenium and tellurium. *Proc. Int. Conf. Phys. Chem. Semicond. Heterojunctions Layer Structures, Budapest, 1970* **5**, 53. Hung. Acad. Sciences, Budapest, Hungary.
3. ABRAHAMS, M. S., WEISBERG, L. R., and TIETJEN, J. J. (1969). Stresses in hetero-epitaxial layers: $GaAs_{1-x}P_x$ on GaAs. *J. Appl. Phys.* **40**, 3754.
4. ABRAHAMS, M. S., BRAUNSTEIN, R., and ROSI, F. D. (1959). Thermal, electrical and optical properties of (In, Ga) as alloys. *Phys. Chem. Solids* **10**, 204.
5. ABRAHAMS, M. S., WEISBERG, L. R., BUIOCCHI, C. J., and BLANC, J. (1969). Dislocation morphology in graded heterojunctions: $GaAs_{1-x}P_x$. *J. Mater. Sci.* **4**, 223.
6. ABRAMOV, B. G., DRUGINKIN, I. F., OKUNEV, V. D., PANTELEEV, J. K., and RAMOSANOV, P. E. (1970). Influence of built-in field on electrical and optical properties of InGaAs–GaAs heterojunctions. *Abstr. Int. Conf. Phys. Chem. Semicond. Heterojunctions Layer Structures, Budapest, 1970,* p. 37. Hung. Acad. Sciences, Budapest, Hungary.
7. ADAMS, M. J., and CROSS, M. (1970). Wave-guiding properties of $GaAs-Al_xGa_{1-x}As$ heterostructure lasers. *Phys. Lett.* **32A**, 207.
8. ADIROVICH, E. I., YUABOV, Yu. M., and YAGUDAEV, G. R. (1969). Photoelectric effects in film diodes with CdS–CdTe heterojunctions. *Sov. Phys.–Semicond.* **3**, 61.
9. ADIROVICH, E. I., YUABOV, YU. M., and YAGUDAEV, G. R. (1970). Thin-film structures with nCdS–pCdTe heterojunction. *Proc. Int. Conf. Phys. Chem. Semicond. Heterojunctions Layer Structures, Budapest, 1970* **2**, 151. Hung. Acad. Sciences, Budapest, Hungary.
10. ADVANI, G. T., GOTTLING, J. G., and OSMAN, M. S. (1962). Thin film triode research. *Proc. IRE* **50**, 1530.
11. AGUSTA, B., and ANDERSON, R. L. (1965). Opto-electric effects in Ge–GaAs $p-n$ heterojunctions. *J. Appl. Phys.* **36**, 206.
12. AGUSTA, B., LOPEZ, A., and ANDERSON, R. L. (1964). Effect of the notch on opto-electrical characteristics of abrupt heterojunctions. *IEEE Trans. Electron Devices* **ED–11**, 533.
13. AHLSTROM, E., and GARTNER, W. W. (1962). Silicon surface–barrier photocells. *J. Appl. Phys.* **33**, 2602.
14. AÏTKHOZHIN, S. A., and SEMILETOV, S. A. (1965). Production of thin films of GaSb by evaporation in vacuo. *Sov. Phys.–Crystallogr.* **9**, 488.
15. ALADINSKII, V. K., and MASLOV, A. A. (1966). Electrical properties of Ge–GaAs ($p-n$ and $n-n$) heterojunctions. *Sov. Phys.–Solid State* **7**, 2789.
16. ALBERS, W. (1961). Diffusion of arsenic in germanium from the vapor phase. *Solid–State Electron.* **2**, 85.
17. ALEKSEEV, YU. A., STULOVA, G. M., and SHALABUTOV, YU. K. (1968). Rectification at a metal–aluminum oxide contact at high temperatures. *Sov. Phys.–Semicond.* **2**, 283.

18. ALFEROV, ZH. I. (1967). Possible development of a rectifier for very high current densities on the basis of a p–i–n (p–n–n^+, n–p–p^+) structure with heterojunctions. *Sov. Phys.–Semicond.* **1**, 358.

19. ALFEROV, ZH. I. (1970a). Electroluminescence of heavily-doped heterojunctions pAl$_x$Ga$_{1-x}$As–nGaAs. *J. Luminescence* **1,2**, 869.

20. ALFEROV, ZH. I. (1970b). Injection luminescence of heterojunctions in A$_3$B$_5$ semi-conductor compounds. *Proc. Int. Conf. Phys. Chem. Semicond. Layered Structures, Budapest, 1970*, **2**, 7. Hung. Acad. Sciences, Budapest, Hungary.

21. ALFEROV, ZH. I. (1970c). Investigation of heterojunctions and p–n junctions in the AlAs–GaAs system using a scanning electron microscope and a probe microanalyzer. *Sov. Phys.–Semicond.* **3**, 1234.

22. ALFEROV, ZH. I., and GABUZOV, D.Z. (1966). Recombination radiation spectrum of GaAs with current excitation via p–n heterojunctions of GaP–GaAs. *Sov. Phys.–Solid State* **7**, 1919.

23. ALFEROV, ZH. I., and KAZARINOV, R. F. (1963). Author's Certificate No. 28448, March 30, 1963.

24. ALFEROV, ZH. I., and NINUA, O. A. (1970a). Electroluminescence of Al$_x$Ga$_{1-x}$As–GaAs heterojunctions under avalanche breakdown conditions. *Sov. Phys.–Semicond.* **4**, 296.

25. ALFEROV, ZH. I., and NINUA, O. A. (1970b). Photoluminescence of epitaxial films on Al$_x$Ga$_{1-x}$As solid solutions. *Sov. Phys.–Semicond.* **4**, 519.

26. ALFEROV, ZH. I., and ZIMOGOROVA, N. S. (1969). Photoelectric properties of p–n junctions in silicon-doped gallium arsenide. *Sov. Phys.–Semicond.* **3**, 385.

27. ALFEROV, ZH. I., *et al.* (1964). Photoelectric properties of heterojunctions in some semiconductors. *Ukr. Fiz. Zh. (USSR)* **9**, 659.

28. ALFEROV, ZH. I., KOROL'KOV, V. I., MIKHAILOVA-MIKHEEVA, I. P., ROMANENKO, V. N., and TUCHKEVICH, V. M. (1965a). Study of growth of GaP and CdTe on GaAs in gas transport reactions. *Sov. Phys.–Solid State* **6**, 1865.

29. ALFEROV, ZH. I., ZIMOGOROVA, N. S., TRUKAN, M. K., and TUCHKEVICH, V. M. (1965b). Some photoelectric properties of gallium phosphide–gallium arsenide p–n heterojunctions. *Sov. Phys.–Solid State* **7**, 990.

30. ALFEROV, ZH. I., KHALFIN, V. B., and KAZARINOV, R. F. (1967a). A characteristic feature of the injection into heterojunctions. *Sov. Phys.–Solid State* **8**, 2480.

31. ALFEROV, ZH. I., KOROL'KOV, V. I., and TRUKAN, M. K. (1967b). Electrical properties of gallium phosphide–gallium arsenide p–n heterojunctions. *Sov. Phys.–Solid State* **8**, 2813.

32. ALFEROV, ZH. I., *et al.* (1967c). Injection luminescence of epitaxial heterojunctions in the GaP–GaAs system. *Sov. Phys.–Solid State* **9**, 208.

33. ALFEROV, ZH. I., GAMAZOV, A. A., and ZIMOGOROVA, N. S. (1968). Photosensitivity spectra of GaP$_x$As$_{1-x}$–GaAs n–n heterojunctions. *Sov. Phys.–Semicond.* **2**, 493.

34. ALFEROV, ZH. I., ANDREEV, V. M., KOROL'KOV, V. I., PORTNOI, E. L., and TRET'YAKOV, D. N. (1969a). Injection properties of n–Al$_x$Ga$_{1-x}$As–pGaAs hetero-junctions. *Sov. Phys.–Semicond.* **2**, 843.

35. ALFEROV, ZH. I., ANDREEV, V. M., KOROL'KOV, V. I., PORTNOI, E. L., and TRET'YAKOV, D. N. (1969b). Coherent radiation of epitaxial heterojunction structures in the AlAs–GaAs system. *Sov. Phys.–Semicond.* **2**, 1289.

36. ALFEROV, ZH. I., ANDREEV, V. M., KOROL'KOV, V. I., PORTNOI, E. L., and YAKOVENKO, A. A. (1969c). Al$_x$Ga$_{1-x}$As solid solutions with a forbidden-band width gradient. *Sov. Phys.–Semicond.* **3**, 460.

37. ALFEROV, ZH. I., GARBUZOV, D. Z., MOROZOV, E. P., and TRET'YAKOV, D. N. (1969d).

Mechanism of radiative recombination in silicon-doped gallium arsenide epitaxial *p–n* structures. *Sov. Phys.–Semicond.* **3**, 471.

38. ALFEROV, ZH. I., GARBUZOV, D. Z., MOROZOV, E. P., and TRET'YAKOV, D. N. (1969e). Radiative recombination in gallium arsenide *p–n* structures with *p*-type regions doped with germanium. *Sov. Phys.–Semicond.* **3**, 600.

39. ALFEROV, ZH. I., ANDREEV, V. M., KOROL'KOV, V. I., PORTNOI, E. L., and YAKOVENKO, A. A. (1969f). Spontaneous radiation sources based on structures with AlAs–GaAs heterojunctions. *Sov. Phys.–Semicond.* **3**, 785.

40. ALFEROV, ZH. I., ANDREEV, V. M., KONNIKOV, S. G., NIKITIN, J. G., and TRET'YAKOV, D. N. (1970a). Heterojunctions on the base of $A^{III}B^V$ semiconducting compounds and of their solid solutions. *Proc. Int. Conf. Phys. Chem. Semicond. Heterojunctions Layer Structures, Budapest, 1970,* **1**, 93. Hung. Acad. Sciences, Budapest, Hungary.

41. ALFEROV, ZH. I., *et al.* (1970b). Effect of the heterostructure parameters on the characteristics of injection lasers in the AlAs–GaAs system. *Proc. Int. Conf. Phys. Chem. Semicond. Heterojunctions Layer Structures, Budapest, 1970,* **2**, 171. Hung. Acad. Sciences, Budapest, Hungary.

42. ALFEROV, ZH. I., ANDREEV, V. M., BORODULIN, V. I., PACK, G. T., PORTNOI, E. L., and SHVEYKIN, V. I. (1970c). Spatial mission characteristics of injection heterolasers in the AlAs–GaAs system. *Proc. Int. Conf. Phys. Chem. Semicond. Heterojunctions Layer Structures, Budapest, 1970,* **2**, 159. Hung. Acad. Sciences, Budapest, Hungary.

43. ALFEROV, ZH. I., KOROL'KOV, V. I., NIKITIN, V. G., TRET'YAKOV, D. D., and YAKOVENKO, A. A. (1970d). S-diodes based on heterojunctions in the GaAs–AlAs system. *Proc. Int. Conf. Phys. Chem. Semicond. Heterojunctions Layer Structures, Budapest, 1970,* **2**, 183. Hung. Acad. Sciences, Budapest, Hungary.

44. ALFEROV, ZH. I., GARBUZOV, D. Z., MOROZOV, E. P., and PORTNOI, E. L. (1970e). Diagonal tunneling and polarization of radiation in $Al_xGa_{1-x}As$–GaAs heterojunctions and in GaAs *p–n* junctions. *Sov. Phys.–Semicond.* **3**, 885.

45. ALFEROV, ZH. I., ANDREEV, V. M., PORTNOI, E. L., and TRUKAN, M. K. (1970f). AlAs–GaAs heterojunction injection lasers with a low room-temperature threshold. *Sov. Phys.–Semicond.* **3**, 1107.

46. ALFEROV, ZH. I., ANDREEV, V. M., PORTNOI, E. L., and PROTASOV, I. I. (1970g). Coordinate-sensitive photocells based on $Al_xGa_{1-x}As$–GaAs heterojunctions. *Sov. Phys.–Semicond.* **3**, 1103.

47. ALFEROV, ZH. I., ANDREEV, V. M., ZIMOGOROVA, N. S., and TRET'YAKOV, D. N. (1970h). Photoelectric properties of $Al_xGa_{1-x}As$–GaAs heterojunctions. *Sov. Phys.–Semicond.* **3**, 1373.

48. ALFEROV, ZH. I., ANDREEV, V. M., KOROL'KOV, V. I., NIKITIN, V. G., and YAKOVENKO, A. A. (1970i). *pnpn* structures based on GaAs and on $Al_xGa_{1-x}As$ solid solutions. *Sov. Phys.–Semicond.* **4**, 481.

49. ALFREY, G. F., and COOKE, I. (1957). Electric contact to ZnS crystals. *Proc. Phys. Soc. London, Sect. B* **70**, 1096.

50. ALLAKHVERDYAN, R. G., ORAEVSKII, A. N., and SUCHKOV, A. F. (1970). Influence of waveguide properties of a *p–n* junction on the coherent emission of gallium arsenide laser diodes. *Sov. Phys.–Semicond.* **4**, 277.

51. ALLEN, F. G., and GOBELI, G. W. (1962). Work function, photoelectric threshold, and surface states of atomically clean silicon. *Phys. Rev.* **127**, 150.

52. ALLEN, F. G., and GOBELI, G. W. (1964). Comparison of the photoelectric properties of cleaved, heated, and sputtered silicon surfaces. *J. Appl. Phys.* **35**, 597.

53. ALLEN, F. G., and GOBELI, G. W. (1966). Energy structure in photoelectric emission from Cs-covered silicon and germanium. *Phys. Rev.* **144**, 558.

54. ALLEN, H. A., and HENDERSON, G. A. (1968). Efficient red-emitting p–n junctions found in GaP by solution growth of thick layers of p-type material on vapor grown n-type substrates. *J. Appl. Phys.* **39**, 2977.

55. ALLEN, J. W. (1959). The reverse characteristics of gallium arsenide p–n junctions. *J. Electron. Contr.* **7**, 254.

56. ALMASI, G. S., and SMITH, A. C. (1968). CdTe–HgTe heterostructures. *J. Appl. Phys.* **39**, 233.

57. AL'TSCHULER, B. L., GOLOVNER, T. M., KAGAN, M. B., and CHERNOV, YA. I. (1967). Structure on the photoelectric properties of diffusion heterojunctions in the GaAs–GaP structure. *Radio Eng. Electron. Phys. (USSR)* **12**, 999.

58. AMICK, J. A. (1963). The growth of single-crystal gallium arsenide layers on germanium and metallic substrates. *RCA Rev.* **24**, 555.

59. AMSTERDAM, M. F. (1970). The anomalous behavior of Schottky barrier diodes made on lightly doped GaAs. *Trans. AIME* **1**, 643.

60. ANAND, Y., and HOWELL, C. (1968). A burnout criterion for Schottky-barrier mixer diodes. *Proc. IEEE.* **56**, 2098.

61. ANANTHA, N. G., DOO, V. Y., SETO, D. K., and PECENEO, G. (1970). Chromium deposition from dicumene–chromium to form metal–semiconductor devices. *J. Electrochem. Soc.* **117**, 107c. (Also *Electrochem. Soc. Extended Abstr., 137th Nat. Meeting, Los Angeles, Spring 1970*, p. 480.)

62. ANDERSON, J. S., and KLEMPERER, D. F. (1960). Photoelectric measurements on nickel and nickel oxide films. *Proc. Roy. Soc.* **258**, 350.

63. ANDERSON, O. L., CHRISTENSEN, H., and ANDREATCH, P. (1957). Technique for connecting electrical leads to semiconductors. *J. Appl. Phys.* **28**, 923.

64. ANDERSON, P. A. (1959). Work function of gold. *Phys. Rev.* **115**, 553.

65. ANDERSON, R. L. (1960a). Germanium Gallium Arsenide Contacts. Ph.D. Thesis, Syracuse Univ., Syracuse, New York.

66. ANDERSON, R. L. (1960b). Junctions between Ge and GaAs. *Proc. Int. Conf. Semicond., Prague, 1960 (Czech. Acad. Sci.)*, p. 563.

67. ANDERSON, R. L. (1960c). Germanium–gallium arsenide heterojunctions. *IBM J. Res. Develop.* **4**, 283.

68. ANDERSON, R. L. (1962). Experiments on Ge–GaAs heterojunctions. *Solid–State Electron.* **5**, 341.

69. ANDERSON, R. L. (1970). The present state of the theory of heterojunctions. *Proc. Int. Conf. Phys. Chem. Semicond. Heterojunctions Layer Structures, Budapest, 1970* **2**, 55. Hung. Acad. Sciences, Budapest, Hungary.

70. ANDERSON, R. L., MARINACE, J. C., and SILVEY, G. A. (1963). Semiconductor Body Formation. U.S. Patent 3,072,507.

71. ANDREEV, V. M., KOROL'KOV, V. I., KUBINTSEVA, Z. M., NOSOV, Y. R., POSTNIKOVA, N. V., and RESHETNYAK, V. G. (1970a). Electrical properties of high-voltage AlAs–GaAs p–n heterostructures. *Proc. Int. Conf. Phys. Chem. Semicond. Heterojunctions Layered Structures, Budapest, 1970* **2**, 201, Hung. Acad. Sciences, Budapest, Hungary.

72. ANDREEV, V. M., *et al.* (1970b). Spontaneous light emission sources of epitaxial structures with AlAs–GaAs heterojunctions. *Proc. Int. Conf. Phys. Chem. Semicond. Heterojunction Layered Structures, Budapest, 1970* **2**, 195. Hung. Acad. Sciences, Budapest, Hungary.

73. ANDRONIK, I. K., *et al.* (1970). Photoelectrical and luminescent properties of some II–VI compound heterojunctions. *Proc. Int. Conf. Phys. Chem. Semicond. Heterojunctions Layer Structures, Budapest, 1970* **2**, 211. Hung. Acad. Sciences, Budapest, Hungary.

74. ANNER, G. E. (1969a). Avalanche luminescence in CdS–Te heterojunctions. *Proc. IEEE* **57**, 1219.

75. ANNER, G. E. (1969b). An infrared sensitive Ge–Te heterojunction. *Proc. IEEE* **57**, 2150.

76. ANSHON, A. V., and KARPOVITCH, I. A. (1969). Some properties of film photocells with CdS–Cu$_2$S heterojunctions. *Sov. Phys.–Semicond.* **3**, 503.

77. ANSTEAD, R. J., and FLOYD, S. R. (1969). Thermal effects on the integrity of aluminum to silicon contacts in silicon integrated circuits. *IEEE Trans. Electron Devices* **ED–16**, 381.

78. ANTELL, G. R. (1961). Investigation of a method of growing crystals of GaP and GaAs from the vapour phase. *Brit. J. Appl. Phys.* **12**, 687.

79. ANTLE, W. K. (1964). Friction technique for optimum thermocompression bonds. *IEEE Trans. Component Parts* **CP–11**, 25.

80. ANTYPAS, G. A., and JAMES, J. W. (1970). Liquid epitaxial growth of GaAsSb and its use as a high-efficiency long-wavelength threshold photoemitter. *J. Appl. Phys.* **41**, 2165.

81. ARCHER, R. J. (1970). Light emitting diodes in III–V alloys. *Electrochem. Soc. Extended Abstr., 137th Nat. Meeting, Los Angeles, Spring 1970*, p. 182.

82. ARCHER, R. J., and ATALLA, M. M. (1963). Metal contacts on cleaved silicon surfaces. *Ann. N.Y. Acad. Sci.* **101**, 697.

83. ARCHER, R. J., and YEP, T. O. (1970). Dependence of Schottky barrier height on donor concentration. *J. Appl. Phys.* **41**, 303.

84. ARCHER, R. J., YEP, T. O., and MURAY, K. (1968). Control of thin film interface barriers. Tech. Rep. Contract No. F33615–68–C–1054, Air Force Avionics Lab., Wright-Patterson AFB, Ohio (October 1967–March 1968), AD 832 050.

85. ARIZUMI, T., and AKASAKI, I. (1963). Thermodynamics of impurity doping reactions in vapor growth of Ge. *Jap. J. Appl. Phys.* **2**, 602.

86. ARIZUMI, T., and AKASAKI, I. (1964). A supplement to thermodynamics of impurity doping reactions in vapor growth of Ge. *Jap. J. Appl. Phys.* **3**, 87.

87. ARIZUMI, T., and NISHINAGA, T. (1966a). Thermodynamical considerations for the preparation of GaAs–Ge heterojunctions through closed tube process. *Jap. J. Appl. Phys.* **5**, 21.

88. ARIZUMI, T., and NISHINAGA, T. (1966b). Thermodynamics of vapour growth of ZnSe–Ge–I$_2$ system in closed tube process. *Jap. J. Appl. Phys.* **5**, 588.

89. ARIZUMI, T., AKASAKI, I., and NISHINAGA, T. (1963). Experimental studies of impurity doping in vapor growth of Ge. *Jap. J. Appl. Phys.* **2**, 757.

90. ARIZUMI, T., HIROSE, M., and ALTAF, N. (1968). Au–Ag alloy–silicon Schottky barriers. *Jap. J. Appl. Phys.* **7**, 870.

91. ARMANTROUT, G. A. (1970). Defect study and identification in Ge(Li) p–n junction radiation detectors. *Met. Trans. AIME* **1**, 659.

92. ARMSTRONG, A. (1958). On junctions with different energy gaps. *Proc. IRE* **46**, 1307.

93. ARMSTRONG, H. (1964). Contact Structure for Large Transistors. U.S. Patent 3,124,640.

94. ARNOLD, S. R., and PRITCHETT, R. L. (1965). Base-contact resistance studies for microwave germanium transistors. *In* Engineering Services on Transistors. Bell Telephone Lab. Rep. #20, June 30, 1965. AD 619 083.

95. ARSENI, K. A., DZHAFAROV, T. D., and DEDEGKAEV, T. T. (1969). Al$_x$Ga$_{1-x}$P–GaP heterojunctions. *Sov. Phys.–Semicond.* **3**, 788.

96. ATALLA, M. M., and KAHNG, D. (1962). A new "hot electron" triode structure with semiconductor metal emitter. *IRE Trans. Electron Devices* **ED–9**, 507.

97. ATALLA, M. M., and SOSHEA, R. W. (1963). Hot-carrier triodes with thin-film metal base. *Solid-State Electron.* **6**, 245.

98. AUBER, F. (1963). *p–n* junction photovoltaic effect in anodically formed oxide films of titanium. *J. Electrochem. Soc.* **110**, 846.

99. AUGUST, R. R. (1968). Monolithic Schottky barrier mixer and tunnel diode structures in heteroepitaxial gallium arsenide on sapphire. *IEEE Trans. Electron Devices* **ED–15**, 688.

100. AVEN, M. (1965). Efficient injection electroluminescence from *p–n* junctions in $ZnSe_xTe_{1-x}$. *Appl. Phys. Lett.* **7**, 146.

101. AVEN, M., and COOK, D. (1961). Some electrical properties of ZnTe–CdS hetero-junctions. *J. Appl. Phys.* **32**, 960.

102. AVEN, M., and CUSANO, D. A. (1964). Injection electroluminescence in ZnS and ZnSe. *J. Appl. Phys.* **35**, 606.

103. AVEN, M., and GARWACKI, W. (1963). Epitaxial growth and properties of ZnTe–CdS heterojunctions. *J. Electrochem. Soc.* **110**, 401.

104. AVEN, M., and GARWACKI, W. (1964). Synthesis and transport properties of ZnSe–ZnTe mixed crystals in *n*- and *p*-type form. *Appl. Phys. Lett.* **5**, 160.

105. AVEN, M., and GARWACKI, W. (1967a). Ohmic electrical contacts to *p*-type ZnTe and $ZnSe_xTe_{1-x}$. *J. Electrochem. Soc.* **114**, 1063.

106. AVEN, M., and GARWACKI, W. (1967b). Mechanism of charge transport and light emission in $ZnSe_xTe_{1-x}$ *p–n* junctions. *J. Appl. Phys.* **38**, 2302.

107. AVEN, M., and KREIGER, E. L. (1970). Diffusion of aluminum in the ZnSe–ZnTe system. *J. Appl. Phys.* **41**, 1930.

108. AVEN, M., and MEAD, C. A. (1965). Electrical transport and contact properties of low resistivity *n*-type zinc sulfide crystals. *Appl. Phys. Lett.* **7**, 8.

109. AVEN, M., and PRENER, J. S. (1967). "Physics and Chemistry of II–VI Compounds." North-Holland Publ., Amsterdam and Wiley, New York.

110. AVEN, M., and SEGALL, B. (1963). Carrier mobility and shallow impurity states in ZnSe and ZnTe. *Phys. Rev.* **130**, 81.

111. AVEN, M., and SWANK, R. K. (1968). Ohmic contacts to wide band-gap semi-conductors. Abstract No. 493, Trans. Electrochem. Soc. Meeting, October 1968.

112. AVEN, M., and SWANK, R. K. (1969). Ohmic contacts to wide-band-gap semi-conductors. *In* "Ohmic Contacts to Semiconductors" (B. Schwartz, ed.). Electrochem. Soc., New York.

113. AVEN, M., HALL, R. B., and GARWACKI, W. (1968). *p–n* junctions as artificial diffusion barriers for native defects. *Appl. Phys. Lett.* **13**, 292.

114. AVILA, A. J. (1964). Metal bonding in semiconductor manufacture. *Semicond. Prod. Solid State Technol.* **7**, No. 11, 22.

115. BACZEWSKI, A. (1965). Epitaxial growth of ZnSe on GaAs. *J. Electrochem. Soc.* **112**, 577.

116. BADALOV, A. Z. (1970). Photoconductivity of gold-doped *n* type silicon. *Sov. Phys.– Semicond.* **3**, 1435.

117. BAERTSCH, R. D., and RICHARDSON, J. R. (1969). An Ag–GaAs Schottky-barrier ultraviolet detector. *J. Appl. Phys.* **40**, 229.

118. BALK, P., and PILKUHN, M. H. (1966). Majority Carrier Channel Device Using Heterojunctions. U.S. Patent 3,273,030.

119. BARDEEN, J. (1947). Surface states and rectification at a metal–semiconductor contact. *Phys. Rev.* **71**, 717.

120. BARDEEN, J. (1961). Tunneling from a many-particle point of view. *Phys. Lett.* **6**, 57.

121. BARDSLEY, W. (1960). The electrical effects of dislocations in semiconductors. *Prog. Semicond.* **4**, 155.

122. BARNS, R. L. (1967). A survey of precision lattice parameter measurements as a too for the characterization of single-crystal materials. *Mater. Res. Bull.* **2**, 273.

123. BARRETT, C. S., and MASSALSKI ,T. B. (1966). "Structure of Metals." McGraw-Hill, New York.

124. BARTELINK, D. J., MOLL, J. L., and MEYER, N. I. (1963). Hot-electron emission from shallow *p–n* junctions in silicon. *Phys. Rev.* **130**, 972.

125. BEADLE, W. E., DABURLOS, K. E., and ECKTON, W. H., JR. (1969). Design, fabrication and characterization of a germanium microwave transistor. *IEEE Trans. Electron Devices* **ED–16**, 125.

126. BEER, S. Z. (1969). The solution of aluminum phosphide in aluminum. *J. Electrochem. Soc.* **116**, 263.

127. BELL, R. L. (1969). Thermionic emission of the GaAs photocathode. *Solid–State Electron.* **12**, 475.

128. BELL, R. L., and SPICER, W. E. (1970). 3–5 compound photocathodes: A new family of photoemitters with greatly improved performance. *Proc. IEEE* **58**, 1788.

129. BELL, R. L., and UEBBING, J. J. (1968). Photoemission from InP–Cs–O. *Appl. Phys. Lett.* **12**, 76.

130. BELOGLAZOV, A. V., BINDEMANN, R., GRACHEV, V. M., and YUNOVICH, A. E. (1970). Structure of the radiation spectra of gallium phosphide, associated with Zn–O and Cd–O impurity complexes. *Sov. Phys.–Semicond.* **3**, 1352.

131. BENDER, B. G. (1962). Semiconductor Device. U.S. Patent 3,036,250.

132. BENDER, B. G., and BERNSTEIN, L. (1965). Semiconductor Electrode Attachment. U.S. Patent 3,141,226.

133. BERCHTOLD, K. (1970). Some properties of InSb–Si heterojunctions. *Proc. Int. Conf. Phys. Chem. Semicond. Heterojunctions Layer Structures, Budapest, 1970* **2**, 221. Hung. Acad. Sciences, Budapest, Hungary.

134. BERGH, A. A., and STRAIN, R. J. (1968). Contacts for gallium phosphide electroluminescent diodes. Abstract No. 499, Trans. of Electrochem. Soc. Meeting, October 1968.

135. BERGH, A. A., and STRAIN, R. J. (1969). A contact for gallium phosphide electroluminescent diodes. *In* "Ohmic Contacts to Semiconductors" (B. Schwartz, ed.). Electrochem. Soc., New York.

136. BERGLUND, C. N. (1966). Electroluminescence using GaAs MIS structures. *Appl. Phys. Lett.* **9**, 441.

137. BERGLUND, C. N., and POWELL, R. J. (1970). The use of photoinjection for studying the interfacial properties of MIS structures. *Proc. Int. Conf. Phys. Chem. Semicond. Heterojunctions Layer Structures, Budapest, 1970* **5**, 71. Hung. Acad. Sciences, Budapest, Hungary.

138. BERKENBLIT, M., REISMAN, A., and LIGHT, T. B. (1968). Epitaxial growth of mirror smooth Ge on GaAs and Ge by the low temperature GeI$_2$ disproportionation reaction. *J. Electrochem. Soc.* **115**, 966.

139. BERMAN, I., and COMER, J. J. (1970). A resistance furnace for the heteroepitaxial growth of single crystal beta SiC through a molten metal intermediate. *Electrochem. Soc. Extended Abstr., 137th Nat. Meeting, Los Angeles, Spring 1970*, p. 427. (Also *J. Electrochem. Soc.* **117**, 104c.)

140. BERNSTEIN, L. J. (1961). Gold alloying to germanium, silicon and aluminum silicon eutectic surfaces. *Semicond. Prod.* **4** (July), 29.

141. BERNSTEIN, L. J. (1962). Alloying to III–V compound surfaces. *J. Electrochem. Soc.* **109**, 270.

142. BERNSTEIN, L. J. (1966). Semiconductor joining by the solid–liquid-interdiffusion

(SLID) process: I. The systems Ag–In, Au–In, and Cu–In. *J. Electrochem. Soc.* **113**, 1282.

143. BERNSTEIN, L. J., and BEALS, R. J. (1961). Thermal expansion and related bonding problems of some III–V compound semiconductors. *J. Appl. Phys.* **32**, 122.

144. BERRY, W. B. (1967). Population inversion in heterojunction structures. *Solid-State Electron.* **10**, 79.

145. BERTOTI, I., FARKAS-JAHNKE, M., LENDVAY, E., and NEMETH, T. (1969). Hetero-epitaxial growth of ZnS on GaP. *J Mater. Sci.* **4**, 699.

146. BERTOTI, I., GORYUNOVA, N. A., VARGA, L. and NEMETH, T. (1970). Heteroepitaxy involving $ZnSiP_2$, $ZnGeP_2$ and $CdSnAs_2$ compounds. *Proc. Int. Conf. Phys. Chem. Semicond. Heterojunctions Layer Structures, Budapest, 1970* **1**, 107. Hung. Acad. Sciences, Budapest, Hungary.

147. BHOLA, S. R., and MAYER, A. (1963). Epitaxial deposition of silicon by thermal decomposition of silane. *RCA Rev.* **24**, 511.

148. BIARD, J. R., and SHAUNFIELD, W. N., JR. (1967). A model of the avalanche photo-diode. *IEEE Trans. Electron Devices* **ED-14**, 233.

149. BILENKO, D. I., ILIN, V. S., GALASHNIKOVA, UN. N., KOSTYUNINA, G. P., KAZANOVA, N. P., and SINTZOVA, G. I. (1970). Electromagnetic radiation reflection from thin semiconductor single-crystal layers and heterojunctions. *Proc. Int. Conf. Phys. Chem. Semicond, Heterojunctions Layer Structures, Budapest, 1970* **2**, 229. Hung. Acad. Sciences, Budapest, Hungary.

150. BIONDI, F. J. (ed.) (1958). "Transistor Technology," Vol. 3, p. 175. Van Nostrand–Reinhold, Princeton, New Jersey.

151. BITER, W. J., and LAUER, R. B. (1970). Epitaxially grown graded heterojunctions of (Zn, Cd)S. *J. Electrochem. Soc.* **117**, 126.

152. BLACK, J., and LUBLIN, P. (1964). Electrical measurements and X-ray lattice para-meter measurements of GaAs doped with Se, Te, Zn and Cd and the stress effects of these elements as diffusants in GaAs. *J. Appl. Phys.* **35**, 2462.

153. BLOUNT, G. H. (1966). Ohmic electrical contacts to high-resistivity ZnS crystals. *J. Electrochem. Soc.* **113**, 690.

154. BLUM, S. E., and CHICOTKA, R. J. (1970). The preparation of homogeneous alloys of III–V compounds. *Electrochem. Soc. Extended Abstr., 137th Nat. Meeting, Los Angeles, Spring 1970*, p. 202.

155. BLUM, W., and HOGABOOM, G. B. (1949). "Principles of Electroplating and Electro-forming." McGraw-Hill, New York.

156. BOBB, L. C., HOLLOWAY, H., MAXWELL, K. H., and ZIMMERMAN, E. (1966a). Epi-taxial growth of bulk-quality gallium arsenide on gallium arsenide and germanium substrates. *J. Appl. Phys.* **37**, 3909.

157. BOBB, L. C., HOLLOWAY, H., MAXWELL, K. H., and ZIMMERMAN, E. (1966b). Oriented growth of semiconductors II. Homoepitaxy of gallium arsenide. *Phys. Chem. Solids* **27**, 1679.

158. BOBB, L. C., HOLLOWAY, H., MAXWELL, K. H., and ZIMMERMAN, E. (1966c). Oriented growth of semiconductors, III. Growth of gallium arsenide on germanium. *J. Appl. Phys.* **37**, 4687.

159. BOER, K. W., and HALL, R. B. (1966). Multilayer ohmic contacts on CdS. *J. Appl. Phys.* **37**, 4739.

160. BOER, K. W., and LUBITZ, K. (1962). Zum Kontaktproblem an CdS-ein Kristallen. *Z. Naturforsch.* **17a**, 397.

161. BOLTAKS, B. I. (1963). "Diffusion in Semiconductors." Academic Press, New York.

162. BOND, W. L. (1960). Precision lattice constant determination. *Acta Crystallogr.* **13**, 814.

163. BONNET, D., and RABENHORST, H. (1970). CdS–CdTe thin film *p–n* heterodiodes with graded energy gaps. *Proc. Int. Conf. Phys. Chem. Semicond. Heterojunctions Layer Structures, Budapest, 1970* **1**, 119. Hung. Acad. Sciences, Budapest, Hungary.

164. BORNEMAN, E. H., SCHWARTZ, R. F., and STICKLER, J. J. (1955). Rectification properties of metal–semiconductor contacts. *J. Appl. Phys.* **26**, 1021.

165. BORTFELD, D. P., and KLEINKNECHT, H. P. (1968). Injection electroluminescence in alloyed ZnTe junctions. *J. Appl. Phys.* **39**, 6104.

166. BOTT, I. B., HILSUM, C., and SMITH, K. C. H. (1967). Construction and performance of epitaxial transferred electron oscillators. *Solid–State Electron.* **10**, 137.

167. BOUCHER, A., and HOLLAN, L. (1970). Thermodynamic and experimental aspects of gallium arsenide vapor growth. *J. Electrochem. Soc.* **117**, 932.

168. BRASLAU, N., GUNN, J. B., and STAPLES, J. L. (1967). Metal–semiconductor contacts for GaAs bulk effect devices. *Solid–State Electron.* **10**, 381.

169. BRATTAIN, W. H., and BARDEEN, J. (1953). Surface properties of germanium. *Bell Syst. Tech. J.* **32**, 1.

170. BRAUNSTEIN, A. I., BRAUNSTEIN, M., and PICUS, G. S. (1966). Voltage dependence of the barrier heights in Al_2O_3 tunnel junctions. *Appl. Phys. Lett.* **8**, 95.

171. BRAUNSTEIN, M., HENDERSON, R. R., and BRAUNSTEIN, A. I. (1968). Moving mask growth of single-crystal silicon films on amorphous quartz substrates. *Appl. Phys. Lett.* **12**, 66.

172. BRAUNSTEIN, R. (1963a). Valence band structure of germanium–silicon alloys. *Phys. Rev.* **130**, 869.

173. BRAUNSTEIN, R. (1963b). Lattice vibration spectra of germanium–silicon alloys. *Phys. Rev.* **130**, 879.

174. BRAUNSTEIN, R., PANKOVE, J. I., and NELSON, H. (1963.) Effect of doping on the emission peak and the absorption edge of GaAs. *Appl. Phys. Lett.* **3**, 31.

175. BREITSCHWERDT, K. G. (1963). Direct and indirect tunneling in germanium at different temperatures. *J. Appl. Phys.* **34**, 2610.

176. BREWER, L., SOMAYAJULU, G. R., and BRACKETT, E. (1963). Thermodynamic properties of gaseous metal dihalides. *Chem. Rev.* **63**, 111.

177. BRIDGERS, H. E., SCAFF, J. H., and SHIVE, J. N. (eds.) (1958). "Transistor Technology," Vol. 1, pp. 331, 343. Van Nostrand–Reinhold, Princeton, New Jersey.

178. BROJDO, S. (1963). Characteristics of the dielectric diode and triode at very high frequencies. *Solid–State Electron.* **6**, 611.

179. BROJDO, S., RILEY, T. J., and WRIGHT, G. T. (1965). The heterojunction transistor and the space-charge-limited triode. *Brit. J. Appl. Phys.* **16**, 133.

180. BROOM, R. F. (1963). Room temperature operation of gallium arsenide lasers. *Phys. Lett.* **4**, 330.

181. BROWN, D. M., and GRAY, P. V. (1968). Si–SiO_2 fast interface state measurements. *J. Electrochem. Soc.* **115**, 760.

182. BROWN, D. M., ENGELER, W. E., GARFINKEL, M., and GRAY, P. V. (1968). Refractory metal–silicon device technology. *Solid–State Electron.* **11**, 1105.

183. BROWNSON, J. (1964). Alloyed Ge–Si heterojunctions. *J. Appl. Phys.* **35**, 1356.

184. BROWNSON, J. (1965). High-speed Ge–Si *n–n* alloyed heterodiodes. *Trans. AIME* **233**, 450.

185. BRYLA, S. M., and FELDMAN, C. (1962). Contact potential in thin metal films. *J. Appl. Phys.* **33**, 774.

186. BUBE, R. H. (1960). "Photoconductivity of Solids," p. 82. Wiley, New York.

187. BUBE, R. H. (1962). Trap density determination by space charge limited currents. *J. Appl. Phys.* **33**, 1733.

188. BUBE, R. H., and LIND, E. L. (1958). Photoconductivity of ZnSe crystals and a correlation of donor and acceptor levels in II–VI photoconductors. *Phys. Rev.* **110**, 1040.

189. BUHANAN, D. (1969). Investigation of current-gain temperature dependence in silicon transistors. *IEEE Trans. Electron Devices* **ED-16**, 117.

190. BULLIS, W. M. (1966). Properties of gold in silicon. *Solid-State Electron.* **9**, 143.

191. BURGER, R. M., and DONOVAN, R. P. (eds.) (1967). Fundamentals of Silicon Integrated Device Technology," Vol. 1, Oxidation, Diffusion and Epitaxy. Prentice-Hall, Englewood Cliffs, New Jersey.

192. BURMEISTER, R. A., JR., and REGEHR, R. W. (1969). Epitaxial growth of $GaAs_{1-x}P_x$ on germanium substrates. *Trans. AIME* **245**, 565.

193. BURNHAM, R. D., HOLONYAK, N., JR., and SCIFRES, D. R. (1970). $Al_xGa_{1-x}As_{1-y}P_y$–$GaAs_{1-y}P_y$ heterostructure laser and lamp junctions. *Appl. Phys. Lett.* **17**, 455.

194. BURNS, J. W. (1968). Epitaxia growth of GaSb from the liquid phase. *Trans. AIME* **242**, 432.

195. BURSTEIN, E. (1954). Anomalous optical absorption limit in InSb. *Phys. Rev.* **93**, 632.

196. BUTLER, J. K., SOMMERS, H. S., JR., and KRESSEL, H. (1970). High-order transverse cavity modes in heterojunction diode lasers. *Appl. Phys. Lett.* **17**, 403.

197. BUTLER, W., and MUSCHEID, W. (1954,1955). Die Bedeutung des electrischen Kontakes bei Untersuchungen an Kadmium Sulfid–Enkristallen I and II. *Ann. Phys.* (*Leipzig*) *Sect. 6* **14**, 215; **15**, 82.

198. BUTORINA, L. N., and TOLOMASOV, V. A. (1966). Preparation of germanium films by vacuum evaporation with gas etching. *Sov. Phys.–Crystallogr.* **10**, 456.

199. BYKOVSKII, YU. A., ELESIN, V. F., and ZUEV, V. V. (1970a). Negative differential conductivity in a semiconductor containing attractive impurity centers. *Sov. Phys.–Semicond.* **3**, 1442.

200. BYKOVSKII, YU. A., ELKHOV, V. A., and LARKIN, A. I. (1970b). Coherence of the radiation emitted by a semiconductor laser and its use in holography. *Sov. Phys.–Semicond.* **4**, 819.

201. CALOW, J. T., DEASLEY, P. J., OWEN, S. J. T., and WEBB, P. W. (1967). Review of semiconductor heterojunctions. *J. Mater. Sci.* **2**, 88.

202. CALOW, J. T., OWEN, S. J. T., and WEBB, P. W. (1968). The growth and electrical characteristics of epitaxial layers of ZnSe on *p*-type Ge. *Phys. Status Solidi* **28**, 295.

203. CALZOLARI, P. U., and GRAFFI, S. (1969). On the beta falloff in junction transistors. *Proc. IEEE* **57**, 1293.

204. CARR, T. G., RICHMOND, J. C., and WAGNER, R. G. (1970). Position-sensitive Schottky barrier photodiodes: Time-dependent signals and background saturation effects. *IEEE Trans. Electron Devices* **ED-17**, 507.

205. CARRUTHERS, T. (1969). Temperature dependence of tunneling structure in GaAs–Au Schottky-barrier junctions. *Bull. Amer. Phys. Soc.* **14**, 413.

206. CASEY, H. C., JR. (1962). Fabrication of gallium arsenide subnano-second switching diodes. Stanford Electron. Labs. Tech. Rep. No. 1802-1, Stanford Univ., Stanford, California.

207. CASEY, H. C., JR., and PANISH, M. B. (1969). Composition dependence of the $Ga_{1-x}Al_xAs$ direct and indirect energy gaps. *J. Appl. Phys.*, **40**, 4910.

208. CASEY, H. C., JR., and SILVERSMITH, D. J. (1969). Radiative tunneling in GaAs abrupt asymmetrical junctions. *J. Appl. Phys.* **40**, 241.

209. CAVE, E. F., and CZORNY, B. R. (1963). Epitaxial deposition of silicon and germanium layers by chloride reduction. *RCA Rev.* **24**, 523.

210. CAYWOOD, J. M., and MEAD, C. A. (1969). Origin of field-dependent collection efficiency in contact limited photoconductors. *Appl. Phys. Lett.* **15**, 14.

211. CERNIGLIA, N. P., TONNER, R. C., BERKOVITS, G., and SOLOMON, A. H. (1968). Beam-lead Schottky-barrier diodes for low noise integrated microwave mixers. *IEEE Trans. Electron Devices* **ED-15**, 674.

212. CHANG, C. C. (1970). Contaminants on chemically etched silicon surfaces: LEED–Auger method. *Surface Sci.* **23**, 283.

213. CHANG, C. C. (1971). Auger electron spectroscopy. *Surface Sci.* **25**, 53–79.

214. CHANG, L. L. (1965a). Field dependence of surface mobility at n–n heterojunction interface. *Solid-State Electron.* **8**, 86.

215. CHANG, L. L. (1965b). The conduction properties of Ge–GaAs$_{1-x}$P$_x$ n–n hetero-junctions. *Solid-State Electron.* **8**, 721.

216. CHANG, L. L. (1966). Comments on junction boundary conditions for heterojunctions. *J. Appl. Phys.* **37**, 3908.

217. CHANG, L. L., ESAKI, L., and STILES, P. J. (1968a). GeTe–GaAs tunnel junction—A negative-resistance Schottky-type barrier. *J. Appl. Phys.* **39**, 6049.

218. CHANG, L. L., ESAKI, L., and STILES, P. J. (1968b). Negative resistance in Schottky-type tunnel heterojunctions. *Bull. Amer. Phys. Soc.* **13**, 456.

219. CHANG, Y. F. (1965). Junction boundary conditions for heterojunctions. *J. Appl. Phys.* **36**, 3350.

220. CHANG, Y. F. (1966). On the junction boundary conditions for heterojunctions. *Proc. IEEE* **54**, 1965.

221. CHANG, Y. F. (1970). The formation of heterojunctions by a gold-solvent alloying technique. *J. Electrochem. Soc.* **117**, 97C. (Also *Electrochem. Soc. Extended Abstr.*, *137th Nat. Meeting, Los Angeles, Spring 1970*, p. 268.)

222. CHAPMAN, R. A. (1964). Thermionic work function of thin-oxide-coated aluminum electrodes in vacuum and in cesium vapor. *J. Appl. Phys.* **35**, 2832.

223. CHAPRA, K. L. (1969). Epitaxial growth of films on substrates coated with amorphous deposits. *J. Appl. Phys.* **40**, 906.

224. CHENETTE, E. R., PEDERSEN, R. A., EDWARDS, R., and KLEIMACK, J. J. (1968). Integrated Schottky-diode clamp for transistor storage time control. *Proc. IEEE* **56**, 232.

225. CHINO, K., and ARIYOSHI, H. (1968). Effect of surface polishing on the stress-sensitivity of a Schottky-barrier diode. *Jap. J. Appl. Phys.* **7**, 1130.

226. CHO, A. Y. (1970). Epitaxial growth of gallium phosphide on cleaved and polished (111) calcium fluoride. *J. Appl. Phys.* **41**, 782.

227. CHOW, C. K. (1963a). On tunneling equations of Holm and Stratton. *J. Appl. Phys.* **34**, 2490.

228. CHOW, C. K. (1963b). Effect of insulating-film-thickness nonuniformity on tunnel characteristics. *J. Appl. Phys.* **34**, 2599.

229. CHOW, C. K. (1967). Temperature dependence of BeO tunneling structures. *J. Appl. Phys.* **34**, 2918.

230. CHRIST, J. G., and RAMSEY, J. N. (1969). Analysis tools for microminiaturized circuits. *IEEE Spectrum* **6**, 109.

231. CHRISTENSEN, H. (1959). Electrical contact with thermo-compression bonds. *Bell Lab. Rec.* **36**, 127.

232. CHRISTIAN, S. M. (1965). Semiconductor Devices. U.S. Patent 3,211,970.

233. CHU, T. L., and KELM, R. W. (1969). The etching of germanium with water vapor and hydrogen sulfide. *J. Electrochem. Soc.* **116**, 1261.

234. CHUDOBIAK, W. J. (1968). On crowding effects and failure mechanisms in high power transistor switches. *Proc. IEEE* **56**, 2176.

235. CHUDOBIAK, W. J. (1969). On the static collector–emitter saturation voltage of a transistor with a lightly doped collector. *Proc. IEEE* **57**, 718.

236. CHUNG, TCHANG-II (1962). Study of aluminum fusion into silicon. *J. Electrochem. Soc.* **109**, 229.

237. CHYNOWETH, A. G., and McKAY, K. G. (1957). Internal field emission in silicon *p–n* junctions. *Phys. Rev.* **106**, 418.

238. CHYNOWETH, A. G., FELDMANN, W. L., and LOGAN, R. A. (1961). Excess tunnel current in silicon Esaki junctions. *Phys. Rev.* **121**, 684.

239. CLAASSEN, R. S. (1961). Excess and hump current in Esaki diodes. *J. Appl. Phys.* **32**, 2372.

240. CLARK, A. H., and ALIBOZEK, R. G. (1968). Low-temperature orientation of silicon films deposited by sputtering through a moving mask. *J. Appl. Phys.* **39**, 2156.

241. CLARK, L. E. (1969). High-current beta falloff. *Proc. IEEE* **57**, 1670.

242. CLARK, L., and WOOK, J. (1965). A method of applying ohmic contacts to cadmium sulphide crystals for Hall measurement. *J. Sci. Instrum.* **42**, 51.

243. CLAWSON, A. R., and WIEDER, H. H. (1967). Electrical and galvanomagnetic properties of single crystal InSb dendrites. *Solid-State Electron.* **10**, 57.

244. COHEN, J. (1962a). Tunnel emission into vacuum. *J. Appl. Phys.* **33**, 1999.

245. COHEN, J. (1962b). Tunnel emission into vacuum, II. *Appl. Phys. Lett.* **1**, 61.

246. COHEN, J. (1965). Platinum–silicon thermocompression bonds. *Solid-State Electron.* **8**, 79.

247. COHEN, M. L., and BERGSTRESSER, T. K. (1966). Band structures and pseudopotential form factors for fourteen semiconductors of the diamond and zinc blend structures. *Phys. Rev.* **141**, 789.

248. COHEN, M. M., and BEDARD, F. D. (1968). Electrical properties of single-crystal gallium phosphide doped with zinc. *J. Appl. Phys.* **39**, 285.

249. COHEN-SOLAL, G., and MARFAING, Y. (1968). Transport of photocarriers in $Cd_xHg_{1-x}Te$ graded-gap structures. *Solid-State Electron.* **11**, 1131.

250. COLMAN, D. (1968). High resistivity Hall effect measurements. *Rev. Sci. Instrum.* **39**, 1946.

251. CONLEY, J. W., and MAHAN, G. D. (1967). Tunneling spectroscopy in GaAs. *Phys. Rev.* **161**, 681.

252. CONRAD, R. W., HOYT, P. L., and MARTIN, D. D. (1967a). Preparation of epitaxial $Ga_xIn_{1-x}As$. *J. Electrochem. Soc.* **114**, 164.

253. CONRAD, R. W., REYNOLDS, R. A., and JEFFCOAT, M. W. (1967b). Preparation of high purity epitaxial gallium arsenide from the elements. *Solid-State Electron.* **10**, 507.

254. CONSTANTINESCU, C., GOLDENBLUM, A., and SOSTARICH, M. (1970). On the light emission from heavily-doped AlGaAs–GaAs heterojunctions. *Proc. Int. Conf. Phys. Chem. Semicond. Heterojunctions Layer Structures, Budapest, 1970* **2**, 249. Hung. Acad. Sciences, Budapest, Hungary.

255. CONWELL, E. M. (1958). Properties of germanium and silicon II. *Proc. IRE* **46**, 1281.

256. CONWELL, E. M. (1964). Relative energy loss to optical and acoustic modes of electrons in avalanche breakdown in Ge. *Phys. Rev.* **135**, A1138.

257. CONWELL, E. M. (1967). High field transport in semiconductors. *Solid State Phys. Suppl.* **9**.

258. COOPER, A. S. (1962). Precise lattice constants of germanium, aluminum, gallium arsenide, uranium, sulphur, quartz and sapphire. *Acta Crystallogr.* **15**, 578.

259. COPPEN, P. J., and MATZEN, W. T. (1962). Distribution of recombination current in emitter-base junctions of Si transistors. *IRE Trans. Electron Devices* **ED-9**, 75.

260. COTTRELL, A. H. (1953). "Dislocations and Plastic Flow in Crystals," p. 39. Oxford Univ. Press (Clarendon), London and New York.

261. COURTENS, E., and CHERNOW, F. (1966). Contact barriers on insulating CdS. *Appl. Phys. Lett.* **8**, 3.

262. COURVOISIER, J. (1963). Method of Manufacturing Semiconductor Bodies. U.S. Patent 3,102,828.

263. COWLEY, A. M. (1966). Depletion capacitance and diffusion potential of gallium phosphide Schottky-barrier diodes. *J. Appl. Phys.* **37**, 3024.

264. COWLEY, A. M., and HEFFNER, H. (1964). Gallium-phosphide–gold surface barrier. *J. Appl. Phys.* **35**, 255.

265. COWLEY, A. M., and SORENSON, H. O. (1966). Quantitative comparison of microwave solid state detectors. *IEEE Trans. Microwave Theory Tech.* **MTT-14**, 588.

266. COWLEY, A. M., and SZE, S. M. (1965). Surface states and barrier height of metal–semiconductor systems. *J. Appl. Phys.* **36**, 3212.

267. COWLEY, A. M., and ZETTLER, R. A. (1968). Shot noise in silicon Schottky barrier diodes. *IEEE Trans. Electron Devices* **ED-15**, 761.

268. COX, R. H., and HASTY, T. E. (1968). Metallurgy of alloyed ohmic contacts for the Gunn oscillator. *Trans. Electrochem. Soc. Meeting, October 1968*, Abstr. No. 496.

269. COX, R. H., and HASTY, T. E. (1969). Metallurgy of alloyed ohmic contacts for the Gunn oscillator. *In* "Ohmic Contacts to Semiconductors" (B. Schwartz, ed.). Electrochem. Soc., New York.

270. COX, R. H., and STRACK, H. (1967). Ohmic contacts for GaAs devices. *Solid-State Electron.* **10**, 1213.

271. CRAFORD, M. G., GROVES, W. O., and FOX, M. J. (1970). GaAs–GaAsP hetero structure injection lasers. *J. Electrochem. Soc.* **117**, 93C; **118**, 355. (Also *Electrochem. Soc. Extended Abstr., 137th Nat. Meeting, Los Angeles, Spring 1970*, p. 199.)

272. CRONIN, G. R., CONRAD, R. W., and BORRELLO, S. R. (1966). Epitaxial InAs on semi-insulating GaAs substrates. *J. Electrochem. Soc.* **113**, 1336.

273. CROWDER, B. L., and HAMMER, W. N. (1966). Shallow acceptor states in ZnTe and CdTe. *Phys. Rev.* **150**, 541.

274. CROWDER, B. L., MOREHEAD, F. F., and WAGNER, P. R. (1966). Efficient injection electroluminescence in ZnTe by avalanche breakdown. *Appl. Phys. Lett.* **8**, 148.

275. CROWELL, C. R. (1965). The Richardson constant for thermionic emission in Schottky barrier diodes. *Solid-State Electron.* **8**, 395.

276. CROWELL, C. R. (1968). Thermionic and zero-bias anomalies in metal–semiconductor (Schottky) barriers. *Bull. Amer. Phys. Soc.* **13**, 456.

277. CROWELL, C. R. (1969a). General normalized current voltage characteristic of quasi-ohmic contacts. *In* "Ohmic Contacts to Semiconductors" (B. Schwartz, ed.). Electrochem. Soc., New York.

278. CROWELL, C. R. (1969b). Richardson constant and tunneling effective mass for thermionic and thermionic-field emission in Schottky barrier diodes. *Solid-State Electron.* **12**, 55.

279. CROWELL, C. R. (1969c). Metal–semiconductor interfaces. *Surface Sci.* **13**, 13.

280. CROWELL, C. R., and RIDEOUT, V. L. (1969a). Normalized thermionic-field emission in metal–semiconductor (Schottky) barriers. *Solid-State Electron.* **12**, 89.

281. CROWELL, C. R., and RIDEOUT, V. L. (1969b). Thermionic-field resistance maxima in metal–semiconductor (Schottky) barriers. *Appl. Phys. Lett.* **14**, 85.

282. CROWELL, C. R., and ROBERTS, G. I. (1969). Surface state and interface effects on the capacitance–voltage relationship in Schottky barriers. *J. Appl. Phys.* **40**, 3726.

283. CROWELL, C. R., and SZE, S. M. (1965a). Electron–phonon collector back-scattering in hot electron transistors. *Solid-State Electron.* **8**, 673.

284. CROWELL, C. R., and SZE, S. M. (1965b). Electron–optical-phonon scattering in the

emitter and collector barriers of semiconductor–metal–semiconductor structures. *Solid-State Electron.* **8**, 979.

285. CROWELL, C. R., and SZE, S. M. (1965c). Ballistic mean free path measurements of hot electrons in Au films. *Phys. Rev. Lett.* **15**, 659.

286. CROWELL, C. R., and SZE, S. M. (1966a). Temperature dependence of avalanche multiplication in semiconductors. *Appl. Phys. Lett.* **9**, 242.

287. CROWELL, C. R., and SZE, S. M. (1966b). Quantum-mechanical reflection of electrons at metal–semiconductor barriers: Electron transport in semiconductor–metal–semiconductor structures. *J. Appl. Phys.* **37**, 2683.

288. CROWELL, C. R., and SZE, S. M. (1966c). Current transport in metal–semiconductor barriers. *Solid-State Electron.* **9**, 1035.

289. CROWELL, C. R., SPITZER, W. G., and WHITE, H. G. (1962a). Range of photoexcited holes in Au. *Appl. Phys. Lett.* **1**, 3.

290. CROWELL, C. R., SPITZER, W. G., HOWARTH, L. E., and LEBATE, E. E. (1962b). Attenuation length measurements of hot electrons in metal films. *Phys. Rev.* **127**, 2006.

291. CROWELL, C. R., SZE, S. M., and SPITZER, W. G. (1964). Equality of the temperature dependence of the gold–silicon surface barrier and the silicon energy gap in Au *n*-type Si diodes. *Appl. Phys. Lett.* **4**, 91.

292. CROWELL, C. R., SARACE, J. C., and SZE, S. M. (1965a). Tungsten–semiconductor Schottky-barrier diodes. *Trans. AIME* **233**, 478.

293. CROWELL, C. R., SHORE, H. B., and LEBATE, E. E. (1965b). Surface-state and interface effects in Schottky barriers at *n*-type silicon surfaces. *J. Appl. Phys.* **36**, 3843.

294. CRUCEANU, E., NICULESCU, D., STAMATESCU, I., NISTOR, N., and IONESCU-BUJOR, S. (1965). Electrical properties of some solid solutions of the $Zn_xHg_{1-x}Te$ system. *Sov. Phys.–Solid State* **7**, 1646.

295. CSERVENY, S. I. (1968). Potential distribution and capacitance of abrupt heterojunctions. *Int. J. Electron.* **25**, 65.

296. CSOKA, L. M. (1970). Metal–semiconductor (Schottky) barriers as heterojunctions. *Proc. Int. Conf. Phys. Chem. Semicond. Heterojunctions Layer Structures, Budapest, 1970*, **5**, 265. Hung. Acad. Sciences, Budapest, Hungary.

297. CULLEN, D. E., and COMPTON, W. D. (1969). Tunneling spectroscopy in highly doped *p*-type silicon. *Bull. Amer. Phys. Soc.* **14**, 414.

298. CUNNELL, F. A., EDMOND, J. T., and HARDING, W. R. (1960). Technology of GaAs. *Solid-State Electron.* **1**, 97.

299. CUNNINGHAM, J. A. (1965). Expanded contacts and interconnexions to monolithic silicon integrated circuits. *Solid-State Electron.* **8**, 735.

300. CUNNINGHAM, J. A., and HARPER, J. G. (1967). Semiconductor reliability: Focus on the contacts. *Electron. Eng.* **26**, 74.

301. CURTIS, B. J., EMMENEGGER, F. P., and NITSCHET, R. (1970). The preparation of ternary and quaternary compounds by vapor-phase growth. *RCA Rev.* **31**, 647.

302. CUSANO, D. A. (1963). CdTe solar cells and photovoltaic heterojunctions in II–VI compounds. *Solid-State Electron.* **6**, 217.

303. CUTTRISS, D. B. (1961). Relation between surface concentration and average conductivity of diffused layers in germanium. *Bell Syst. Tech. J.* **40**, 509.

304. DACENKO, L. I., KLOCHKOV, Y. A. and TKHORIK, Y. A. (1970). Investigation of the crystallographic perfection of hetero-structures. *Abstr. Int. Conf. Phys. Chem. Semicond. Heterojunctions Layer Structures, Budapest, 1970*. Hung. Acad. Sciences, Budapest, Hungary.

305. DALE, J. R. (1966). Alloyed semiconductor heterojunctions. *Phys. Status Solidi* **16**, 351.

306. DALE, J. R., and JOSH, M. J. (1964). Alloys for GaAs devices. *Solid-State Electron.* **7**, 177.

307. DALE, J. R., and JOSH, M. J. (1965). Heterojunctions by alloying. *Solid-State Electron.* **8**, 1.

308. DALE, J. R., and TURNER, R. G. (1963). Simple ohmic contacts on gallium arsenide. *Solid-State Electron.* **6**, 388.

309. DANAHY, E. L. (1970). The real world of silicon photo-diodes. *Electro-Optical Systems Design*, May, p. 36.

310. DANILYUK, S. A., and KOT, M. V. (1965). Structure and electric properties of the HgTe–ZnTe system. *Bull. Acad. Sci. USSR Phys. Ser.* **28**, 973.

311. DAS, M. B., and BOOTHROYD, A. R. (1961). Determination of physical parameters of diffusion and drift transistors. *IRE Trans. Electron Devices* **ED-8**, 15.

312. DAVEY, J. E. (1962). Epitaxy of germanium films on germanium by vacuum evaporation. *J. Appl. Phys.* **33**, 1015.

313. DAVEY, J. E. (1966). Electrical characteristics of epitaxial germanium films vacuum deposited on semi-insulating GaAs up to thicknesses of 10^6 Å. *Appl. Phys. Lett.* **8**, 164.

314. DAVEY, J. E., and PANKEY, T. (1964). Structural and optical characteristics of thin GaAs films. *J. Appl. Phys.* **35**, 2203.

315. DAVEY, J. E., and PANKEY, T. (1968). Epitaxial GaAs films deposited by vacuum evaporation. *J. Appl. Phys.* **39**, 1941.

316. DAVEY, J. E., and PANKEY, T. (1969). Structural and optical evaluation of vacuum-deposited GaP films. *J. Appl. Phys.* **40**, 212.

317. DAVIS, J. L., and NORR, M. K. (1966). Ge-epitaxial PbS heterojunctions. *J. Appl. Phys.* **37**, 1670.

318. DAVIS, L. C., and STEINRISSER, F. (1970). One-electron and phonon-assisted tunneling in *n*-Ge Schottky barriers. *Phys. Rev. B* **1**, 614.

319. DAVIS, M. E. (1970). *n*–*n* GaAs–GaP heterojunction. *Proc. Int. Conf. Phys. Chem. Semicond. Heterojunctions Layer Structures, Budapest, 1970* **2**, 259. Hung. Acad. Sciences, Budapest, Hungary.

320. DAVIS, M. E., ZEIDENBERGS, G., and ANDERSON, R. L. (1969). GaAs–GaP heterojunctions. *Phys. Status Solidi* **34**, 385.

321. DAVIS, R. H., and HOSACK, H. H. (1963). Double barrier in thin-film triodes. *J. Appl. Phys.* **34**, 864.

322. DAW, A. N., and MITRA, R. N. (1965). Electroless plating on III–V compound semiconductors. *Solid-State Electron.* **8**, 697.

323. DAW, A. N., MITRA, R. N., and CHOUDHURY, N. K. D. (1967). Cut-off frequency of a drift transistor. *Solid-State Electron.* **10**, 359.

324. DAWSON, L. R., and WHELAN, J. M. (1968). Growth of thin GaAs epitaxial films from Ga solution. *Bull. Amer. Phys. Soc.* **13**, 375.

325. DEAL, B. E., SNOW, E. N., and MEAD, C. A. (1966). Barrier energies in metal–silicon dioxide–silicon structures. *Phys. Chem. Solids* **27**, 1873.

326. DEALBA, E., and WARMAN, J. (1968). Electron mobility in polar semiconductors at intermediate and high electric fields. *Phys. Chem. Solids* **29**, 69.

327. DEKKER, A. J. (1957). "Solid State Physics," Chap. 17. Prentice-Hall, Englewood Cliffs, New Jersey.

328. DEKKER, A. J. (1958). Secondary electron emission. *Solid State Phys.* **6**, 251.

329. DEMOULIN, E., and VAN DE WIELE, F. (1970). Inversion layers in abrupt Si–Ge heterojunctions. *Proc. Int. Conf. Phys. Chem. Semicond. Heterojunctions Layer Structures, Budapest, 1970* **2**, 267. Hung. Acad. Sciences, Budapest, Hungary.

330. DENOBEL, D. (1958). Method of Making Electrical Connection to Semi-Conductive Selenide or Telluride. U.S. Patent 2,865,793.

331. DeNobel, D., and Kock, H. G. (1969). A silicon Schottky barrier avalanche transit time diode. *Proc. IEEE* **57**, 2088.

332. Deslatted, R. D. (1969). Optical and X-ray interferometry of a silicon lattice spacing. *Appl. Phys. Lett.* **15**, 386.

333. Deuling, H. J. (1969). Influence of acoustic phonons on tunneling through metal–semiconductor contacts. *Bull. Amer. Phys. Soc.* **14**, 414.

334. Diedrich, H., and Jotten, K. (1961). Experimental investigations of Ge transistors with wide gap emitters. *Int. Symp. Semicond. Devices, Paris, February 20-25, 1961.*

335. Dierschke, E. G., and Pearson, G. L. (1970). Forward and reverse tunnel currents in gallium phosphide diffused *p–n* junctions. *J. Appl. Phys.* **41**, 329.

336. Dijk, H. V., and Goorissen, J. (1967). Cadmium sulphide single crystals by epitaxial growth on germanium substrates. *In J. Phys. Chem. Suppl.* No. 1, *Proc. Int. Conf. Crystal Growth, Boston, 1966* (H. S. Peiser, ed.), p. 531.

337. DiLorenzo, J. V., Moore, G. E., Jr., and Machala, A. E. (1970). Dependence of the electrical properties of vapor grown epitaxial GaAs on the AsCl$_3$ vapor concentration. *Electrochem. Soc. Extended Abstr., 137th Nat. Meeting, Los Angeles, Spring 1970,* p. 370.

338. Dismukes, J. P., Yim, W. M., Tietjen, J. J., and Novak, R. E. (1970). Vapor deposition of semiconducting mononitrides of scandium, yttrium, and the rare-earth elements. *RCA Rev.* **31**, 680.

339. Ditrick, N. H., and Nelson, H. (1961). Design and fabrication of germanium tunnel diodes. *RCA Eng.* **7**, 19.

340. Dobrynin, V. A. (1969). Analysis of the band diagrams of degenerate heterojunctions. *Sov. Phys.–Semicond.* **3**, 11.

341. Dolega, U. (1963). Theory of the *p–n* heterojunction between semiconductors with different crystal lattices (in German). *Z. Naturforsch.* **18**, 653.

342. Donnelly, J. P. (1965). Studies of Ge–GaAs and Ge–Si Heterojunctions. Ph.D. Thesis, Carnegie Inst. of Technol., Pittsburgh, Pennsylvania.

343. Donnelly, J. P., and Milnes, A. G. (1965). The capacitance of double saturation *n*Ge–*n*Si heterojunctions. *Proc. IEEE* **53**, 2109.

344. Donnelly, J. P., and Milnes, A. G. (1966a). Photovoltaic characteristics of *p–n* Ge–Si and Ge–GaAs heterojunctions. *Int. J. Electron.* **20**, 295.

345. Donnelly, J. P., and Milnes, A. G. (1966b). The epitaxial growth of Ge on Si by solution growth techniques. *J. Electrochem. Soc.* **113**, 297.

346. Donnelly, J. P., and Milnes, A. G. (1966c). Current voltage characteristics of *p–n* Ge–Si and Ge–GaAs heterojunctions. *Proc. IEE (London)* **113**, 1468.

347. Donnelly, J. P., and Milnes, A. G. (1966d). The photovoltaic response of *n*Ge–*n*Si heterodiodes. *Solid-State Electron.* **9**, 174.

348. Donnelly, J. P., and Milnes, A. G. (1967). The capacitance of *p–n* heterojunctions including the effects of interface states. *IEEE Trans. Electron Devices* **ED-14**, 63.

349. Dousmanis, G. C., and Duncan, R. C. (1958). Calculations on the shape and extent of space charge regions in semiconductor surfaces. *J. Appl. Phys.* **29**, 1627.

350. Dousmanis, G. E., Mueller, C. W., and Nelson, H. (1963). Effect of doping on frequency of stimulated and incoherent emission in GaAs diodes. *Appl. Phys. Lett.* **3**, 133.

351. Drangeid, K. E., and Sommerhalder, R. (1970). Dynamic performance of Schottky-barrier field-effect transistors. *IBM J. Res. Develop.* **14**, 82.

352. Drapak, I. T. (1968). Visible luminescence of a ZnO–Cu$_2$O heterojunction. *Sov. Phys.–Semicond.* **2**, 513.

353. Dubenskii, K. K., Rumyantseva, A. V., Ryzhkin, Yu. S., and Trapeznikov, V. I.

(1970). Recombination radiation emitted by ZnSe–ZnTe heterostructures. *Sov. Phys.–Semicond.* **4**, 845.

354. DUMIN, D. J., and ROBINSON, P. H. (1968). Electrically and optically active defects in silicon-on-sapphire films. *J. Cryst. Growth* **3/4**, 214.

355. DUMIN, D. J., ROBINSON, P. H., CULLEN, G. W., and GOTTLIEB, G. E. (1970). Heteroepitaxial growth of germanium and silicon on insulating substrates. *RCA Rev.* **31**, 620.

356. DUNLAP, W. C., JR. (1954). Diffusion of impurities in germanium. *Phys. Rev.* **94**, 1531.

357. DURUPT, P., RAYNAUD, J.-P., and MESNARD, G. (1969). Hétérojonctions Ge *p*–Si *n* obtenues par épitaxie sous vide. *Solid-State Electron.* **12**, 469.

358. DUTTON, R. W., and MULLER, R. S. (1968). Thin film CdS–CdTe diodes. *Solid-State Electron.* **11**, 749.

359. DYMENT, J. C., and D'ASARO, L. A. (1967). Continuous operations of GaAs junction lasers on diamond heat sinks at 200° K. *Appl. Phys. Lett.* **11**, 292.

360. DYMENT, J. C., RIPPER, J. E., and ROLDAN, R. H. R. (1969a). Spiking in light pulses from GaAs Q-switched junction lasers. *IEEE J. Quantum Electron.* **QE-5**, 415.

361. DYMENT, J. C., RIPPER, J. E., and ZACHOS, T. H. (1969b). Optimum stripe width for continuous operation of GaAs junction lasers. *J. Appl. Phys.* **40**, 1802.

362. DZHAFAROV, T. D. (1970). The drift of zinc ions in the $GaAs_{1-y}P_y$–GaAs heterojunctions. *Proc. Int. Conf. Phys. Chem. Semicond. Heterojunctions Layer Structures, Budapest, 1970* **1**, 131. Hung. Acad. Sciences, Budapest, Hungary.

363. DZHAFAROV, T. D., STEPANOVA, M. I., and SUBASHIEV, V. K. (1967). Preparation of alloyed AlAs–GaAs heterojunctions and investigations of their electrical properties. *Sov. Phys.–Semicond.* **1**, 1087.

364. EARLY, J. M. (1958). Structure-determined gain-band product of junction triode transistors. *Proc. IRE* **46**, 1924.

365. EASTMAN, P. C., HAERING, R. R., and BARNES, P. A. (1964). Injection electroluminescence in metal–semiconductor tunnel diodes. *Solid-State Electron.* **7**, 879.

366. EDEN, R. C., MOLL, J. L., and SPICER, W. E. (1967). Experimental evidence for optical population of the X minima in GaAs. *Phys. Rev. Lett.* **18**, 597.

367. EFFER, D. (1965). Epitaxial growth of doped and pure GaAs in an open flow system. *J. Electrochem. Soc.* **112**, 1020.

368. EFFER, D., and ANTELL, G. R. (1960). Preparation of InP, GaAs, and GaP by chemical methods. *J. Electrochem. Soc.* **107**, 252.

369. EGIAZARYAN, G. A., MURYGIN, V. I., RUBIN, V. S., and STAFEEV, V. I. (1970). Some investigations of S type diodes made of semi-insulating gallium arsenide. *Sov. Phys.–Semicond.* **3**, 1389.

370. EHRENREICH, H. (1961). Band structure and transport properties of some 3–5 compounds. *J. Appl. Phys.* **32**, 2155.

371. ELLIS, B., and MOSS, T. S. (1970). Calculated efficiencies of practical GaAs and Si solar cells including the effect of built-in electric fields. *Solid-State Electron.* **13**, 1.

372. ELLIS, W. C., FROSCH, C. J., and ZETTERSTROM, R. B. (1968). Morphology of gallium phosphide crystals grown by VLS mechanism with gallium as liquid-forming agent *J. Cryst. Growth*, **2**, 61.

373. EMTAGE, P. R. (1962). Electrical conduction and the photovoltaic effect in semiconductors with position-dependent band gaps. *J. Appl. Phys.* **33**, 1950.

374. EMTAGE, P. R., and TANTRAPORN, W. (1962). Schottky emission through thin insulating films. *Phys. Rev. Lett.* **8**, 267.

375. ENGSTROM, R. W. (1947). Multiplier photo-tube characteristics: Application to low light levels. *J. Opt. Soc. Amer.* **37**, 420.

376. EPSTEIN, A. S., and DEBAETS, M. C. (1966). Efficiency of a gallium arsenide phosphide solar cell at high light intensities. *Solid-State Electron.* **9**, 1019.
377. ESAKI, L. (1970). A superlattice—periodic array of heterojunctions. *Proc. Int. Conf. Phys. Chem. Semicond. Heterojunctions Layer Structures, Budapest, 1970* **1**, 13. Hung. Acad. Sciences, Budapest, Hungary.
378. ESAKI, L., HOWARD, W. E., and HEER, J. (1964a). The interface transport properties of Ge–GaAs heterojunctions. *Surface Sci.* **2**, 127.
379. ESAKI, L., HOWARD, W. E., and HEER, J. (1964b). The field effect interface conductance in Ge–GaAs *n–n* heterojunctions. *Appl. Phys. Lett.* **4**, 3.
380. ETTENBERG, M., and GILBERT, S. (1970). Vapor phase transport of high purity AlAs. *Electrochem. Soc. Extended Abstr., 137th Nat. Meeting, Los Angeles, Spring 1970*, p. 270.
381. EVERSTEYN, F. C., SEVERIN, P. J. W., BREKEL, C. H. J. v.d., and PEEK, H. L. (1970). A stagnant layer model for the epitaxial growth of silicon from silane in a horizontal reactor. *J. Electrochem. Soc.* **117**, 925.
382. EWING, R. E., and SMITH, D. K. (1968). Compositional inhomogeneities in GaAs$_{1-x}$P$_x$ alloy epitaxial layers. *J. Appl. Phys.* **39**, 5943.
383. FAINER, M. SH., OBUKHOVSKII, YA. A., SYSOEV, L. A., and GAISINSKII, V. B. (1970). Preparation of ohmic contacts on cadmium sulfide single crystals by the electrolytic deposition of indium. *Sov. Phys.–Semicond.* **3**, 1465.
384. FAN, H. Y., NAVON, D., and GEBBIE, H. (1954). Recombination and trapping of carriers in germanium. *Physica (Utrecht)* **20**, 855.
385. FANG, F. F., and HOWARD, W. E. (1964). Effect of crystal orientation on Ge–GaAs heterojunctions. *J. Appl. Phys.* **35**, 612.
386. FANG, F. F., and YU, H. N. (1963). Experiments on optical coupling between GaAs *p–n* junction and heterojunction. *Proc. IEEE* **51**, 860.
387. FAUST, J. W., JR., JOHN, H. F., and RUBENSTEIN, M. S. (1969). Formation of Heterojunction Devices by Epitaxial Growth from Solutions. U.S. Patent 3,447,976.
388. FEDER, R., and LIGHT, T. (1968). Precision thermal expansion measurements of semi-insulating GaAs. *J. Appl. Phys.* **39**, 4870.
389. FEDOTOV, YA. A., MATSON, E. A., and CHERKAS, K. V. (1967). Electrical properties of germanium–gallium arsenide heterojunctions. *Sov. Phys.–Semicond.* **1**, 106.
390. FEDOTOV, YA. A., GRATSERSHTEIN, A. I., and ZIMOGOROVA, N. S. (1970). Photosensitivity spectra and electro-luminescence of GaAs–In$_x$Ga$_{1-x}$As *p–n* heterojunctions. *Sov. Phys.–Semicond.* **4**, 838.
391. FEDOTOV, YA. A., GRUZDEVA, G. A., KOVALEV, A. N., and SUPALOV, V. A. (1970). Germanium–silicon *n–n* heterojunctions. *Sov. Phys.–Semicond.* **4**, 699.
392. FEDOTOV, YA. A., *et al.* (1970a). Electrophysical properties of heterojunctions on the base of A^{III}B^V, A^{II}B^V and Si–SiO$_2$ semiconductors. *Abstr. Int. Conf. Phys. Chem. Semicond. Heterojunctions Layer Structures, Budapest, 1970*, p. 8. Hung. Acad. Sciences, Budapest, Hungary.
393. FEDOTOV, YA. A., *et al.* (1970b). Methods for producing Ge–GaAs heterojunctions and investigation of their electrophysical properties. *Abstr. Int. Conf. Phys. Chem. Semicond. Heterojunctions Layer Structures, Budapest, 1970*, p. 9. Hung. Acad. Sciences, Budapest, Hungary.
394. FELTINSH, I., and FREIBERGA, L. (1965). Some experiments on Si–SiC and Ge–SiC heterojunctions. *Latv. PSR Zinat. Akad. Vestis Fiz. Teh. Zinat. Ser.* p. 123.
395. FERGUSSON, R. R., and GABOR, T. (1964). The transport of gallium arsenide in the vapor phase by chemical reaction. *J. Electrochem. Soc.* **111**, 585.
396. FERTIN, J., LACH, B., MEULEMAN, J., DUPUIS, J., L'HERMITE, and PETIT, R. (1968).

Reverse epitaxial silicon diode for hybrid photomultiplier tube. *IEEE Trans. Nucl. Sci.* **NS-15**, 179.

397. FEUCHT, D. L. (1970). Preparation and properties of Ge–Ge$_x$Si$_{1-x}$, Ge–GaAs and Si–GaP heterojunctions. *Proc. Int. Conf. Phys. Chem. Semicond. Heterojunctions Layer Structures, Budapest, 1970,* **1**, 39. Hung. Acad. Sciences, Budapest, Hungary.

398. FILLARD, J. P., and MANIFACIER, J. C. (1970). Ge–SnO$_2$ heterojunctions. *Proc. Int. Conf. Phys. Chem. Semicond. Heterojunctions Layer Structures, Budapest, 1970* **1**, 139. Hung. Acad. Sciences, Budapest, Hungary.

399. FINNE, R. M., and KLEIN, D. L. (1967). A water–amine-complexing agent system for etching silicon. *J. Electrochem. Soc.* **114**, 965.

400. FIRESTER, A. H., and HELLER, M. E. (1970). Use of diode lasers to recover holographically stored information. *IEEE J. Quantum Electron.* **QE-6**, 572.

401. FISCHER, A. G. (1970). Electroluminescent II–VI heterojunctions. *Proc. Int. Conf. Phys. Chem. Semicond. Heterojunctions Layer Structures, Budapest, 1970* **2**, 71. Hung. Acad. Sciences, Budapest, Hungary.

402. FISCHER, A. G., and Moss, H. I. (1963). Tunnel-injection electroluminescence. *J. Appl. Phys.* **34**, 2112.

403. FISHER, J. C., and GIAEVER, I. (1961). Tunneling through thin insulating layers. *J. Appl. Phys.* **32**, 172.

404. FLORES, J. M. (1964). Simple technique for making an electric contact on silicon. *Rev. Sci. Instrum.* **35**, 112.

405. FOGIEL, M. (1968). "Microelectronics." Research and Education Assoc., New York.

406. FOSS, N. A. (1968). Formation of Hg$_{1-x}$CdTe by Hg ion bombardment of CdTe single crystals. *J. Appl. Phys.* **39**, 6029.

407. FOSTER, L. M., and SCARDEFIELD, J. E. (1970). The solidus boundary in the GaP–InP pseudobinary system. *J. Electrochem. Soc.* **117**, 534. (Also *Electrochem. Soc. Extended Abstr., Los Angeles, Spring 1970,* p. 227.)

408. FOWLER, R. H. (1931). The analysis of photoelectric sensitivity curves for clean metals at various temperatures. *Phys. Rev.* **38**, 45.

409. FOWLER, R. H., and NORDHEIM, D. L. (1928). Electron emission in intense electric fields. *Proc. Roy. Soc., Ser. A* **119**, 173.

410. FOXHALL, G. F., and MILLER, L. E. (1966). Diffusion of arsenic in germanium from a germanium arsenide source. *J. Electrochem. Soc.* **113**, 698.

411. FOYT, A. G. (1963). A proposed UHF transistor using an electroluminescent emitter and collector-base heterojunction. *Proc. IEEE* **51**, 852.

412. FRANCOMBE, M. H. (1963). Preparation and properties of sputtered films. *Trans. Nat. Vacuum Symp. Amer. Vacuum Soc. 10th* (E. Bancroft, ed.), p. 316. Macmillan, New York.

413. FRANKL, D. R. (1967). Preparation of Surface, "Electrical Properties of Semiconductor Surfaces," Chap. 5. Pergamon, Oxford.

414. FRANTSEVICH, I. N. (ed.) (1970). "Silicon Carbide; Structure, Properties and Uses." Plenum Press, New York.

415. FRANZ, W. (1956). "Handbuch der Physik" (S. Flügge, ed.), Vol. **17**, p. 155 (in German). Springer, Berlin.

416. FREISER, R. G. (1968). Low temperature silicon epitaxy. *J. Electrochem. Soc.* **115**, 401.

417. FROMHOLD, A. T., JR. (1963). Kinetics of oxide film growth on metal crystals—II homogeneous field approximations. *Phys. Chem. Solids* **24**, 1309.

418. FROSCH, C. J., and FOY, P. W. (1961). An open tube carrier gas method for the epitaxial growth of gallium phosphide (Abstract). *J. Electrochem. Soc.* **108**, 177c.

419. FROSCH, C. J., and THURMOND, C. D. (1962). Vapor growth of single crystals of GaP

and GaAs by a Ga$_2$O vapor transport mechanism (Abstract). *J. Electrochem. Soc.* **109**, 301c.

420. FUJIMOTO, T., and YAMAGUCHI, J. (1968). X-ray analysis of interface layer in Ge–Si *n–n* heterojunction. *Jap. J. Appl. Phys.* **7**, 562.

421. FULLER, C. S., and ALLISTON, H. W. (1962). A polishing etchant for III–V compounds. *J. Electrochem. Soc.* **109**, 880.

422. FURUKAWA, Y. (1967a). Trap levels in gallium arsenide. *Jap. J. Appl. Phys.* **6**, 675.

423. FURUKAWA, Y. (1967b). Epitaxial growth of gallium arsenide by using silicon tetrachloride. *Jap. J. Appl. Phys.* **6**, 1344.

424. FURUKAWA, Y., and ISHIBASHI, Y. (1967a). Transient phenomena in capacitance and reverse current in a GaAs Schottky barrier diode. *Jap. J. Appl. Phys.* **6**, 13.

425. FURUKAWA, Y., and ISHIBASHI, Y. (1967b). Trapping effects in Au–*n*-type GaAs Schottky barrier diodes. *Jap. J. Appl. Phys.* **6**, 503.

426. FURUKAWA, Y., and ISHIBASHI, Y. (1967c). Vapor plating of tin onto GaAs. *Jap. J. Appl. Phys.* **6**, 787.

427. GALLI, G., and MORRITZ, F. L. (1966). The epitaxy of some chalcogenides on substrates of Ge and GaAs. *Electrochem. Soc. Extended Abstr. Electron. Div. 15, Cleveland, Ohio Meeting, 1966.*

428. GANS, F. F. (1962). Heterojunction Transistor Manufacturing Process. U.S. Patent 3,057,762.

429. GARBE, S., and FRANK, G. (1970). Photoemission from silicon-doped *p*-type gallium arsenide. *Int. Symp. Gallium Arsenide Related Compounds, 3rd, Aachen, Germany, October 1970.*

430. GARRETT, C. G., and BRATTAIN, W. H. (1955). Physical theory of semiconductor surfaces. *Phys. Rev.* **99**, 376.

431. GAVRISHCHAK, I. V., TSYUTSYURA, D. I., and SHNEIDER, A. D. (1968). Electrical properties of the HgTe–ZnTe system near a composition with maximum mobility. *Sov. Phys.–Semicond.* **1**, 1198.

432. GEPPERT, D. V. (1962a). A metal base transistor. *Proc. IRE* **50**, 1967.

433. GEPPERT, D. V. (1962b). Space-charge-limited tunnel emission into an insulating film. *J. Appl. Phys.* **33**, 2993.

434. GEPPERT, D. V., COWLEY, A. M., and DOU, B. V. (1966). Correlation of metal–semiconductor barrier height and metal work function—effect of surface states. *J. Appl. Phys.* **37**, 2458.

435. GERGELY, G., JANOSSY, I., MENYHARD, M., PEISNER, J., and SZEKELY, C. (1970). Some surface properties of heteroepitaxial Ge$_x$Si$_{1-x}$ layers on Si. *Proc. Int. Conf. Phys. Chem. Semicond. Heterojunctions Layer Structures, Budapest, 1970* **1**, 147. Hung. Acad. Sciences, Budapest, Hungary.

436. GERSHENZON, M., and MIKULYAK, R. M. (1961). Vapor phase preparation of gallium phosphide crystals. *J. Electrochem. Soc.* **108**, 548.

437. GERSHENZON, M., LOGAN, R. A., and NELSON, D. F. (1966). Electrical and electroluminescent properties of gallium phosphide diffused *p–n* junctions. *Phys. Rev.* **149**, 580.

438. GILL, R. B. (1970). Room-temperature close-confinement GaAs laser with overall external quantum efficiency of 40 per cent. *Proc. IEEE* **58**, 949.

439. GILL, W. D., and BUBE, R. H. (1969). Mechanism for the photovoltaic effect in heat-treated Cu$_2$S–CdS heterojunctions. *Proc. 3rd Int. Conf. Photoconduct., Stanford Univ., 1969.*

440. GILL, W. D., and BUBE, R. H. (1970). Light microprobe investigation of Cu$_2$S–CdS heterojunctions. *J. Appl. Phys.* **41**, 1694.

441. GIVARGIZOV, E. I. (1963). Mechanism underlying the epitaxial growth of germanium films from the gaseous phase. *Sov. Phys.–Solid State* **5**, 840.

442. GLANG, R., HOLMWOOD, R. A., and ROSENFELD, R. L. (1965). Determination of stress in films on single crystalline silicon substrates. *Rev. Sci. Instrum.* **36**, 7.

443. GLAUBERMAN, A. E., DROZDOV, V. A., and KURMASCHEV, SH.D. (1970). Optically stimulated relaxation of electrical properties of wide gap heterojunctions. *Abstr. Int. Conf. Phys. Chem. Semicond. Heterojunctions Layer Structures, Budapest, 1970*, p. 12. Hung. Acad. Sciences, Budapest, Hungary.

444. GLICKSMAN, R. (1970). Technology and design of GaAs laser and non-coherent IR-emitting diodes. *Solid State Technol.* **13**, 39.

445. GOBAT, A. R., LAMORTE, M. F., and McIVER, G. W. (1962). Characteristics of high conversion efficiency gallium arsenide solar cells. *IRE Trans. Military Electron.* **MIL-6**, 20.

446. GOBELI, G. W., and ALLEN, F. G. (1960). Surface measurements on freshly cleaved silicon *p–n* junctions. *Phys. Chem. Solids* **14**, 23.

447. GOBELI, G. W., and ALLEN, F. G. (1962). Direct and indirect excitation processes in photoelectric emission from silicon. *Phys. Rev.* **127**, 141.

448. GOBELI, G. W., and ALLEN, F. G. (1963). Photoelectric yield and work function of cleaved GaAs. *Bull. Amer. Phys. Soc.* **8**, 198.

449. GOETZBERGER A., and SHOCKLEY, W. (1960). Metal precipitates in silicon *p–n* junctions. *J. Appl. Phys.* **31**, 1821.

450. GOLDBERG, N., and POLLACK, S. R. (1963). Oscillatory tunneling current through thin-film insulating barriers in a magnetic field. *J. Appl. Phys.* **34**, 3556.

451. GOLDBERG, P., and NICKERSON, J. W. (1963). DC electroluminescence in thin films of zinc sulfide. *J. Appl. Phys.* **34**, 1601.

452. GOL'DBERG, YU. A., and TSARENKOV, B. V. (1970). Dependence of the resistance of metal–gallium arsenide ohmic contacts on the carrier density. *Sov. Phys.–Semicond.* **3**, 1447.

453. GOL'DBERG, YU. A., NASLEDOV, D. N., and TSARENKOV, B. V. (1967a). A GaAs–indium ohmic contact. *Instrum. Exp. Tech.* **4**, 969.

454. GOL'DBERG, YU. A., NASLEDOV, D. N., and TSARENKOV, B. V. (1967b). Thin multilayer GaAs–metal contacts. *Instrum. Exp. Tech.* **6**, 1472.

455. GOLDSTEIN, B., and PERLMAN, S. S. (1966). Electrical and optical properties of high-resistivity gallium phosphide. *Phys. Rev.* **148**, 715.

456. GONTAR', V. M., EGIAZARYAN, G. A., RUBIN, V. S., MURYGIN, V. L., and STAFEEV, V. I. (1970). Some properties of diode structures made of semi-insulating gallium arsenide. *Sov. Phys.–Semicond.* **3**, 1460.

457. GOOCH, C. H. (ed.) (1969). "Gallium Arsenide Lasers." Wiley (Interscience), New York.

458. GOODMAN, A. M. (1963). Metal–semiconductor barrier height measurement by the differential capacitance method–one carrier system. *J. Appl. Phys.* **34**, 329.

459. GOODMAN, A. M. (1964a). Evaporated metallic contacts to conducting cadmium sulfide single crystals. *J. Appl. Phys.* **35**, 573.

460. GOODMAN, A. M. (1964b). Electroplated gold and copper contacts to cadmium sulfide. *Surface Sci.* **1**, 54.

461. GOODMAN, A. M. (1966a). Photoemission of electrons from silicon and gold into silicon dioxide. *Phys. Rev.* **144**, 588.

462. GOODMAN, A. M. (1966b). Photoemission of holes from silicon into silicon dioxide. *Phys. Rev.* **152**, 780.

463. GOODMAN, A. M. (1968). Photoemission of electrons and holes into silicon nitride. *Appl. Phys. Lett.* **13**, 275.

464. GOODMAN, A. M. (1969). Electrically conducting photoluminescent ZnSe films. *J. Electrochem. Soc.* **116**, 364.

465. GOODMAN, A. M. (1970). Photoemission of holes and electrons from aluminum into aluminum oxide. *J. Appl. Phys.* **41**, 2176.

466. GOODMAN, A. M., and PERKINS, D. M. (1964). Metal–semiconductor barrier-height measurement by the differential capacitance method—degenerate one-carrier system. *J. Appl. Phys.* **35**, 3351.

467. GOODWIN, A. R., and SELWAY, P. R. (1970). Gain and loss processes in AlGaAs–GaAs heterojunction lasers. *IEEE J. Quantum Electron.* **QE-6**, 285.

468. GORA, T., and WILLIAMS, F. (1969). Theory of electronic states and transport in graded mixed semiconductors. *Phys. Rev.* **177**, 1179.

469. GORADIA, C. P. (1968). Surface states and the gold–*n*-silicon surface barrier. *Diss. Abstr.* **29**, 1148-B.

470. GORDY, W., and THOMAS, W. J. O. (1956). Electronegativities of the elements. *J. Chem. Phys.* **24**, 439.

471. GORODETSKY, A. F., MEL'NIK, V. G., and MEL'NIK, I. G. (1959). A method of producing ohmic contacts with silicon. *Sov. Phys.–Solid State* **1**, 153.

472. GORYUNOVA, N. A., KARPOVITCH, I. A., ANSHON, A. V., LEONOV, E. I., and ORLOV, V. M. (1970). Photoelectrical properties of CdSnP$_2$–Cu$_2$S heterojunctions. *Proc. Int. Conf. Phys. Chem. Semicond. Heterojunctions Layer Structures, Budapest, 1970*, **2**, 275. Hung. Acad. Sciences, Budapest, Hungary.

473. GOSS, A. J., BENSON, K. E., and PFANN, W. G. (1956). Dislocations at compositional fluctuations in germanium–silicon alloys. *Acta Met.* **4**, 332.

474. GOTTLIEB, G. E., and CORBOY, J. F. (1963). Epitaxial growth of GaAs using water vapor. *RCA Rev.* **24**, 585.

475. GRANT, J. T., and HAAS, T. W. (1970). Auger electron spectroscopy of Si. *Surface Sci.* **23**, 347–62.

476. GRAY, P. E. (1969). The saturated photovoltage of a *p–n* junction. *IEEE Trans. Electron Devices* **ED-16**, 424.

477. GRAY, P. V., and BROWN, D. M. (1966). Density of SiO$_2$–Si interface states. *Appl. Phys. Lett.* **8**, 31.

478. GREEN, J. M. (1970). Fast growing protrusions from epitaxial semiconductor surfaces. *Metallurgical Trans.* **1**, 647.

479. GREEN, M. (1965). Effects of temperature and injection level on the photovoltage generated at a rhodium to *p*-silicon contact. *Solid-State Electron.* **8**, 855.

480. GREEN, M., GREENBURG, I. N., BANGHOF, J., SASS, A., WHITE, H. S., and McKNIGHT, R. V. (1961). Electrical characteristics of reduced-area, rhodium-plated, silicon end contact. *Solid-State Electron.* **3**, 1.

481. GRIBNIKOV, Z. S., and MELNIKOV, V. I. (1966a). Diffusion of "hot" electrons in *n–n* heterojunctions. *Sov. Phys.–Solid State* **7**, 1612.

482. GRIBNIKOV, Z. S., and MELNIKOV, V. I. (1966b). Injection and extraction of hot electrons in *n–n* heterojunctions with rapid Maxwellization of the electron gas. *Sov. Phys.–Solid State* **7**, 2364.

483. GRIGOROVICI, R., CROITORU, N., MARINA, M., and NASTASE, L. (1968). Heterojunctions between amorphous silicon and silicon single crystals. *Rev. Roum. Phys.* **13**, 317.

484. GRIMMEISS, H. G., and MEMMING, R. (1962). *p–n* photovoltaic effect in CdS. *J. Appl. Phys.* **33**, 2217.

485. GROSSMAN, J. (1963). A kinetic theory for autodoping for vapor phase epitaxial growth of germanium. *J. Electrochem. Soc.* **110**, 1065.

486. GROSVALET, J., NOTSCH, C., and TRIBES, R. (1963). Physical phenomena responsible for saturation current in field effect devices. *Solid-State Electron.* **6**, 65.

487. GRUZDEVA, G. A., KOVALEV, A. N., SUPALOV, V. A., and FEDOTOV, Ya. A. (1970). Ge–Si heterojunctions. Fabrication and properties. *Proc. Int. Conf. Phys. Chem. Semicond. Heterojunctions Layer Structures, Budapest, 1970* **1**, 155. Hung. Acad. Sciences, Budapest, Hungary.

488. GUBANOV, A. I. (1950). Theory of the contact between two semiconductors with different types of conduction. *Zh. Tekh. Fiz.* **20**, 1287.

489. GUBANOV, A. I. (1951a). Theory of the contact of two semiconductors of the same type of conductivity. *Zh. Tekh. Fiz.* **21**, 304.

490. GUBANOV, A. I. (1951b). Theory of the contact of two semiconductors with mixed conductivity. *Zh. Eksp. Teor. Fiz.* **21**, 721.

491. GUBANOV, A. I. (1952). Theory of the contact phenomena in semiconductors. *Zh. Tekh. Fiz.* **22**, 729.

492. GUBANOV, A. I., and SHARAPOV, B. N. (1970). Threshold current of an injection laser with a p–n heterojunction. *Sov. Phys.–Semicond.* **4**, 367.

493. GUIZZETTI, G., REGUZZONI, E., and SAMOGGIA, G. (1970). Experiments on PbS–Ge heterojunctions. *Proc. Int. Conf. Phys. Chem. Semicond. Heterojunctions Layer Structures, Budapest, 1970* **2**, 293. Hung. Acad. Sciences, Budapest, Hungary.

494. GUNN, J. B. (1954). The theory of rectification and injection of a metal–semiconductor contact. *Proc. Phys. Soc., London, Sect. B* **67**, 575.

495. GUNN, J. B. (1956). The field dependence of electron mobility in Ge. *J. Electron. Contr.* **2**, 87.

496. GUNN, J. B. (1964). Instabilities of current in III–V semiconductors. *IBM J. Res. Develop.* **8**, 141.

497. GUNTHER, K. G. (1967). Vaporization and reaction of the elements. *In* "Compound Semiconductors" (R. K. Willardson and H. L. Goering, eds.), Vol. 1, Preparation of III–V Compounds, p. 313. Van Nostrand–Reinhold, Princeton, New Jersey.

498. GUPTA, D. C., and YEE, R. (1969). Silicon epitaxial layers with abrupt interface impurity profiles. *J. Electrochem. Soc.* **116**, 1561.

499. GUTAI, L., PFEIFER, J., and MARKO, I. (1970). Some properties of Si–Ge heterojunctions obtained by vacuum evaporation. *Proc. Int. Conf. Phys. Chem. Semicond. Heterojunctions Layer Structures, Budapest, 1970* **2**, 301. Hung. Acad. Sciences, Budapest, Hungary.

500. GUTIERREZ, W. A., and WILSON, H. L. (1956). CdSe–ZnSe thin-film rectifier. *Proc. IEEE* **53**, 749.

501. GUTKIN, A. A., NASLEDOV, D. N., and SEDOV, V. E. (1965). Spectral characteristics of gallium arsenide photocells. *Sov. Phys.–Solid State* **7**, 58.

502. GUTKIN, A. A., KAGAN, M. B., MAGERRAMOV, E. M., and CHERNOV, YA. I. (1967). Spectral characteristics of GaP–GaAs photocells in the photon energy range up to 5.4 eV. *Sov. Phys.–Solid State* **8**, 2474.

503. HAANSTRA, J. H. (1967). Some electrical and optical properties in ZnSe due to the incorporation of Cu and Fe. *In* "II–VI Semiconducting Compounds" (D. G. Thomas, ed.) (*Proc. Int. Conf. Brown Univ., 1967*), p. 207. Benjamin, New York.

504. HABERECHT, R. R., and KERN, E. L. (eds.) (1969). "Semiconductor Silicon/1969." Electrochem. Soc., New York.

505. HACKETT, W. H., JR., and SAUL, R. H. (1970). High efficiency red-emitting GaP diodes grown by liquid-phase epitaxy. *Electrochem. Soc. Extended Abstr., 137th Nat. Meeting, Los Angeles, Spring 1970*, p. 223.

506. HAGEN, S. N. (1968). Surface-barrier diodes on silicon carbide. *J. Appl. Phys.* **39**, 1458.

507. HAGSTRUM, H. D. (1960). Studies of Auger electrons ejected from germanium by slowly moving positive ions. *Phys. Chem. Solids* **14**, 33.
508. HAHN, L. A. (1969). The effect of collector resistance upon the high current capability of $n-p-v-n$ transistors. *IEEE Trans. Electron Devices* **ED-16**, 654.
509. HAKKI, B. W., and KNIGHT, S. (1966). Microwave phenomena in bulk GaAs. *IEEE Trans. Electron Devices* **ED-13**, 94.
510. HALE, A. P. (1963). Preparation and evaluation of epitaxial silicon films prepared by vacuum evaporation. *Vacuum* **13**, 93.
511. HALE, J. A. (1970). Germanium–gallium arsenide heterojunctions: Effect of oxygen on the barrier height. *Abstr. Int. Conf. Phys. Chem. Semicond. Heterojunctions Layer Structures, Budapest, 1970*, p. 54. Hung. Acad. Sciences, Budapest, Hungary.
512. HALL, R. N. (1958). Electrical contacts to silicon carbide. *J. Appl. Phys.* **29**, 914.
513. HALL, R. N. (1963). Coherent light emission from $p-n$ junctions. *Solid-State Electron.* **6**, 405.
514. HALPERN, J., and REDIKER, R. H. (1960). Low reverse leakage GaAs diodes. *Proc. IRE* **48**, 1780.
515. HALSTED, R. E., and AVEN, M. (1961). Correlation of ZnTe and other II–VI compound band edge emission spectra. *Bull. Amer. Phys. Soc.* **6**, 312.
516. HAMAKAWA, Y., NISHINO, T., and IKIDA, K. (1969). Interface barrier electroreflectance in semiconductor heterojunction (*Ann. Physical Electron. Conf., 29th, Yale, March 1969*). *Bull. Amer. Phys. Soc.* **14**, 790.
517. HAMAKAWA, Y., WU, T. S., and KITAMURA, H. (1970). Interface barrier and optoelectronic properties of vapor grown GaP- and Ge-heterojunctions. *Proc. Int. Conf. Phys. Chem. Semicond. Heterojunctions Layer Structures, Budapest, 1970* **2**, 313. Hung. Acad. Sciences, Budapest, Hungary.
518. HAMAKER, R. W., and WHITE, W. B. (1969). Growth and characterization of $Ga_xIn_{1-x}Sb$ solid solutions using temperature-gradient zone melting. *J. Electrochem. Soc.* **116**, 478.
519. HAMILTON, P. M. (1964). Advances in III–V and II–VI semiconductor compounds. *Semicond. Prod. Solid State Technol.* **7**, 15.
520. HAMPSHIRE, M. J. (1970). Private communication.
521. HAMPSHIRE, M. J., and TOMLINSON, R. D. (1970a). Small signal equivalent circuit of an isotype heterojunction dominated by traps. *Solid-State Electron.* **13**, 41.
522. HAMPSHIRE, M. J., and TOMLINSON, R. D. (1970b). Optical properties of isotype heterojunctions. *Proc. Int. Conf. Phys. Chem. Semicond. Heterojunctions Layer Structures, Budapest, 1970* **2**, 323. Hung. Acad. Sciences, Budapest, Hungary.
523. HAMPSHIRE, M. J., and WRIGHT, G. T. (1964). The Si–Ge heterojunction. *Brit. J. Appl. Phys.* **15**, 1331.
524. HAMPSHIRE, M. J., PRITCHARD, T. I., TOMLINSON, R. D., and HACKNEY, C. (1970a). An isotype heterojunction detector for use in pyrometry. *J. Phys. E: Sci. Instrum.* **3**, 185.
525. HAMPSHIRE, M. J., PRITCHARD, T. I., and TOMLINSON, R. D. (1970b). A null detecting frequency meter operating in the near infra-red using a CdSe–Ge isotype heterojunction. *Solid-State Electron.* **13**, 1073.
526. HANDELMAN, E. T., and POVILONIS, E. I. (1964). Epitaxial growth of silicon by vacuum sublimation. *J. Electrochem. Soc.* **111**, 201.
527. HANDLER, P. (1960). Section I. Cleaned surfaces of germanium and silicon: Energy level diagrams for germanium and silicon surfaces. *Phys. Chem. Solids* **14**, 1.
528. HANDY, R. J. (1967). Theoretical analysis of the series resistance of a solar cell. *Solid-State Electron.* **10**, 765.

529. HANDY, R. M. (1962). Electrode effects on aluminum oxide tunnel junctions. *Phys. Rev.* **126**, 1968.

530. HANEMAN, D. (1959). Photoelectric emission and work functions of InSb, GaAs, Bi_2Te_3, and germanium. *Phys. Chem. Solids* **11**, 205.

531. HANNAY, N. B. (ed.) (1959). "Semiconductors." Van Nostrand–Reinhold, Princeton, New Jersey.

532. HANNEMAN, R. C., OGILVIE, R. E., and MODRZEJEWSKI, A. (1962). Kossel line studies of irradiated nickel crystals. *J. Appl. Phys.* **33**, 1429.

533. HANSEN, M. (1958). "Constitution of Binary Alloys." McGraw-Hill, New York.

534. HANUS, W., and OSZWALDOWSKI, M. (1970). Structure and electrical properties of GaSb thin films evaporated in vacuo. *Proc. Int. Conf. Phys. Chem. Semicond. Heterojunctions Layer Structures, Budapest, 1970* **3**, 115. Hung. Acad. Sciences, Budapest, Hungary.

535. HAQ, K. E. (1965). Deposition of germanium films by sputtering. *J. Electrochem. Soc.* **112**, 500.

536. HARA, T., and SKASAKO, J. (1968). Electrical properties of sulfur-doped gallium phosphide. *J. Appl. Phys.* **39**, 285.

537. HARMAN, G. G. (1960). Hard gallium alloys for use as low contact resistance electrodes and for bonding thermocouples into samples. *Rev. Sci. Instrum.* **31**, 717.

538. HARMAN, G. G., and HIGIER, T. (1962). Some properties of dirty contacts on semiconductors and resistivity measurement by a two-terminal method. *J. Appl. Phys.* **33**, 2198.

539. HARRICK, N. J. (1961). Determination of the semiconductor surface potential under a metal contact. *J. Appl. Phys.* **32**, 568.

540. HARRIS, J. S., and SNYDER, W. L. (1969). Homogeneous solution grown epitaxial GaAs by tin doping. *Solid-State Electron.* **12**, 337.

541. HARRIS, J. S., NANNICHI, Y., PEARSON, G. L., and DAY, G. F. (1969). Ohmic contacts to solution-grown gallium arsenide. *J. Appl. Phys.* **40**, 4575.

542. HARRIS, L. A. (1968a). Analysis of materials by electron-excited Auger electrons. *J. Appl. Phys.* **39**, 1419.

543. HARRIS, L. A. (1968b). Some observations of surface segregation by Auger electron emission. *J. Appl. Phys.* **39**, 1428.

544. HARRIS, L. A. (1969). Angular dependencies in electron-excited Auger emission. *Surface Sci.* **15**, 77.

545. HÁRSY, M., GERGELY, G., SCHANDA, J., SOMOGYI, M., SVISZT, P., and SZIGETI, G. (1967). ZnS single crystals grown from gallium and indium melts. *In* "II–VI Semiconducting Compounds" (D. G. Thomas, ed.) (*Proc. Int. Conf., Brown Univ., 1967*), p. 413. Benjamin, New York.

546. HÁRSY, M., VARGHA, N., VARGA, L., and LENDVAY, E. (1970). Preparation and structure of ZnS–GaS heterojunctions. *Proc. Int. Conf. Phys. Chem. Semicond. Heterojunctions Layer Structures, Budapest, 1970* **1**, 179. Hung. Acad. Sciences, Budapest, Hungary.

547. HASEGAWA, F., and SAITO, T. (1968). Impurity transfer in GaAs vapor growth and carrier-concentration profiles on the grown films. *Jap. J. Appl. Phys.* **7**, 1342.

548. HASHIMOTO, N., and KOGA, Y. (1968). Observation of vanadium silicides on silicon. *J. Appl. Phys.* **39**, 5798.

549. HAYASHI, I., and PANISH, M. B. (1970). GaAs–$Ga_xAl_{1-x}As$ heterostructure injection lasers which exhibit low thresholds at room temperature. *J. Appl. Phys.* **41**, 150.

550. HAYASHI, I., PANISH, M. B., and FOY, P. W. (1969). A low threshold room temperature injection laser. *IEEE J. Quantum Electron.* **QE-5**, 211.

551. HAYASHI, I., PANISH, M. B., FOY, P. W., and SUMSKI, S. (1970a). GaAs–$Al_xGa_{1-x}As$

double-heterostructure injection lasers: Heat sinking and continuous operation at room temperature. *Proc. Int. Conf. Semicond. Devices, Tokyo, September 1970.*

552. HAYASHI, I., PANISH, M. B., FOY, P. W., and SUMSKI, S. (1970b). Junction lasers which operate continuously at room temperature. *Appl. Phys. Lett.* **17**, 109.

553. HAYASHI, I., PANISH, M. B., and REINHART, F. K. (1971). GaAs–Al$_x$Ga$_{1-x}$As double heterostructure injection lasers. *J. Appl. Phys.* **42**, 1929.

554. HECK, L. D., and LOONEY, J. C. (1965). Percussive welding of metal–semiconductor contacts. *Semicond. Prod. Solid State Technol.* June 1965, p. 11.

555. HEINE, M. A., and PRYOR, M. J. (1963). The distribution of A-C resistance in oxide films on aluminum. *J. Electrochem. Soc.* **110**, 1205.

556. HEINE, V. (1965). Theory of surface states. *Phys. Rev.* **138**, A1689.

557. HEINZ, D. M., HEBERT, H. J., and SHARP, W. N. (1966). Solid state image intensifier panel. Nat. Tech. Information Service, Springfield, Virginia, Document AD 479 258.

558. HEISE, B. H. (1962). Precision determination of the lattice constant by the Kossel line technique. *J. Appl. Phys.* **33**, 938.

559. HEMMENT, P. L. F. (1966). The preparation of a non-rectifying contact to *n* type silicon. *J. Sci. Instrum.* **43**, 389.

560. HENISCH, H. K. (1957). "Rectifying Semi-Conductor Contacts." Oxford Univ. Press (Clarendon), London and New York.

561. HERMAN, F. (1955). Speculations on the energy band structure of Ge–Si alloys. *Phys. Rev.* **95**, 847.

562. HERRING, C. (1955). Transport properties of a many-valley semiconductor. *Bell Syst. Tech. J.* **34**, 237.

563. HERRMANN, F. P., and RAAB, S. (1970). The effect of boron on the crystal growth of GaAs/Ga$_{1-x}$Al$_x$As-heterojunctions. *Proc. Int. Conf. Phys. Chem. Semicond. Heterojunctions Layer Structures, Budapest, 1970* **1**, 187. Hung. Acad. Sciences, Budapest, Hungary.

564. HERSHINGER, L. W. (1966). Solid-State Thin-Film Triode with a Graded Energy Band Gap. U.S. Patent 3,290,568.

565. HERZOG, D. G., and KRESSEL, H. (1970). Thermoelectrically cooled GaAlAs laser illuminator. *Appl. Opt.* **9**, 2249.

566. HEYMANN, H., and PETRUZELLA, J. (1969). The development and fabrication of a power Schottky diode. NASA Scientific and Technical Information Facility, College Park, Maryland. 20740, Document N70-11411.

567. Hibberd, R. G. (1960). Semi-Conductor Devices. U.S. Patent 2,966,434.

568. HICKMOTT, T. W. (1962). Low-frequency negative resistance in thin anodic oxide films. *J. Appl. Phys.* **33**, 2669.

569. HILL, J. S., RAWLINS, T. G. R., and SIMPSON, G. N. (1970). Si–ZnS heterojunctions II. Single crystal films of zinc sulphide on silicon. Document AD707343. Nat. Tech. Information Service, Springfield, Virginia.

570. HILLEGAS, W. J., JR., and SCHNABLE, G. L. (1963). Plating of metals on semiconductors. *Electrochem. Technol.* **1**, 228.

571. HILSUM, C., and ROSE-INNES, A. C. (1961). "Semiconducting III–V Compounds." Pergamon Press, Oxford.

572. HINKLEY, E. D., and REDIKER, R. H. (1967). GaAs–InSb graded-gap heterojunction. *Solid-State Electron.* **10**, 671.

573. HINKLEY, E. D., REDIKER, R. H., and LAVINE, M. C. (1964). Inversion of (111) surfaces in single crystal regrowth during interface alloying of intermetallic compounds. *Appl. Phys. Lett.* **5**, 110.

574. HINKLEY, E. D., REDIKER, R. H., and JADUS, D. K. (1965). GaAs–InSb *n*–*n* heterojunctions: A single crystal "Schottky" barrier. *Appl. Phys. Lett.* **6**, 144.

575. Hirayama, C. (1964). Thermodynamic properties of the solid monoxides, mono-sulfides, monoselenides, and monotellurides of Ge, Sn, and Pb. *J. Chem. Eng. Data* **9**, 65.

576. Hofstein, S. R., and Zaininger, K. H. (1964). Physical limitations on the frequency response and performance of the surface inversion layers in MOS devices. *IEEE Trans. Electron Devices* (Abstract) **ED-11**, 530.

577. Holland, L. (1960). "Vacuum Deposition of Thin Films." Wiley, New York.

578. Holloway, H. (1966). Some diffraction methods for the study of epitaxial layers. In "The Use of Thin Films in Physical Investigations" (J. C. Anderson, ed.). Academic Press, New York.

579. Holloway, H., and Bobb, L. C. (1967). Oriented growth of semiconductors. V. Surface features and twins in epitaxial gallium arsenide. *J. Appl. Phys.* **38**, 2893.

580. Holloway, H., and Bobb, L. C. (1968). Slip in epitaxial Ge–GaAs combinations. *J. Appl. Phys.* **39**, 2467.

581. Holloway, H., and Wilkes, E. (1968). Epitaxial growth of cubic CdS. *J. Appl. Phys.* **39**, 5807.

582. Holloway, H., Wollmann, K., and Joseph, A. S. (1965). Oriented growth of semi-conductors, I. Orientations in gallium arsenide grown epitaxially on germanium. *Phil. Mag.* **11**, 263.

583. Holloway, H., Richards, J. L., Bobb, L. C., Perry, J., Jr., and Zimmerman, E. (1966). Oriented growth of semiconductors. IV. Vacuum deposition of epitaxial indium antimonide. *J. Appl. Phys.* **37**, 4694.

584. Holonyak, N., Jr. (1961). Halogen vapor transport and growth of epitaxial layers of intermetallic compounds and compound mixtures. *AIME Tech. Conf., Metall. Semicond. Mater., Los Angeles, 1961.*

585. Holonyak, N., Jr., and Lesk, I. A. (1960). Gallium arsenide tunnel diodes. *Proc. IRE* **48**, 1405.

586. Holonyak, N., Jr., Jillson, D. C., and Bevacqua, S. F. (1962). Halogen vapor transport and growth of epitaxial layers of intermetallic compounds and compound mixtures. In "Metallurgy of Semiconductor Materials" (J. B. Schroeder, ed.). Wiley (Interscience), New York.

587. Holt, D. B. (1966a). Defects in epitaxial films of semiconducting compounds with the sphalerite structure. *J. Mater. Sci.* **1**, 280.

588. Holt, D. B. (1966b). Misfit dislocations in semiconductors. *Phys. Chem. Solids* **27**, 1053.

589. Honig, R. E. (1957). Vapor pressure data for the more common elements. *RCA Rev.* **18**, 195.

590. Hooper, R. C., Cunningham, J. A., and Harper, J. G. (1965). Electrical contacts to silicon. *Solid-State Electron.* **8**, 831.

591. Hoshina, H., and Okazaki, S. (1968). Au–As alloy–Si Schottky barrier. *Jap. J. Appl. Phys.* **7**, 686.

592. Hovel, H. J. (1968). ZnSe–Ge Heterojunctions and Heterojunction Transistors. Ph.D. Thesis, Carnegie-Mellon Univ., Pittsburgh, Pennsylvania.

593. Hovel, H. J. (1970). Switching and memory in ZnSe–Ge heterojunctions. *Appl. Phys. Lett.* **17**, 141.

594. Hovel, H. J., and Milnes, A. G. (1967). ZnSe–Ge heterojunction transistors. *Electron Devices Meeting, Washington, D.C., October 1967.* Late News Suppl. [See also *IEEE Trans. Electron Devices* **ED-16**, 766 (1969).]

595. Hovel, H. J., and Milnes, A. G. (1968). Electrical characteristics of nZnSe–pGe heterodiodes. *Int. J. Electron.* **25**, 201.

596. HOVEL, H. J., and MILNES, A. G. (1969). The epitaxy of ZnSe on Ge, GaAs and ZnSe by an HCl close-spaced transport process. *J. Electrochem. Soc.* **116**, 843.

597. HOWARD, W. E., FOWLER, A. B., and McLEOD, D. (1968). The pressure dependence of barrier height in Ge–GaAs n–n heterojunctions. *J. Appl. Phys.* **39**, 1533.

598. HOWSON, R. P., and MALINA, V. (1970). Getter evaporation of thin films of III–V semiconductors. *Proc. Int. Conf. Phys. Chem. Semicond. Heterojunctions Layer Structures, Budapest, 1970* **3**, 141. Hung. Acad. Sciences, Budapest, Hungary.

599. HSU, S. T. (1970). Low-frequency excess noise in metal–silicon Schottky barrier diodes. *IEEE Trans. Electron Devices* **ED-17**, 496.

600. HUBER, D. (1970). Calculation of III–V ternary phase diagrams Ga–As–P and Ga–Al–As. *Proc. Int. Conf. Phys. Chem. Semicond. Heterojunctions Layer Structures, Budapest, 1970* **1**, 195. Hung. Acad. Sciences, Budapest, Hungary.

601. HUBER, H. H., JR. (1966). The effect of mercury contamination on the work function of gold. *Appl. Phys. Lett.* **8**, 169.

602. HUBNER, K. (1965). Photoresponsive Semiconductor Device. U.S. Patent 3,196,285.

603. HUGHES, A. L., and DuBRIDGE, L. A. (1932). "Photoelectric Phenomena." McGraw-Hill, New York.

604. HUNTER, L. P. (1956). "Handbook of Semiconductor Electronics." McGraw-Hill, New York (2nd ed., 1962).

605. IDO, T., OSHIMA, S., and SAJI, M. (1968). Preparation of ZnTe epitaxial films on CdS by vapor transport method. *Jap. J. Appl. Phys.* **7**, 1141.

606. IDO, T., HIROSE, M., and ARIZUMI, T. (1970). Electrical properties of pInSb–nSi heterojunctions. *Proc. Int. Conf. Phys. Chem. Semicond. Heterojunctions Layer Structures, Budapest, 1970* **2**, 335. Hung. Acad. Sciences, Budapest, Hungary.

607. IGNATKINA, R. S., LIBOV, L. D., and MESKIN, S. S. (1965). Alloyed contacts on gallium phosphide. *Instrum. Exp. Tech. (USSR)* No. 3, 726.

608. IIDA, S. (1966). Epitaxial growth of germanium by the hydrogen reduction of GeI₄. *Jap. J. Appl. Phys.* **5**, 138.

609. IIZUKA, H., and KITAOKA, S. (1970). Low-noise GaAs Schottky-barrier beam-lead mixer diodes. *Proc. IEEE* **58**, 1372.

610. ING, D. W., McAVOY, B. R., and URE, R. W. (1968). Alloying contacts to GaAs by hot hydrogen and HCl gases. *Solid-State Electron.* **11**, 469.

611. ING, S. W., JR., and MINDEN, H. T. (1962). Open tube epitaxial synthesis of GaAs and GaP. *J. Electrochem. Soc.* **109**, 995.

612. IOFFE, A. V. (1948). Rectification at the boundary of two semiconductors. *Zh. Tekh. Fiz.* **18**, 1498.

613. ISAWA, N. (1968). Diffusion of arsenic near the interface of GaAs–Ge epitaxy. *Jap. J. Appl. Phys.* **7**, 81.

614. ISHII, H., and YAMADA, K. (1967). Effects of traps on thin film transistors. *Solid-State Electron.* **10**, 1201.

615. ISHII, M., SUSAKI, W., SOGO, T., and OKU, T. (1970). Luminescent properties of Te-doped Ga₁₋ₓAlₓAs epitaxial layers grown from Ge solution. *Electrochem. Soc. Extended Abstr., 137th Nat. Meeting, Los Angeles, Spring 1970*, p. 210.

616. ITO, K., and FUJISAWA, H. (1970). Effect of crystal orientation on n–InSb–n–GaAs heterojunctions. *Abstr. Int. Conf. Phys. Chem. Semicond. Heterojunctions Layer Structures, Budapest, 1970*, p. 57. Hung. Acad. Sciences, Budapest, Hungary.

617. ITO, K., and TAKAHASHI, K. (1968). Epitaxial growth of Ge layers on Si substrates by vacuum evaporation. *Jap. J. Appl. Phys.* **7**, 821.

618. ITOH, T. (1970). Energy distribution of electrons emitted from silicon surface-barrier diodes. *J. Appl. Phys.* **41**, 1951.

619. Iтон, T., Hasegawa, S., and Kaminaka, N. (1968). Electrical properties of *n*-type epitaxial films of silicon on sapphire formed by vacuum evaporation. *J. Appl. Phys.* **39**, 5310.

620. Iтон, T., Hasegawa, S., and Kaminaka, N. (1969). Epitaxial films on silicon on spinel by vacuum evaporation. *J. Appl. Phys.* **40**, 2597.

621. Iтон, T., Matsuda, I., and Hasegawa, K. (1970). Field-induced photoelectron emission from *p*-type silicon aluminum surface-barrier diodes. *J. Appl. Phys.* **41**, 1945.

622. Iтон, Y. (1969). Silicon surface effect on Schottky barriers. *J. Electrochem. Soc.* **116**, 146C.

623. Iтон, Y., and Hashimoto, N. (1967). Photoelectric measurement of tungsten silicide and *n*-type silicon barriers. *J. Appl. Phys.* **38**, 899.

624. Iтон, Y., and Hashimoto, N. (1969). Reaction-process dependence of barrier height between tungsten silicide and *n*-type silicon. *J. Appl. Phys.* **40**, 425.

625. Iwasa, H., Yokozawa, M., and Teramoto, I. (1968). Electroless nickel plating on silicon. *J. Electrochem. Soc.* **115**, 485.

626. Iwerson, J. E., Bray, A. R., and Kleimack, J. J. (1962). Low current alpha in Si transistors. *Trans. IRE Electron Devices* **ED-9**, 474.

627. Jackson, D. M., and Howard, R. W. (1965). Fabrication of epitaxial SiC films on Si. *Trans. AIME* **233**, 468.

628. Jadus, D. K. (1967). The Realization of a Wide Band Gap Emitter Transistor. Ph.D. Thesis, Carnegie-Mellon Univ., Pittsburgh, Pennsylvania.

629. Jadus, D. K., and Feucht, D. L. (1969). The realization of a wide band gap emitter transistor. *IEEE Trans. Electron Devices* **ED-16**, 102.

630. Jadus, D. K., Ladd, G. O., and Feucht, D. L. (1965). Heterojunction formation by the low-temperature vapor growth of GaAs. *J. Electrochem. Soc.* **112**, 66c.

631. Jadus, D. K., Reedy, H. E., and Feucht, D. L. (1967). Ohmic contacts to GaAs by a simple low temperature alloying process. *J. Electrochem. Soc.* **114**, 408.

632. Jain, V. K., and Sharma, S. K. (1970). On the preparation of epitaxial films of III–V compounds. *Solid-State Electron.* **13**, 1145.

633. Jaklevic, R. C., Donald, D. K., Lambe, J., and Vassell, W. C. (1963). Injection electroluminescence in cadmium sulfide by tunneling films. *Appl. Phys. Lett.* **2**, 7.

634. James, H. M., and Ginzbarg, A. S. (1953). Band structure in disordered alloys and impurity semiconductors. *J. Phys. Chem.* **57**, 840.

635. James, L. W., and Uebbing, J. J. (1970). Long-wavelength threshold of Cs_2O-coated photoemitters. *Appl. Phys. Lett.* **16**, 370.

636. James, L. W., Moll, J. L., and Spicer, W. E. (1968). The GaAs photocathode. *Proc. 1968 Symp. GaAs*, pp. 230–237. Brit. Inst. of Phys. and Phys. Soc., London.

637. James, L. W., Antypas, G. A., Uebbing, J. J., Edgecumbe, J., and Bell, R. L. (1970). III–V ternary photocathodes. *Int. Symp. Gallium Arsenide Related Compounds, 3rd, Aachen, Germany, October 1970.*

638. James, R. W. (1958). "The Optical Principles of the Diffraction of X-Rays," Chap. 8. Bell, London.

639. Janik, E. (1965). Ge–Si heterojunctions. *Przeglad Electron. (Poland)* **6**, 65.

640. Jazwinski, S. T., and Szummer, A. (1970). The methodology of the investigation of anomalies occurring during formation of metal–semiconductor contacts. *Proc. Int. Conf. Phys. Chem. Semicond. Heterojunctions Layer Structures, Budapest, 1970* **5**, 303. Hung. Acad. Sciences, Budapest, Hungary.

641. Jeffes, J. H. E. (1968). The physical chemistry of transport processes. *J. Cryst. Growth* **3/4**, 13.

642. JENNY, D. A. (1958). The status of transistor research in compound semiconductors. *Proc. IRE* **46**, 959.

643. JESSER, W. A. (1970). Misfit accommodation by imperfect dislocations in epitaxial fcc films. *J. Appl. Phys.* **41**, 39.

644. JOHNSON, P. S. (1954). The solar constant. *J. Meteorol.* **11**, 431.

645. JOHNSON, R. T., and DARSEY, D. M. (1968). Resistive properties of indium and indium–gallium contacts of CdS. *Solid-State Electron.* **11**, 1015.

646. JONA, F. (1966). Study of the early stages of the epitaxy of silicon on silicon. *Appl. Phys. Lett.* **9**, 235.

647. JONES, M. E. (1962). Ohmic Connection for Silicon Semiconductor Devices. U.S. Patent 3,021,462.

648. JONES, M. E., MILLER, D. P., and WURST, E. C. (1961). Contact for Gallium Arsenide. U.S. Patent 3,012,175.

649. JONES, P. L., LITTING, C. N. W., MASON, D. E., and WILLIAMS, V. A. (1968). The epitaxial growth of zinc sulphide on silicon by vacuum evaporation. *Brit. J. Appl. Phys. (J. Phys. D)* **1**, 283.

650. JONSCHER, A. K. (1962). Physics of semiconductor switching devices. *Prog. Semicond.* **6**, 143–197

651. JOYCE, B. A. (1968). Growth and perfection of chemically-deposited epitaxial layers of Si and GaAs. *J. Cryst. Growth* **3/4**, 43.

652. JOYCE, B. A., and BRADLEY, R. R. (1963). Epitaxial growth of silicon from the pyrolysis of monosilane on silicon substrates. *J. Electrochem. Soc.* **110**, 1235.

653. JUNGBLUTH, E. D. (1970). A review of bulk and process-induced defects in GaAs semiconductors. *Metallurgical Trans. AIME* **1**, 575.

654. JUNGBLUTH, E. D., and CHIAO, H. C. (1968). Intense interjunction strain in phosphorus-diffused silicon. *J. Electrochem. Soc.* **115**, 429.

655. KAGAN, M. B., and LYUBASHEVSKAYA, T. L. (1967). Selection of the optimum combination of semiconductor materials for a two-stage photocell. *Sov. Phys.–Semicond.* **1**, 1091.

656. KAGAN, M. B., LANDSMAN, A. P., and CHERNOV, YA. I. (1965). Certain photoelectric properties of p–n junctions in the GaP–GaAs system. *Sov. Phys.–Solid State* **6**, 2149.

657. KAHNG, D. (1962). A hot electron transistor. *Proc. IEEE* **50**, 1534.

658. KAHNG, D. (1963). Conduction properties of the Au–n-type-Si Schottky barrier. *Solid-State Electron.* **6**, 281.

659. KAHNG, D. (1964). Au–n type GaAs Schottky barrier and its varactor application. *Bell Syst. Tech. J.* **43**, 215.

660. KAHNG, D., and D'ASARO, L. A. (1964). Gold–epitaxial silicon high-frequency diodes. *Bell Syst. Tech. J.* **43**, 225.

661. KAHNG, D., and LEPSELTER, M. P. (1965). Planar epitaxial silicon Schottky barrier diodes. *Bell Syst. Tech. J.* **44**, 1525.

662. KAHNG, D., and WEMPLE, S. H. (1965). Measurement of nonlinear polarization of $KTaO_3$ using Schottky diodes. *J. Appl. Phys.* **36**, 2925.

663. KAMANDJIEV, P., MLADJOV, L., PRAMATAROVA, L., SVIRACHEV, D., and NICOLOVA, L. (1970). Electrical and photoelectrical properties of epitaxial heterojunctions of zinc selenide on germanium. *Abstr. Int. Conf. Phys. Chem. Semicond. Heterojunctions Layer Structures, Budapest, 1970*, p. 58. Hung. Acad. Sciences, Budapest, Hungary.

664. KAMATH, G. S., and BOWMAN, D. (1967). Preparation and properties of epitaxial gallium phosphide. *J. Electrochem. Soc.* **114**, 192.

665. KANDA, Y., TOKAI, T., and KOZUKA, H. (1965). Uniaxial stress effect on Ge–Si alloyed heterojunction. *Jap. J. Appl. Phys.* **4**, 701.

666. KANDILAROV, B. (1970). The S-matrix method in the theory of heterojunctions. *Proc. Int. Conf. Phys. Chem. Semicond. Heterojunctions Layer Structures, Budapest, 1970* **2**, 345. Hung. Acad. Sciences, Budapest, Hungary.

667. KANDILAROV, B., and ANDREYTCHIN, R. (1965). Photovoltaic effects in CdS–CdSe heterojunctions. *Phys. Status Solidi* **8**, 897.

668. KANE, E. O. (1962). Theory of photoelectric emission from semiconductors. *Phys. Rev.* **127**, 131.

669. KANE, P. F., and LARRABEE, G. B. (1970). "Characterization of Semiconductor Materials," McGraw-Hill, New York.

670. KANG, C. S., and GREENE, P. E. (1967). Preparation and properties of high-purity epitaxial GaAs grown from Ga solutions. *Appl. Phys. Lett.* **11**, 171.

671. KANO, G., and TAKAYANAGI, S. (1967). Reverse characteristics of Mo–Si epitaxial Schottky diodes. *IEEE Trans. Electron. Devices* **ED-14**, 822.

672. KANO, G., NAKAI, J., and YASUOKA, M. (1965). An effect of pinholes on the CdSe–Au–Ge metal-base-transistors. *Jap. J. Appl. Phys.* **4**, 822.

673. KANO, G., INOUE, M., MATSUNO, J.-I., and TAKAYANAGI, S. (1966). Molybdenum–silicon Schottky barrier. *J. Appl. Phys.* **37**, 2985.

674. KANO, G., FUJIWARA, S., IIZUKA, M., HASEGAWA, H., and SAWAKI, T. (1969). Improvements of reverse I–V characteristics in epitaxial planar Schottky barrier diodes. *Appl. Phys. Lett.* **15**, 138.

675. KANTER, H. (1963). Slow electron transfer through evaporated Au films. *J. Appl. Phys.* **34**, 3629.

676. KARPINSKAS, S., SMILGA, A., VAITKUS, J., and VISCAKAS, J. (1970). Homo- and hetero-junctions with cadmium selenide. *Proc. Int. Conf. Phys. Chem. Semicond. Heterojunctions Layer Structures, Budapest, 1970* **2**, 355. Hung. Acad. Sciences, Budapest, Hungary.

677. KASANO, H., and IIDA, S. (1967). Preparation of GaAs–Ge and InAs–GaAs hetero-junctions in a closed tube system using iodine process. *Jap. J. Appl. Phys.* **6**, 1038.

678. KASTNING, J. A. (1968). A study of electroluminescent cadmium sulfide diodes. Nat. Tech. Information Service, Springfield, Virginia, Document AD 681 705.

679. KAZARINOV, R. F., SURIS, R. A., and KHOLODNOV, V. P. (1970). Lux-ampere characteristic of a germanium *pnp* phototransistor with a gold-compensated base. *Sov. Phys.–Semicond.* **3**, 1363.

680. KEATING, P. N. (1963). Hole injection into CdS from Cu_2S. *Phys. Chem. Solids* **24**, 1101.

681. KENDALL, D. L. (1964). Simultaneous indium and cadmium diffusion in gallium arsenide. *Appl. Phys. Lett.* **4**, 67.

682. KENNEDY, D. I., and RUSS, M. J. (1968). Electroluminescence in polycrystalline ZnTe. *Solid-State Electron.* **11**, 513.

683. KENNEDY, D. I., KOTELES, E. S., and WEBB, W. A. (1969). Voltage dependence of electroluminescence from GaP diodes prepared by liquid epitaxial techniques. *J. Appl. Phys.* **40**, 875.

684. KERR, D. R. (1962). Theory and Application of Piezoresistance of Diffused Layers in Germanium. Ph.D. Thesis, Carnegie Inst. Technol., Pittsburgh, Pennsylvania.

685. KERRIGAN, J. V. (1963). Studies on the transport and deposition of alpha-aluminum oxide. *J. Appl. Phys.* **34**, 3408.

686. KESPERIS, J. S., YATSKO, R. S., and NEWMAN, P. A. (1964). Research on hetero-junctions (growth of gallium phosphide, silicon heterojunctions). Tech. Rep. ECOM2471, Nat. Tech. Information Service, Springfield, Virginia, Document AD 611 039.

687. KEYES, R. J., and QUIST, T. M. (1962). Recombination radiation emitted by gallium arsenide. *Proc. IRE* **50**, 1822.

688. KEYES, R. W. (1966). Mobility Anisotropic Semiconductor Device. U.S. Patent 3,283,220.

689. KHAN, I. H., and LEARN, A. J. (1969). Formation of epitaxial beta SiC films on sapphire. *Appl. Phys. Lett.* **15**, 410.

690. KILGORE, B. F., and ROBERTS, R. W. (1963). Preparation of evaporated silicon films. *Rev. Sci. Instrum.* **34**, 11.

691. KIM, C. W., and SCHWARTZ, R. J. (1969). A p–i–n thermo-photovoltaic diode. *IEEE Trans. Electron Devices* **ED-16**, 657.

692. KIMURA, T., NUNOSHITA, M., and YAMAGUCHI, J. (1966). Electrical and optical properties of Ge–Si n–n heterojunctions. *Jap. J. Appl. Phys.* **5**, 639.

693. KING, V. J. (1961). Liquid alloy for making contacts to metallic and nonmetallic surfaces. *Rev. Sci. Instrum.* **32**, 1407.

694. KINGSTON, R. H. (1956). Review of germanium surface phenomena. *J. Appl. Phys.* **27**, 101.

695. KINGSTON, R. H., and NEUSTADTER, S. F. (1955). Calculation of the surface charge, field and carrier concentrations at the surface of a semiconductor. *J. Appl. Phys.* **26**, 718.

696. KIRBY, R. K. (1963). Thermal expansion. *In* "American Institute of Physics Handbook" (D. Gray, ed.), 2nd ed. McGraw-Hill, New York.

697. KIRCHER, C. J. (1970). Properties of palladium silicide–silicon contacts. *J. Electrochem. Soc.* **117**, 100C. (Also *Electrochem. Soc. Extended Abstr., 137th Nat. Meeting, Los Angeles, Spring 1970*, p. 335.)

698. KIRVALIDZE, I. D., and ZHUKOV, V. F. (1960). The possibility of creating ohmic contact with silicon by rubbing the metal onto the semiconductor using dry friction. *Sov. Phys.–Solid State* **1**, 1446.

699. KISELEVA, N. K. (1964a). Production and some properties of InSb–GaSb heterojunctions. *Sov. Phys.–Crystallogr.* **9**, 365.

700. KISELEVA, N. K. (1964b). The electrical properties of InSb–GaSb heterojunctions. *Radio Eng. Electron. Phys.* (*U.S.S.R.*) **9**, 1574.

701. KISELEVA, N. K., and KOLOMIETS, B. T. (1970). Heterocrystals based on InSb and GaSb. *Proc. Int. Conf. Phys. Chem. Semicond. Heterojunctions Layer Structures, Budapest, 1970* **1**, 203. Hung. Acad. Sciences, Budapest, Hungary.

702. KLEIN, W. (1969). Photoemission from cesium oxide covered GaInAs. *J. Appl. Phys.* **40**, 4384.

703. KLEINMAN, D. A. (1961). Considerations on the solar cell. *Bell Syst. Tech. J.* **40**, 85.

704. KLOHN, K. L., and WANDINGER, L. (1969). Variation of contact resistance of metal–GaAs contacts with impurity concentration and its device implication. *J. Electrochem. Soc.* **116**, 507.

705. KNIGHT, S., and PAOLA, C. (1969). Ohmic contacts to gallium arsenide. *In* "Ohmic Contacts to Semiconductors" (B. Schwartz, ed.). Electrochem. Soc., New York.

706. KOBAYASHI, A., ODA, Z., KAWAJI, S., ARATA, H., and SUGIYAMA, K. (1960). Impurity conduction of cleaned germanium surfaces at low temperatures. *Phys. Chem. Solids* **14**, 37.

707. KOC, S., *et al.* (1970). Electrical and optical properties of amorphous germanium layers prepared by various technique. *Proc. Int. Conf. Phys. Chem. Semicond. Heterojunctions Layer Structures, Budapest, 1970* **3**, 163. Hung. Acad. Sciences, Budapest, Hungary.

708. KODERA, H., SHIRAFUJI, J., and KURATA, K. (1966). A method of producing heterojunctions between compound semiconductors by alloying and substitution reaction. *Jap. J. Appl. Phys.* **5**, 743.

709. KOENIG, H. R., and MAISSEL, L. I. (1970). Application of RF Discharges to sputtering. *IBM J. Res. Develop.* **14**, 168.

710. KOENIG, S. H. (1959). Hot and warm electrons—a review. *Phys. Chem. Solids* **8**, 227.
711. KOHN, E. S. (1970). Investigation of solid-state cold cathodes. Rep. AFCRL-70-0711. Air Force Cambridge Res. Labs., Bedford, Massachusetts.
712. KOLM, C., KULIN, S. A., and AVERBACH, B. L. (1957). Studies on group III–V intermetallic compounds. *Phys. Rev.* **108**, 965.
713. KOLOMIETS, B. T., and MALKOVA, A. A. (1958). The properties and structure of ternary semiconductor systems IV. Electrical and photoelectric properties of substitutional solid solutions in the ZnTe–CdTe system. *Sov. Phys.–Tech. Phys.* **3**, 1532.
714. KOMASHCHENKO, V. N., and FEDORUS, G. A. (1970). Electrical properties of $Cu_{2-x}Se$–$CdSe$ pn heterojunctions. *Sov. Phys.–Semicond.* **3**, 1001.
715. KOMASHCHENKO, V. N., LUKYANCHIKOVA, N. B., FEDORUS, G. A., and SHEINKMAN, M. K. (1970). Mechanism of current passing through the $p(Cu_{2-x}Se)$–$n(CdSe)$ thin film heterojunction in dark and under illumination. *Proc. Int. Conf. Phys. Chem. Semicond. Heterojunctions Layer Structures, Budapest, 1970* **2**, 213. Hung. Acad. Sciences, Budapest, Hungary.
716. KONAKA, M., ABE, T., and SATO, K. (1968). Non-destructive determination of impurity concentration in silicon epitaxial layer using metal–silicon Schottky barrier. *Jap. J. Appl. Phys.* **7**, 790.
717. KONAPLYA, L. N., KRAVCHENKO, A. F., and SIROTKINA, U. P. (1965). Nonrectifying contacts on gallium arsenide. *Instrum. Exp. Tech. (USSR)*, No. 3, 728.
718. KONNERTH, K. L., MARINACE, J. C., and TOPALIAN, J. C. (1970). Zinc-diffused GaAs electroluminescent diodes with long operating life. *J. Appl. Phys.* **41**, 2060.
719. KONOZENKO, I. D., KUSHNEIR, S. CH., MUSALEVSKY, E. A., OSTRANITZA, A. P., PETRENCO, N. S., and PAVLENKO, A. A. (1970). Growth and investigations of crystal structure of the graded CdS CdSe heterojunctions. *Proc. Int. Conf. Phys. Chem. Semicond. Heterojunctions Layer Structures, Budapest, 1970* **1**, 221. Hung. Acad. Sciences, Budapest, Hungary.
720. KORZO, V. F., and RYABOVA, L. A. (1969). Switching effect in $Ge–In_2O_3$ and $Ge–Al_2O_3$ heterojunctions. *Sov. Phys.–Semicond.* **3**, 517.
721. KOSOGOV, O. V., and MARAMZINA, M. A. (1970). Photodiodes with an n type filter layer. *Sov. Phys.–Semicond.* **3**, 1467.
722. KOSONOCKY, W. F., and CORNELY, R. H. (1968). GaAs laser amplifiers. *IEEE J. Quantum Electron.* **QE-4**, 125.
723. KOT, M. V., PANASYUK, L. M., SIMASHKEVICH, A. V., TSURKAN, A. E., and SHERBAN, D. A. (1965). Intrinsic recombination radiation of ZnSe–ZnTe heterojunctions. *Sov. Phys.–Solid State* **7**, 1001.
724. KOTELYANSKIY, I. M., MITYAGIN, A. YU., and ORLOV, V. P. (1970). Growth and crystal structures in heteroepitaxial CdS deposited onto GaP. *Proc. Int. Conf. Phys. Chem. Semicond. Heterojunctions Layer Structures, Budapest, 1970* **1**, 231. Hung. Acad. Sciences, Budapest, Hungary.
725. KOVALENKO, P. A., KOROTKOV, V. A., and PANASYUK, L. M. (1970). Transversal and longitudinal photoeffect in heterojunction between elemental semiconductors and II–VI compounds. *Proc. Int. Conf. Phys. Chem. Semicond. Heterojunctions Layer Structures, Budapest, 1970* **2**, 363. Hung. Acad. Sciences, Budapest, Hungary.
726. KRAUSE, G. O., and TEAGUE, E. C. (1967). Observation of misfit dislocations in GaAs–Ge heterojunctions. *Appl. Phys. Lett.* **10**, 251.
727. KRENZKE, G., LEMKE, H., and MULLER, G. O. (1968). Properties of gold contacts on clean n-silicon surfaces. *Phys. Status Solidi* **25**, K131.
728. KRESSEL, H. (1970). The small, economy size laser. *Laser Focus*, November 1970, p. 45.

729. KRESSEL, H., and LADANY, I. (1968). Electroluminescence in $Al_xGa_{1-x}P$ diodes prepared by liquid-phase epitaxy. *J. Appl. Phys.* **39**, 5339.

730. KRESSEL, H., and NELSON, H. (1969a). Close-confinement gallium arsenide *PN* junction lasers with reduced optical loss at room temperature. *RCA Rev.* **30**, 106.

731. KRESSEL, H., and NELSON, H. (1969b). Improved red and infrared light-emitting $Al_xGa_{1-x}As$ laser diodes using the close-confinement structure. *Appl. Phys. Lett.* **15**, 7.

732. KRESSEL, H., NELSON, H., McFARLANE, S. H., ABRAHAMS, M. S., LeFUR, P., and BUIOCCHI, C. J. (1969). Effect of substrate imperfections on GaAs injection lasers prepared by liquid-phase epitaxy. *J. Appl. Phys.* **40**, 3587.

733. KRESSEL, H., and NELSON, H. (1970a). Low-threshold $Al_xGa_{1-x}As$ visible and IR light-emitting diode lasers. *IEEE J. Quantum Electron.* **QE-6**, 278.

734. KRESSEL, H., NELSON, H., and HAWRYLO, F. Z. (1970b). Control of optical losses in *p–n* junction lasers by use of a heterojunction: Theory and experiment. *J. Appl. Phys.* **41**, 2019.

735. KRESSEL, H., KOHN, E. S., NELSON, H., and TIETJEN, J. J. (1970c). An optoelectronic cold cathode using an $Al_xGa_{1-x}As$ heterojunction structure. *Appl. Phys. Lett.* **16**, 359.

736. KRESSEL, H., BUTLER, J. K., and SOMMERS, H. S., JR. (1970d). High-order transverse cavity modes in heterojunction diode lasers. *Appl. Phys. Lett.* **17**, 403.

737. KRESSEL, H., SOMMERS, H. S., JR., LOCKWOOD, H. F., ETTENBERG, M., and NELSON, H. (1970e). Catastrophic damage in "close-confined" AlGaAs–GaAs heterojunction injection lasers. *Proc. 1970 IEEE Reliability Conf., Las Vegas.*

738. KRESSEL, H., LOCKWOOD, H. F., and HAWRYLO, F. Z. (1971). Low threshold LOC GaAs injection lasers. *Appl. Phys. Lett.* **18**, 43.

739. KRIKORIAN, E., and SNEED, R. J. (1966). Epitaxial deposition of germanium by both sputtering and evaporation. *J. Appl. Phys.* **37**, 3665.

740. KRÖGER, F. A., DIEMER, G., and KLASENS, H. A. (1956). Nature of an ohmic metal–semiconductor contact. *Phys. Rev.* **103**, 279.

741. KROEMER, H. (1956a). Band structure of semiconductor alloys with locally varying composition. *Bull. Amer. Phys. Soc.* **1**, 143.

742. KROEMER, H. (1956b). The drift transistor. *Arch. Elektrotech. Ubert.* **8**, 223.

743. KROEMER, H. (1956c). The apparent contact potential of a pseudo-abrupt *p–n* junction. *RCA Rev.* **17**, 515.

744. KROEMER, H. (1957a). Theory of a wide gap emitter for transistors. *Proc. IRE* **45**, 1535.

745. KROEMER, H. (1957b). Quasi-electric and quasi-magnetic fields in a non-uniform semiconductor. *RCA Rev.* **28**, 332.

746. KROEMER, H. (1963). A proposed class of heterojunction injection lasers. *Proc. IEEE* **51**, 1782.

747. KROEMER, H. (1965). Heterojunction device concepts. Tech. Rep. No. AFAL-TR-65-243. Document AD 471 872. Nat. Tech. Information Service, Springfield, Virginia.

748. KROGER, F. A., and DeNOBEL, K. (1958). Semiconductor Device with Telluride Containing Ohmic Contact and Method of Forming the Same. U.S. Patent 2,865,794.

749. KRUSE, P. W., and SCHULZE, R. G. (1969). Photoeffects in *n*-Ge–*p*-GaAs heterojunctions. *Proc. Int. Conf. Photoconduct., 3rd, Stanford Univ., 1969.*

750. KRUSE, P. W., PRIBBLE, F. C., and SCHULZE, R. G. (1967). Solid-state infrared-wavelength converter employing high-quantum-efficiency Ge–GaAs heterojunctions. *J. Appl. Phys.* **38**, 1718.

751. KRUSE, P. W., LIU, S. T., SCHULZE, R. G., and PETERSON, S. R. (1969). Avalanche breakdown in *n*-Ge, *p*-GaAs heterojunctions. *J. Appl. Phys.* **40**, 5401.

752. KU, S. M., and BLACK, J. F. (1966). Injection electroluminescence in $(Al_xGa_{1-x})As$ diodes of graded energy gap. *J. Appl. Phys.* **37**, 3733.

753. KUCERA, L. (1970). The behaviour of selenium heterojunctions prepared by means of different technologies. *Proc. Int. Conf. Phys. Chem. Semicond. Heterojunctions Layer Structures, Budapest, 1970* **5**, 329. Hung. Acad. Sciences, Budapest, Hungary.

754. KUKHARSKII, A. A., SUBASHIEV, V. K., and USHAKOVA, M. B. (1967). Degenerate *n*-type germanium layer in heterojunctions. *Sov. Phys.–Semicond.* **1**, 161.

755. KUMAR, R. C. (1968). Current transport in isotype heterojunctions. *Int. J. Electron.* **25**, 239.

756. KUMAR, R. C. (1969). Diffusion theory of current transport in anisotype heterojunctions. *Int. J. Electron.* **27**, 185.

757. KUNIOKA, A., and SAKAI, Y. (1968). Electrical and optical properties of CdO–Si junctions. *Jap. J. Appl. Phys.* **7**, 1138.

758. KURATA, K., and HIRAI, T. (1968). A study on making abrupt heterojunctions by solution growth. *J. Electrochem. Soc.* **115**, 869.

759. KURTIN, S., and MEAD, C. A. (1968a). Surface barriers on layer semiconductors: GaSe. *Phys. Chem. Solids* **29**, 1865.

760. KURTIN, S., and MEAD, C. A. (1968b). GaSe Schottky barrier gate FET. *Proc. IEEE* **56**, 1594.

761. KURTIN, S., McGILL, T. C., and MEAD, C. A. (1969). Fundamental transition in the electronic nature of solids. *Phys. Rev. Lett.* **22**, 1433.

762. KUROV, G. A., SEMILETOV, S. A., and PINSKER, Z. G. (1957). Electrical properties and real structure of single-crystal germanium films produced by evaporation in vacuum. *Sov. Phys.–Crystallogr.* **2**, 53.

763. LADD, G. O. (1969). Growth and Electrical Characteristics of *p*Ge–*n*GaAs Heterojunctions. Ph.D. Thesis, Carnegie-Mellon Univ., Pittsburgh, Pennsylvania.

764. LADD, G. O., and FEUCHT, D. L. (1970a). Autodoping effects at the interface of GaAs–Ge heterojunctions. *Metallurgical Trans. AIME* **1**, 609.

765. LADD, G. O., and FEUCHT, D. L. (1970b). Performance potential of high frequency heterojunction transistors. *IEEE Trans. Electron Devices* **ED-17**, 413.

766. LADE, R. W. (1964). Alloyed and diffused high–low junctions in silicon. *J. Electron. Contr.* **17**, 415.

767. LADE, R. W., and JORDAN, A. G. (1962). On the static characteristics of high–low junction devices. *J. Electron. Contr.* **13**, 23.

768. LAKATOS, A. I., and ROBERTS, G. G. (1968). Double Schottky-barrier capacitance on trigonal selenium. *J. Appl. Phys.* **39**, 5308.

769. LAKSHMANAN, T. K. (1960). *p–n* junctions between semiconductors having different energy gaps. *Proc. IRE* **48**, 1646.

770. LAMMING, J. S., and FOXELL, C. A. P. (1965). GaAs semiconductor devices. *Mullard Tech. Comm.* **8**, 118.

771. LAMPERT, M. A. (1956). Simplified theory of space charge limited currents in an insulator with traps. *Phys. Rev.* **103**, 1648.

772. LAMPERT, M. A. (1959). Transient behavior of the ohmic contact. *Phys. Rev.* **113**, 1236.

773. LAMPERT, M. A. (1962). Injection currents in insulators. *Proc. IRE* **50**, 1781.

774. LANDER, J. J., GOBELI, G. W., and MORRISON, J. (1963). Structural properties of cleaved silicon and germanium surfaces. *J. Appl. Phys.* **34**, 2298.

775. LANDIS, H. M. (1965). Electrodes for ceramic barium titanate type semiconductors. *J. Appl. Phys.* **36**, 2000.

776. LANDSBERG, P. T. (1951a). The theory of direct-current characteristics of rectifiers. *Proc. Roy. Soc., Ser. A* **206**, 463.

777. LANDSBERG, P. T. (1951b). Contributions to the theory of heterogeneous barrier layer rectifiers. *Proc. Roy. Soc., Ser. A* **206**, 477.

778. LANDSMAN, A. P., SHVETZ, U. I., ALTSHULER, B. L., and TYKVENKO, R. N. (1970). Photoelectric properties of Cu_2S–CdTe heterojunctions. *Proc. Int. Conf. Phys. Chem. Semicond. Heterojunctions Layer Structures, Budapest, 1970* **2**, 373. Hung. Acad. Sciences Budapest, Hungary.

779. LANE, C. H. (1968). Stress at the Si–SiO_2 interface and its relationship to interface states. *IEEE Trans. Electron Devices* **ED-15**, 998.

780. LANE, C. H. (1970). Aluminum metallization and contacts for integrated circuits. *Metallurgical Trans. AIME* **1**, 713.

781. LANG, J., KISPETER, J., and GOMBAY, L. (1970). Dependence of electric properties of Se–CdSe heterojunctions on doping. *Abstr. Int. Conf. Phys. Chem. Semicond. Heterojunctions Layer Structures, Budapest, 1970*, p. 63. Hung. Acad. Sciences, Budapest, Hungary.

782. LARACH, S., SHRADER, R. E., and STOCKER, C. F. (1957). Anomalous variation of band gap with composition in zinc sulfo- and seleno-tellurides. *Phys. Rev.* **108**, 587.

783. LASHER, G. J. (1963). Threshold relations and diffraction loss for injection lasers. *IBM J. Res. Develop.* **7**, 58.

784. LASHER, G. J., and STERN, F. (1964). Spontaneous and stimulated recombination in semiconductors. *Phys. Rev.* **133**, A553.

785. LAUGIER, A. (1970). Phonon spectroscopy in Ge–GaAs tunnel heterojunctions. *Proc. Int. Conf. Phys. Chem. Semicond. Heterojunctions Layer Structures, Budapest, 1970* **2**, 375. Hung. Acad. Sciences, Budapest, Hungary.

786. LAUGIER, A., DURUPT, P., VICARIO, E., and PITAVAL, M. (1970). Pseudo-Kikuchi-patterns: A new method for orientation, structure and quality control of epitaxial layers. *Proc. Int. Conf. Phys. Chem. Semicond. Heterojunctions Layer Structures, Budapest, 1970* **1**, 239. Hung. Acad. Sciences, Budapest, Hungary.

787. LAVINE, J. M. (1963). GaAs injection lasers. *IBM J. Res. Develop.* **7**, 17.

788. LAVINE, J. M. (1964). A survey of hot-electron and thin-film transistors. *Semicond. Prod. Solid State Technol.* **7**, 17.

789. LAVRENTJEVA, L. G., and ZAKHAROV, I. S. (1970). Properties of the metallurgical interface in Ge–GaAs. *Proc. Int. Conf. Phys. Chem. Semicond. Heterojunctions Layer Structures, Budapest, 1970* **1**, 253. Hung. Acad. Sciences, Budapest, Hungary.

790. LAW, J. T. (1960). Surface properties of vacuum cleaned silicon. *Phys. Chem. Solids* **14**, 9.

791. LAWLEY, K. L., HEILIG, J. A., and KLEIN, D. L. (1967). Preparation of ohmic contacts for *n*-type GaAs. *Electrochem. Techn.* **5**, 374.

792. LAX, B., and NEUSTADTER, S. F. (1954). Transient response of a *p–n* junction. *J. Appl. Phys.* **25**, 1148.

793. LEARN, A. J., and GAG, K. E. (1969). Reactive deposition of cubic silicon carbide. *J. Appl. Phys.* **40**, 430.

794. LEARN, A. J., and SCOTT-MONCK, J. A. (1968). Thin-film CdS diodes heat treated in various ambients. *J. Appl. Phys.* **39**, 2480.

795. LEE, C. A. (1956). A high frequency diffused base germanium transistor. *Bell Syst. Tech. J.* **35**, 23.

796. LEE, C. A., LOGAN, R. A., BATDORF, R. L., KLEIMACK, J. J., and WIEGMANN, W. (1964). Ionization rates of holes and electrons in silicon. *Phys. Rev.* **134**, A761.

797. LEE, D. H., and NICOLET, M. A. (1965). Space charge limited currents in solids for various geometries and field dependent mobility. *Solid-State Electron.* **8**, 182.

798. LEE, Y. S., and KIM, C. K. (1970). Two-watt cw GaAs Schottky-barrier IMPATT diodes. *Proc. IEEE* **58**, 1153.

799. LEHMANN, W. (1968). *N–i* and *P–i* heterojunctions in II–VI compounds. *Bull. Amer. Phys. Soc.* **13**, 456.

800. LEHOVEC, K. (1952). New photoelectric devices utilizing carrier injection. *Proc. IRE* **40**, 1407.

801. LEHOVEC, K., and SLOBODSKOY, A. (1961). Diffusion of charged particles into a semiconductor under consideration of the built-in field. *Solid-State Electron.* **3**, 45.

802. LEHOVEC, K., SLOBODSKOY, A., and SPRAGUE, J. L. (1963). Field effect–capacitance analysis of surface states on silicon. *Phys. Status Solidi* **3**, 447.

803. LEIDERMAN, A. YU. (1970). Influence of trapping centers on current characteristics of a semiconductor *pnn*+ structure. *Sov. Phys.–Semicond.* **3**, 1251.

804. LEIGHTON, R. B. (1959). "Principles of Modern Physics." McGraw-Hill, New York.

805. LEIGHTON, W. H., JR. (1970). Monolithic X Band Microstrip Line Mixers on Semi-insulating Gallium Arsenide. Ph.D. Thesis, Carnegie-Mellon Univ., Pittsburgh, Pennsylvania.

806. LEITE, R. C. C., SARACE, J. C., OLSON, D. H., COHEN, B. G., SHELAN, J. M., and YARIV, A. (1965). Injection mechanisms in GaAs diffused electroluminescent junctions. *Phys. Rev.* **137**, A1583.

807. LEITZ, P., MARCHAL, G., and PALZ, W. (1970a). Research on some physical aspects of the CdS–Cu₂S thin film solar cell. *Proc. Int. Conf. Phys. Chem. Semicond. Heterojunctions Layer Structures, Budapest, 1970* **2**, 385. Hung. Acad. Sciences, Budapest, Hungary.

808. LEITZ, P., MARCHAL, G., and PALZ, W., (1970b). Physical model for the CdTe–Cu₂Te thin film solar cell. *Abstr. Int. Conf. Phys. Chem. Semicond. Heterojunctions Layer Structures, Budapest, 1970,* p. 64. Hung. Acad. Sciences, Budapest, Hungary.

809. LENDVAY, D., BALAZS, J., GAL, M., GERGELY, G., and SCHANDA, J. (1970). Preparation and some properties of Si–ZnS heterojunctions. *Proc. Int. Conf. Phys. Chem. Semicond. Heterojunctions Layer Structures, Budapest, 1970* **1**, 263. Hung. Acad. Sciences, Budapest, Hungary.

810. LENZLINGER, M., and SNOW, E. H. (1969). Fowler–Nordheim tunneling into thermally grown SiO₂. *J. Appl. Phys.* **40**, 278.

811. LEPSELTER, M. P. (1960). Semiconductor Device Fabrication. U.S. Patent 3,106,489.

812. LEPSELTER, M. P. (1966). Beam-lead technology. *Bell Syst. Tech. J.* **45**, 233.

813. LEPSELTER, M. P., and ANDREWS, J. M. (1969). Ohmic contacts to silicon. *In* "Ohmic Contacts to Semiconductors" (B. Schwartz, ed.). Electrochem. Soc., New York.

814. LEPSELTER, M. P., and SZE, S. M. (1968a). SB-IGFET: An insulated-gate field effect transistor using Schottky barrier contacts for source and drain. *Proc. IEEE* **56**, 1400.

815. LEPSELTER, M. P., and SZE, S. M. (1968b). Near ideal metal–semiconductor barriers. *IEEE Trans. Electron Devices* **ED-15**, 434.

816. LEPSELTER, M. P., MACRAE, A. U., and MACDONALD, R. W. (1969). SB-IGFET, II: An ion implanted IGFET using Schottky barriers. *Proc. IEEE* **57**, 812.

817. LESK, I. A., HOLONYAK, N., Jr., ALDRICH, R. W., BROUILLETTE, J. W., and GHANDHI, S. K. (1960). A categorization of the solid-state device aspects of microsystems electronics. *Proc. IRE* **48**, 1833.

818. LEVERENZ, D. J., and GADDY, O. L. (1970). Subnanosecond gating properties of the dynamic cross-field photomultiplier. *Proc. IEEE* **58**, 1487.

819. LEVITAS, A., WANG, C. C., and ALEXANDER, B. H. (1954). Energy gap of germanium–silicon alloys. *Phys. Rev.* **95**, 846.

820. LEWICKI, G. W., and MEAD, C. A. (1966). Voltage dependence of barrier height in AlN tunnel junctions. *Appl. Phys. Lett.* **8**, 98.

821. LIBICKY, A. (1967). Synthesis and crystal growth of CdSe, ZnTe and ZnSe. *In* "II–VI Semiconducting Compounds" (D. G. Thomas, ed.). Benjamin, New York.

822. LIBOV, L. D., MESKIN, S. S., NASLEDOV, D. N., SEDOV, V. E., and TSARENKOV, B. V.

(1965). Ohmic contacts of metals with gallium arsenide (review). *Instrum. Exp. Tech.* (*USSR*), 746.

823. LIDORENKO, N. S., LANDSMAN, A. P., KAGAN, M. B., and LJUBASHEVSKAJA, T. L. (1970). Some properties of heterostructure photocells as radiant energy converters. *Abstr. Int. Conf. Phys. Chem. Semicond. Heterojunctions Layer Structures, Budapest, 1970*, p. 66. Hung. Acad. Sciences, Budapest, Hungary.

824. LIECHTI, C. A. (1970). Down-converters using Schottky-barrier diodes. *IEEE Trans. Electron. Devices* **ED-17**, 975.

825. LIGHT, T. B., HULL, E. M., and GERETH, R. (1968). Formation of thick semi-insulating GaAs films by flash evaporation. *J. Electrochem. Soc.* **115**, 857.

826. LIN, L. Y. (1957). Study of injecting and extracting contacts on germanium single crystals. *Rev. Sci. Instrum.* **28**, 187.

827. LINDEN, K. J. (1969). Injection electroluminescence from diffused gallium–aluminum arsenide diodes. *J. Appl. Phys.* **40**, 2325.

828. LINDLEY, W. T., PHELAN, R. J., JR., WOLFE, C. M., and FOYT, A. G. (1969). GaAs Schottky barrier avalanche photo-diodes. *Appl. Phys. Lett.* **14**, 197.

829. LINDMAYER, J. (1964). The metal-gate transistor. *Proc. IEEE* **52**, 1751.

830. LINDMAYER, J. (1965). Heterojunction properties of the oxidized semiconductor. *Solid-State Electron.* **8**, 523.

831. LINDMAYER, J., and BUSEN, K. M. (1965). The semiconductor–oxide interface as a heterojunction. *Trans. AIME* **233**, 530.

832. LINDMAYER, J., and REVESZ, A. G. (1970). Electronic processes in heterojunction photovoltaic cells. *Proc. Int. Conf. Phys. Chem. Semicond. Heterojunctions Layer Structures, Budapest, 1970* **2**, 397. Hung. Acad. Sciences, Budapest, Hungary.

833. LINDMAYER, J., REYNOLDS, J., and WRIGLEY, C. (1963). One carrier space charge limited current in solids. *J. Appl. Phys.* **34**, 809.

834. LIU, Y.-Z., MOLL, J. L., and SPICER, W. E. (1969). Effects of heat cleaning on the photoemission properties of GaAs surfaces. *Appl. Phys. Lett.* **14**, 275.

835. LIU, Y.-Z., MOLL, J. L., and SPICER, W. E. (1970). Quantum yield of GaAs semi-transparent photocathode. *Appl. Phys. Lett.* **17**, 60.

836. LOCKWOOD, H. F., KRESSEL, H., SOMMERS, H. S., JR., and HAWRYLO, F. Z. (1970). An efficient large optical cavity injection laser. *Appl. Phys. Lett.* **17**, 499.

837. LOEBNER, E. E. (1957). Electroluminescent Device. U.S. Patent 2,817,783.

838. LOEBNER, E. E. (1963). Hewlett Packard Associates Interim Eng. Rep. No. 3, March–May 1963, Palo Alto, California (A.F. Contract No. AF33(657)-9772, BPSN: 3-6799 760 E 415906).

839. LOFERSKI, J. J. (1963). Recent research on photovoltaic solar energy converters. *Proc. IEEE* **51**, 667.

840. LOGAN, R. A., and SZE, S. M. (1966). Avalanche multiplication in Ge and GaAs *p–n* junctions. *Proc. Int. Conf. Phys. Semicond., Kyoto.* Physical Soc. of Japan, Tokyo, Japan.

841. LOGAN, R. A., CHYNOWETH, A. G., and COHEN, B. G. (1962). Avalanche breakdown in gallium arsenide *p–n* junctions. *Phys. Rev.* **128**, 2518.

842. LONGINI, R. L., and FEUCHT, D. L. (1965). Semiconductor heterojunctions. *Trans. AIME* **233**, 433.

843. LONGINI, R. L., and GREENE, R. F. (1956). Ionization interaction between impurities in semiconductors and insulators. *Phys. Rev.* **102**, 992.

844. LONSDALE, K. (1947). Divergent-beam X-ray photography of crystals. *Phil. Trans. Roy. Soc. London* **240**, 219.

845. LOONEY, D. H. (1958). Contact Structures. U.S. Patent 2,820,932.

846. LOPEZ, A., and ANDERSON, R. L. (1964). Photocurrent spectra of Ge–GaAs hetero-junctions. *Solid-State Electron.* **7**, 695.

847. LORENZ, M. R., and PILKUHN, M. (1966). Preparation and properties of solution-grown epitaxial p–n junctions in GaP. *J. Appl. Phys.* **37**, 4094.

848. LORENZ, M. R., AVEN, M., and WOODBURY, H. H. (1963). Correlation between irradiation and thermally induced defects in II–VI compounds. *Phys. Rev.* **132**, 143.

849. LOSEE, D. L., and WOLF, E. L. (1969). Tunneling spectroscopy of CdS Schottky-barrier junctions. *Phys. Rev.* **187**, 925.

850. LOWEN, J., and REDIKER, R. H. (1960). Gallium arsenide diffused diodes. *J. Electrochem. Soc.* **107**, 26.

851. LOWENHEIM, F. A., Ed. (1963). "Modern Electroplating." Wiley, New York.

852. LUCOVSKY, G., and CHOLET, P. H. (1960). Gallium arsenide, a sensitive photodiode for the visible spectrum. *J. Opt. Soc. Amer.* **50**, 979.

853. LUDEKE, R., and PAUL, W. (1967). Growth and optical properties of epitaxial thin films of some II–VI compounds. *In* "II–VI Semiconducting Compounds" (D. G. Thomas, ed.), p. 123. Benjamin, New York.

854. LUDWIG, G. W. (1967). Gunn effect in ZnSe and CdTe. *In* "II–VI Semiconducting Compounds" (D. G. Thomas, ed.), p. 1287. Benjamin, New York.

855. LUDWIG, G. W., and AVEN, M. (1967). Gunn effect in ZnSe. *J. Appl. Phys.* **38**, 5326.

856. LUTHER, L. C. (1969). Growth of zinc-doped gallium phosphide by the water vapor transport method. *J. Electrochem. Soc.* **116**, 374.

857. LUTHER, L. C. (1970). Bulk growth of GaP by halogen vapor transport. *Metallurgical Trans. AIME* **1**, 593.

858. LUZHNAYA, N. P. (1968). Growth from metal solutions. *J. Cryst. Growth* **3/4**, 97.

859. MACDONALD, J. R. (1962). Accurate solution of an idealized one-carrier metal–semiconductor contact. *Solid-State Electron.* **5**, 11.

860. MACH, R., LUDWIG, W., EICHHORN, G., and ARNOLD, H. (1970). Epitaxial growth and the conduction properties of ZnSe–GaAs heterojunctions. *Phys. Status Solidi* **2**, 701.

861. MAHMOUD, A. A. (1968). Analysis of small-area metal–semiconductor contacts. *Bull. Amer. Phys. Soc.* **13**, 1676.

862. MALININ, A. U., PAPKOV, V. S., and SUROVIKOV, M. V. (1970). The electric properties of layers grown on insulator substrates near the heterojunction interface. *Proc. Int. Conf. Phys. Chem. Semicond. Heterojunctions Layer Structures, Budapest, 1970* **3**, 173. Hung. Acad. Sciences, Budapest, Hungary.

863. MAMMANA, C. I. Z., and ANDERSON, R. L. (1970). Tunneling in isotype heterojunctions. *Proc. Int. Conf. Phys. Chem. Semicond. Heterojunctions Layer Structures, Budapest, 1970* **1**, 279. Hung. Acad. Sciences, Budapest, Hungary.

864. MANASEVIT, H. M., and SIMPSON, W. I. (1969). The use of metal-organics in the preparation of semiconductor materials. 1. Epitaxial gallium–V compounds. *J. Electrochem. Soc.* **116**, 1725.

865. MANASEVIT, H. M., and THORSEN, A. C. (1970). Heteroepitaxial GaAs on aluminum oxide: 1. Early growth studies. *Metallurgical Trans. AIME* **1**, 623.

866. MARINACE, J. C. (1960a). Epitaxial vapor growth of Ge single crystals in a closed cycle process. *IBM J. Res. Develop.* **4**, 248.

867. MARINACE, J. C. (1960b). Tunnel diodes by vapor growth of Ge on Ge and GeAs. *IBM J. Res. Develop.* **4**, 280.

868. MARPLE, D. T. F. (1964). Electron effective mass in ZnSe. *J. Appl. Phys.* **35**, 1879.

869. MARTIN, D. D., and STRATTON, R. (1966). The operation of graded band-gap base transistors at high currents. *Solid-State Electron.* **9**, 237.

870. MARTINELLI, R. U. (1970). Infrared emission from silicon. *Appl. Phys. Lett.* **16**, 261.

871. MARTINUZZI, S., DAVID, J. P., CABANE-BROUTY, F., and SORBIER, J. P. (1970). On physical properties and analysis of CdS–Cu₂S heterojunctions. *Proc. Int. Conf. Phys. Chem. Semicond. Heterojunctions Layer Structures, Budapest, 1970* **1**, 287. Hung. Acad. Sciences, Budapest, Hungary.

872. MARUSKA, H. P., and TIETJEN, J. J. (1969). The preparation and properties of vapor-deposited single crystaline GaN. *Appl. Phys. Lett.* **15**, 327.

873. MARUYAMA, M., KIKUCHI, S., and MIZUNO, O. (1969). Preparation and properties of epitaxial gallium arsenide. *J. Electrochem. Soc.* **116**, 413.

874. MATARE, H. F. (1959). Dislocation planes in semiconductors. *J. Appl. Phys.* **30**, 581.

875. MATARE, H. F. (1960). Anisotropy of carrier transport in semiconductor bi-crystals. *Solid State Phys.* **1**, 73.

876. MATARE, H. F. (1969). Heteroepitaxy of silicon on insulator crystal substrates. *J. Electrochem. Soc.* **116**, 146C.

877. MATINO, H., and TOKUNAGA, M. (1969). Contact resistances of several metals and alloys to GaAs. *J. Electrochem. Soc.* **116**, 709.

878. MATLOW, S. L., and RALPH, E. L. (1959). Ohmic aluminum–*n* type silicon contact. *J. Appl. Phys.* **30**, 541.

879. MATLOW, S. L., and RALPH, E. L. (1961). A low-resistance ohmic contact for silicon semiconductor devices. *Solid-State Electron.* **2**, 202.

880. MATSUKURA, Y., and ODA, J. (1966). Multiple Hetero-Layer Composite Semiconductor Device. U.S. Patent 3,275,906.

881. MATSUURA, E., MATSUI, K., and HASIGUTI, R. R. (1962). Technics for ohmic connecting leads to silicon. *J. Appl. Phys.* **33**, 1610.

882. MAY, G. A. (1968). The Schottky-barrier-collector transistor. *Solid-State Electron.* **11**, 613.

883. MAY, J. E. (1969). Tungsten–aluminum ohmic contact. *In* "Ohmic Contacts to Semiconductors" (B. Schwartz, ed.). Electrochem. Soc., New York.

884. MAY, J. E. (1968). Alloy barrier ohmic contacts for aluminum to *n*-type silicon. *Electrochem. Soc. Extended Abstr., 134th Nat. Meeting, Montreal, Fall 1968*, Abstract 505.

885. MAYBURG, S., and SMITH, B. (1962). High frequency base transport factor and transit time of graded base transistors. *IRE Trans. Electron Devices* **ED-9**, 161.

886. MAYER, H. (1970). New results in the problems of epitaxy. *Abstr. Int. Conf. Phys. Chem. Semicond. Heterojunctions Layer Structures, Budapest, 1970*, p. 93. Hung. Acad. Sciences, Budapest, Hungary.

887. MAYER, S. E. (1959). Low Resistance Contacts to Germanium. U.S. Patent 2,914,449.

888. McAFEE, K. B., RYDER, E. J., SHOCKLEY, W., and SPARKS, M. (1951). Observation of Zener current in germanium *p–n* junctions. *Phys. Rev.* **83**, 650.

889. McCALDIN, J. O., and HARADA, R. (1960). Influence of arsenic pressure on the doping of gallium arsenide with germanium. *J. Appl. Phys.* **31**, 2065.

890. McCOMBS, A. E., JR., and MILNES, A. G. (1968). Calculation of drift velocity in silicon at high electric fields. *Int. J. Electron.* **24**, 573.

891. McKAY, K. G. (1954). Avalanche breakdown in silicon. *Phys. Rev.* **94**, 877.

892. McKELVEY, J. P. (1957). Experimental determination of injected carrier recombination rates at dislocations in semiconductors. *Phys. Rev.* **106**, 910.

893. McKELVEY, J. P., and LONGINI, R. L. (1955). Recombination of injected carriers at dislocation edges in semiconductors. *Phys. Rev.* **99**, 1227.

894. McWHORTER, A. L. (1963). Electromagnetic theory of the semiconductor junction laser. *Solid-State Electron.* **6**, 417.

895. MEAD, C. A. (1960a). The tunnel-emission amplifier. *Proc. IRE* **48**, 359.

896. MEAD, C. A. (1960b). A note on tunnel emission. *Proc. IRE* **48**, 1478.

897. MEAD, C. A. (1961a). Operation of tunnel-emission devices. *J. Appl. Phys.* **32**, 646.
898. MEAD, C. A. (1961b). Anomalous capacitance of thin dielectric structures. *Phys. Rev.* **6**, 545.
899. MEAD, C. A. (1962a). Electron transport mechanisms in thin insulating films. *Phys. Rev.* **128**, 2088.
900. MEAD, C. A. (1962b). Transport of hot electrons in thin gold films. *Phys. Rev. Lett.* **8**, 56.
901. MEAD, C. A. (1966a). Schottky barrier gate field effect transistor. *Proc. IEEE* **54**, 307.
902. MEAD, C. A. (1966b). Metal–semiconductor surface barriers. *Solid-State Electron.* **9**, 1023.
903. MEAD, C. A. (1968). Physics of interfaces. *Electrochem. Soc. Extended Abstr., 134th Nat. Meeting, Montreal, Fall 1968*, Abstract 491.
904. MEAD, C. A. (1969a). Physics of interfaces. *In* "Ohmic Contacts to Semiconductors" (B. Schwartz, ed.). Electrochem. Soc., New York.
905. MEAD, C. A. (1969b). Some properties of exponentially damped wave functions. *In* "Tunneling Phenomena in Solids" (E. Burstein and S. Lundquist, eds.), Chap. 9. Plenum Press, New York.
906. MEAD, C. A., and SPITZER, W. G. (1963a). Photoemission from Au and Cu into CdS. *Appl. Phys. Lett.* **2**, 74.
907. MEAD, C. A., and SPITZER, W. G. (1963b). Fermi level position at semiconductor surfaces. *Phys. Rev. Lett.* **10**, 471.
908. MEAD, C. A., and SPITZER, W. G. (1964). Fermi level position at metal–semiconductor interfaces. *Phys. Rev.* **134**, A713.
909. MEHL, W., GOSSENBERGER, H. F., and HELPERT, E. (1963). A method for the preparation of low-temperature alloyed gold contacts to silicon and germanium. *J. Electrochem. Soc.* **110**, 239.
910. MEIERAN, E. S. (1967). Reflection X-ray topography of gallium arsenide deposited on germanium. *J. Electrochem. Soc.* **114**, 292.
911. MELCHIOR, H., LEPSELTER, M. P., and SZE, S. M. (1968). Schottky-barrier avalanche photodiodes. *IEEE Trans. Electron Devices* **ED–15**, 687.
912. MELEHY, M. A. (1970). New aspects of transport in heterojunctions and metal-semiconductor diodes. *Abstr. Int. Conf. Phys. Chem. Semicond. Heterojunctions Layer Structures, Budapest, 1970*, p. 26. Hung. Acad. Sciences, Budapest, Hungary.
913. MESNARD, G. (1970). Influence of temperature on tunnel current in heterojunctions. *Proc. Int. Conf. Phys. Chem. Semicond. Heterojunctions Layer Structures, Budapest, 1970* **1**, 309. Hung. Acad. Sciences, Budapest, Hungary.
914. MEYER, D. E. (1969). Improving aluminum-silicon ohmic contact without high temperature alloy. *In* "Ohmic Contacts to Semiconductors" (B. Schwartz, ed.). Electrochem. Soc., New York.
915. MEYERHOF, W. E. (1947). Contact potential difference in silicon crystal rectifiers. *Phys. Rev.* **71**, 727.
916. MEYERHOFER, D., and OCHS, S. A. (1963). Current flow in very thin films of Al_2O_3 and BeO. *J. Appl. Phys.* **34**, 2535.
917. MICHAELSON, H. B. (1950). Work functions of the elements. *J. Appl. Phys.* **21**, 536.
918. MICHEL, A. E., BLUM, J. M., and HOEKSTRA, J. P. (1969). Base contacts for high-speed germanium transistors. *In* "Ohmic Contacts to Semiconductors" (B. Schwartz, ed.). Electrochem. Soc., New York.
919. MIDDELHOEK, S. (1970). Metallization processes in fabrication of Schottky-barrier FET's. *IBM J. Res. Develop.* **14**, 148.
920. MIKSIC, M. G., MANDEL, G., MOREHEAD, F. F., ONTON, A. A., and SCHLIG, E. S. (1964a). Injection electroluminescence in *p*-type ZnTe. *Phys. Lett.* **11**, 202.

921. MIKSIC, M. G., SCHLIG, E. S., and HAERING, R. R. (1964b). Behavior of CdS thin film transistors. *Solid-State Electron.* **7**, 39.

922. MILES, J. L., and SMITH, P. H. (1963). The formation of metal oxide films using gaseous and solid electrolytes. *J. Electrochem. Soc.* **110**, 1240.

923. MILLEA, M. F., McCOLL, M., and MEAD, C. A. (1969). Schottky barriers on GaAs. *Phys. Rev.* **177**, 1164.

924. MILLER, K. J., and GRIECO, M. J. (1962). Epitaxial silicon–germanium alloy films on silicon substrates. *J. Electrochem. Soc.* **109**, 70.

925. MILLER, K. J., GRIECO, M. J., and SZE, S. M. (1966). Growth of vanadium on silicon substrates. *J. Electrochem. Soc.* **113**, 902.

926. MINDEN, H. T. (1963). Gallium arsenide electroluminescent and laser diodes. *Semicond. Prod. Solid State Technol.* **6**, 34.

927. MINDEN, H. T. (1965). The preparation and properties of GaAs–InAs mixed crystals. *J. Electrochem. Soc.* **112**, 300.

928. MINDEN, H. T. (1970). Some optical properties of aluminum arsenide. *Appl. Phys. Lett.* **17**, 358.

929. MIRONOV, I. A., OKSMAN, YA. A., and RYZHKIN, YU. S. (1969). Electroluminescence of low-resistivity single crystal of zinc selenide. *Sov. Phys.–Semicond.* **2**, 1155.

930. MLAVSKI, A. J., and WEINSTEIN, M. (1963). Crystal growth of GaAs by a traveling solvent method. *J. Appl. Phys.* **34**, 2885.

931. MOEST, R. R., and SHUPP, B. R. (1961). Vapor phase growth of GaAs and GaP. *J. Electrochem. Soc.* **108**, 178c.

932. MOEST, R. R., and SHUPP, B. R. (1962). Preparation of epitaxial GaAs and GaP films by vapor phase reaction. *J. Electrochem. Soc.* **109**, 1061.

933. MOLL, J. L. (1964). "Physics of Semiconductors." McGraw-Hill, New York.

934. MOLL, J. L., MEYER, N. I., and BARTELINK, D. J. (1961). Hot-electron emission from silicon pn junctions parallel to the surface. *Phys. Rev. Lett.* **7**, 87.

935. MOLNAR, B., FLOOD, J. J., and FRANCOMBE, M. H. (1964). Fibered and epitaxial growth in sputtered films of GaAs. *J. Appl. Phys.* **35**, 3554.

936. MONTGOMERY, H. C., and BROWN, W. L. (1956). Field-induced conductivity changes in germanium. *Phys. Rev.* **103**, 865.

937. MOON, P. (1940). Proposed standard-radiation curves for engineering use. *J. Franklin Inst.* **230**, 583.

938. MOORE, R. M. (1969). The effect of a field-independent polarization discontinuity on heterojunction characteristic. *IEEE Trans. Electron Devices* **ED–16**, 186.

939. MOORE, R. M., and BUSANOVICH, C. J. (1969a). An evaporated heterojunction diode strain sensor. *Proc. IEEE* **57**, 735.

940. MOORE, R. M., and BUSANOVICH, C. J. (1969b). The heterode strain sensor: An evaporated heterojunction device. *IEEE Trans. Electron Devices* **ED–16**, 850.

941. MOORE, R. M., and BUSANOVICH, C. J. (1969c). The feasibility of a heterode pressure transducer. *Proc. IEEE* **57**, 1433.

942. MOORE, R. M., and BUSANOVICH, C. J. (1970). Vacuum-deposited thin-film p-Se n-CdSe heterojunction diodes. *IEEE Trans. Electron. Devices* **ED–17**, 305.

943. MOREHEAD, F. F., and MANDEL, G. (1964). Efficient, visible electro-luminescence from p–n junctions in $Zn_xCd_{1-x}Te$. *Appl. Phys. Lett.* **5**, 53.

944. MOREHEAD, F. F., and MANDEL, G. (1965). Self-compensation-limited conductivity in binary semiconductors, n-$Zn_xCd_{1-x}Te$. *Phys. Rev.* **137**, 924.

945. MORIIZUMI, T., and TAKAHASHI, K. (1969). Si- and Ge-doped GaAs p–n junctions. *Jap. J. Appl. Phys.* **8**, 348.

946. MORIIZUMI, T., and TAKAHASHI, K. (1970). Epitaxial vapor growth of ZnTe on InAs. *Jap. J. Appl. Phys.* **9**, 849.

947. MORTON, G. A. (1949). Photomultipliers for scintillation counting. *RCA Rev.* **10**, 525.

948. MORTON, G. A. (1956). Recent developments in the scintillation counter field. *IRE Trans. Nucl. Sci.* **NS–3**, 122.

949. MORTON, G. A., SMITH, H. M., and KRAIL, H. R. (1968). Pulse height resolution of high gain first dynode photomultipliers. *Appl. Phys. Lett.* **13**, 356.

950. MOSS, H. I. (1961). Large-area thin-film photovoltaic cells. *RCA Rev.* **22**, 29.

951. MOSS, T. S. (1959). "Optical Properties of Semiconductors." Butterworth, London.

952. MOSS, T. S. (1961). The potentialities of silicon and gallium arsenide solar batteries. *Solid-State Electron.* **2**, 222.

953. MOSS, T. S., and HAWKINS, R. D. F. (1961). Infrared absorption in gallium arsenides. *Infrared Phys.* **1**, 111.

954. MOULTON, C. (1962). Sputtered III–V intermetallic films. *Nature* **193**, 793.

955. MROCZKOWSKI, R. S., LAVINE, M. C., and GATOS, H. C. (1965). Metallurgical aspect of interface-alloyed GaAs–Ge heterojunctions. *Trans. AIME* **233**, 456.

956. MÜLLER, E. K. (1964). Structure of oriented, vapor-deposited GaAs films, studied by electron diffraction. *J. Appl. Phys.* **35**, 580.

957. MULLER, R. S. (1963). Behavior of gold blocking contacts to CdS at high impressed fields. *J. Appl. Phys.* **34**, 2401.

958. MULLER, R. S., and ZULEEG, R. (1964). Vapour-deposited, thin-film heterojunction diodes. *J. Appl. Phys.* **35**, 1550.

959. MURAVYEVA, D. D., KALINKIN, I. P., and ALESKOVSKY, V. B. (1970). Thermodynamic conditions of synthesis of highly oriented $A^{II}B^{VI}$ films under vacuum condensation. *Proc. Int. Conf. Phys. Chem. Semicond. Heterojunctions Layer Structures, Budapest, 1970* **3**, 183. Hung. Acad. Sciences, Budapest, Hungary.

960. MURPHY, E. L., and GOOD, R. H., JR. (1956). Thermionic emission, field emission, and the transition region. *Phys. Rev.* **102**, 1464.

961. MUZALEVSKY, E. A. (1966). Some properties of CdS–CdSe heterojunctions. *Ukr. Fiz. Zh.* **11**, 436.

962. NAKAI, J., MITSUSADA, K., and YASUOKA, A. (1964). Characteristics of the contact between vacuum-deposited cadmium selenide films and gold films. *Jap. J. Appl. Phys.* **3**, 490.

963. NAKAI, J., YASUOKA, A., OKUMURA, T., and KANO, G. (1965). CdSe–Ge heterojunctions. *Jap. J. Appl. Phys.* **4**, 545.

964. NAKAI, J., KAMURO, S., FUKUSHIMA, M., and HAMAGUCHI, C. (1970). ZnTe–GaSb heterojunctions. *Proc. Int. Conf. Phys. Chem. Semicond. Heterojunctions Layer Structures, Budapest, 1970* **1**, 319. Hung. Acad. Sciences, Budapest, Hungary.

965. NAKAI, Y., WATANABE, M., OI, T., and AKATSUKA, M. (1969). Strain effect at insulator–GaAs interface and characteristics of planar p–n GaAs diode. *Electrochem. Soc. Extended Abstr., 136th Nat. Meeting, Detroit, Fall 1969*, p. 269.

966. NAKAMURA, S., and FUKAI, M. (1967). Preparation of thin films of ZnSe, ZnTe and $ZnSe_xTe_{1-x}$ by flash evaporation. *Jap. J. Appl. Phys.* **6**, 1473.

967. NAKANO, T. (1967). Preparation and properties of GaAs–Si heterojunctions by solution growth method. *Jap. J. Appl. Phys.* **6**, 854.

968. NANAVATI, R. P. (1963). "An Introduction to Semiconductor Electronics." McGraw-Hill, New York.

969. NANNICHI, Y. (1963). Epitaxial growth of silicon by vacuum sublimation. *Nature* **200**, 1087.

970. NANNICHI, Y., and PEARSON, G. L. (1969). Properties of GaP Schottky-barrier diodes at elevated temperatures. *Solid-State Electron.* **12**, 341.

971. NASLEDOV, D. N., PROTASOV, YU. V., and RUMYANTSEV, A. P. (1970). Thin-film transistor controlled by an Al–CdS barrier layer. *Sov. Phys.–Semicond.* **3**, 968.

972. NATHAN, M. I. (1962a). Current voltage characteristics of germanium tunnel diodes. *J. Appl. Phys.* **33**, 1460.

973. NATHAN, M. I. (1962b). Semiconductor lasers. *Appl. Opt.* **5**, 1514.

974. NATHAN, M. I. (1963). Recombination radiation and stimulated emission in GaAs. *Solid-State Electron.* **6**, 425.

975. NATHAN, M. I., and MARINACE, J. C. (1962). Phonon and polaron interaction in Ge–GaAs tunnel heterojunctions. *Phys. Rev.* **128**, 2149.

976. NATHAN, M. I., DUMKE, W. P., BURNS, G., DILL, F. H., JR., and LASHER, G. (1962). Stimulated emission of radiation from GaAs $p–n$ junctions. *Appl. Phys. Lett.* **1**, 62.

977. NATHANSON, H. C., and JORDAN, A. G. (1962). Characteristics of an exponentially retrograded variable capacitance diode, Part I. *Semicond. Prod. Solid State Technol.* **5**, 38.

978. NELSON, D. F., and McKENNA, J. (1967). Electromagnetic modes of anisotropic dielectric waveguides at $p–n$ junctions. *J. Appl. Phys.* **38**, 4057.

979. NELSON, H. (1963). Epitaxial growth from the liquid state and its application to the fabrication of tunnel and laser diodes. *RCA Rev.* **24**, 603.

980. NELSON, H., and KRESSEL, H. (1969). Improved red and infrared light emitting $Al_xGa_{1-x}As$ laser diodes using the close-confinement structure. *Appl. Phys Lett.* **15**, 7.

981. NELSON, H., PANKOVE, J. I., HAWRYLO, F., DOUSMANIS, G. C., and RENO, C. (1964). High-efficiency injection laser at room temperature. *Proc. IEEE* **52**, 1360.

982. NELSON, O. L., and ANDERSON, D. E. (1966). Hot-electron transfer through thin-film $Al–Al_2O_3$ triodes. *J. Appl. Phys.* **37**, 66.

983. NEUDECK, G. W. (1969). An Investigation of Noise in $N–N$ Germanium–Silicon Heterojunctions. Ph.D. Thesis, Purdue Univ., Lafayette, Indiana. (Also University Microfilms, Ann Arbor, Michigan, Order No. 69-17,230.)

984. NEUDECK, G. W., THOMPSON, H. W., JR., and SCHWARTZ, R. J. (1969). High-frequency noise in germanium–silicon $n–n$ heterojunctions. *J. Appl. Phys.* **40**, 4108.

985. NEVILLE, R. C., and MEAD, C. A. (1970). Surface barriers on zinc oxide. *J. Appl. Phys.* **41**, 3795.

986. NEWKIRK, J. B., and WERNICK, J. H., Ed. (1962). "Direct Observation of Imperfections in Crystals." Wiley (Interscience), New York.

987. NEWMAN, P. C. (1965). Forward characteristics of heterojunctions. *Electron. Lett.* **1**, 265.

988. NEWMAN, R. C. (1964). A review of the growth and structure of thin films of germanium and silicon. *Microelectronics and Reliability*, **3**, 121.

989. NEWMAN, R. C., and WAKEFIELD, J. (1960). The formation of thin films of germanium by the disproportionation of germanium di-iodide. *In* "Solid State Physics in Electronics and Telecommunications" (M. Désirant and J. L. Michiels, eds.), Vol. 1. Academic Press, New York.

990. NEWMAN, R. C., and WAKEFIELD, J. (1962). Vapor growth of germanium–silicon alloy films on germanium substrates. *J. Electrochem. Soc.* **109**, 201c.

991. NEWMAN, R. L., and GOLDSMITH, N. (1961). Vapor growth of gallium arsenide. *J. Electrochem. Soc.* **108**, 1127.

992. NICOLET, M. A. (1966). Unipolar space charge limited current in solids with non-uniform spatial distribution of shallow traps. *J. Appl. Phys.* **37**, 4224.

993. NICOLL, F. H. (1963). The use of close spacing in chemical-transport systems for growing epitaxial layers of semiconductors. *J. Electrochem. Soc.* **110**, 1165.

994. NIEMYSKI, T., and WEYDMAN, Z. (1970). Growth of heavily doped SiC layers on SiC

single crystals. *Electrochem. Soc. Extended Abstr., 137th Nat. Meeting, Los Angeles, Spring 1970*, p. 485.

995. NISHINO, T., and HAMAKAWA, Y. (1970). Opto-electronic properties of Si–SnO$_2$ heterojunctions. *Proc. Int. Conf. Phys. Chem. Semicond. Heterojunctions Layer Structures, Budapest, 1970* **2**, 409. Hung. Acad. Sciences, Budapest, Hungary.

996. NISHIZAWA, J., TERASAKI, T., SHIMBO, M., and SUNAMI, H. (1970). Dependence of surface defects on growing process in silicon epitaxial growth. *Proc. Int. Conf. Phys. Chem. Semicond. Heterojunctions Layer Structures, Budapest, 1970* **3**, 201. Hung. Acad. Sciences, Budapest, Hungary.

997. NOHAVICA, D. (1970). Epitaxial growth of GaP from gaseous and liquid phase. *Proc. Int. Conf. Phys. Chem. Semicond. Heterojunctions Layer Structures, Budapest, 1970* **3**, 209. Hung. Acad. Sciences, Budapest, Hungary.

998. NOJIMA, K., and IBUKI, S. (1966). Preparation and some properties of ZnSe diodes. *Jap. J. Appl. Phys.* **5**, 253.

999. NOJIMA, K., KOMIYA, H., and IBUKI, S. (1968). Visible electroluminescence in Au–Si$_3$N$_4$–ZnSe diode. *Jap. J. Appl. Phys.* **7**, 559.

1000. NOVIKOVA, S. I. (1961a). Thermal expansion of α-Sn, InSb and CdTe. *Sov. Phys.–Solid State* **2**, 2087.

1001. NOVIKOVA, S. I. (1961b). Investigation of thermal expansion of GaAs and ZnSe. *Sov. Phys.–Solid State* **3**, 129.

1002. NOYCE, R. N., BOHN, R. E., and CHUA, H. T. (1969). Schottky diodes make IC scene. *Electronics* **42**, July 21, p. 74.

1003. NUESE, C. J., and GANNON, J. J. (1968) Silver–manganese evaporated ohmic contacts to *p*-type gallium arsenide. *J. Electrochem. Soc.* **115**, 327.

1004. NUESE, C. J., TIETJEN, J. J., GANNON, J. J., and GOSSENGERGER, H. F. (1969). Optimization of electroluminescent efficiencies for vapor-grown GaAs$_{1-x}$P$_x$ diodes. *J. Electrochem. Soc.* **116**, 248.

1005. NUESE, C. J., SIGAI, A. G., ETTENBERG, M., GANNON, J. J., and GILBERT, S. L. (1970). Preparation of visible-light-emitting *p–n* junctions in AlAs. *Appl. Phys. Lett.* **17**, 90.

1006. ODA, J. (1962). Vapor growth of silicon–germanium crystals. *Jap. J. Appl. Phys.* **1**, 131.

1007. OGURA, S., and VISWANATHAN, C. R. (1970). Photoemission studies in MNS structures. *Electrochem. Soc. Extended Abstr., 137th Nat. Meeting, Los Angeles, Spring 1970*, p. 275.

1008. OKADA, J., and MATINO, H. (1964). Continuous oscillations of acousto-electric current in cadmium sulfide. *Jap. J. Appl. Phys.* **3**, 698.

1009. OKADA, J., KANO, T., and SASAKI, Y. (1961). Epitaxial vapour growth of GaAs on Ge single crystals. *J. Phys. Soc. Jap.* **16**, 2591.

1010. OKAZAKI, S., and OTAKI, E. (1966). Properties of intersurface in Au–insulator–CdS measured by capacitance method. *Jap. J. Appl. Phys.* **5**, 181.

1011. OKIMURA, H. (1968). Anomalous breakdown in CdS(CdSe)–*n*Ge junctions. *Jap. J. Appl. Phys.* **7**, 1297.

1012. OKIMURA, H., and KONDO, R. (1970). Electrical and photovoltaic properties of CdS–S junctions. *Jap. J. Appl. Phys.* **9**, 274.

1013. OKIMURA, H., KAWAKAMI, M., and SAKAI, Y. (1967). Photovoltaic properties of CdS–*p*Si heterojunction cells. *Jap. J. Appl. Phys.* **6**, 908.

1014. OLDHAM, W. G. (1963). Semiconductor Heterojunctions. Ph.D. Thesis, Carnegie Inst. of Technol., Pittsburgh, Pennsylvania.

1015. OLDHAM, W. G. (1965). Vapor growth of GaP on GaAs substrates. *J. Appl. Phys.* **36**, 2887.

1016. OLDHAM, W. G., and BAHRAMAN, A. (1967). Electro-optic junction modulators. *IEEE J. Quantum Electron.* **QE–3**, 278.

1017. OLDHAM, W. G., and MILNES, A. G. (1962). Heterojunctions in InP–GaAs. *IRE Trans. Electron Devices* **ED–9**, 509.

1018. OLDHAM, W. G., and MILNES, A. G. (1963). *n–n* semiconductor heterojunctions. *Solid-State Electron.* **6**, 121.

1019. OLDHAM, W. G., and MILNES, A. G. (1964). Interface states in abrupt semiconductor heterojunctions. *Solid-State Electron.* **7**, 153.

1020. OLDHAM, W. G., RIBEN, A. R., FEUCHT, D. L., and MILNES, A. G. (1963). Epitaxial growth of Ge on Si. *J. Electrochem. Soc.* **110**, 53c.

1021. O'REILLY, T. J. (1965). The transient response of insulated gate field effect transistors. *Solid-State Electron.* **8**, 947.

1022. OSTROBORODOVA, V. V., and DIAS, P. (1970). Ionization energy of zinc in gallium phosphide. *Sov. Phys.–Semicond.* **3**, 1319.

1023. OXLEY, T. H., and SUMMERS, J. G. (1966). Metal–gallium arsenide diodes as mixers. *1966 Symp. GaAs, Reading, England.* Inst. of Phys. and the Phys. Soc., London, p. 138.

1024. PADOVANI, F. A. (1966). Graphical determination of the barrier height and excess temperature of a Schottky barrier. *J. Appl. Phys.* **37**, 921.

1025. PADOVANI, F. A. (1967). Forward voltage–current characteristics of metal–silicon Schottky barriers. *J. Appl. Phys.* **38**, 891.

1026. PADOVANI, F. A. (1968). Thermionic emission in Au–GaAs Schottky barriers. *Solid-State Electron.* **11**, 193.

1027. PADOVANI, F. A., and STRATTON, R. (1966a). Field and thermionic-field emission in Schottky barriers. *Solid-State Electron.* **9**, 695.

1028. PADOVANI, F. A., and STRATTON, R. (1966b). Experimental energy–momentum relationship determination using Schottky barriers. *Phys. Rev. Lett.* **16**, 1202.

1029. PADOVANI, F. A., and STRATTON, R. (1968). The accuracy of the WKB approximation for tunneling in metal–semiconductor junctions. *Appl. Phys. Lett.* **13**, 167.

1030. PADOVANI, F. A., and SUMNER, G. G. (1965). Experimental study of gold gallium arsenide Schottky barriers. *J. Appl. Phys.* **36**, 3744.

1031. PAGE, D. J. (1965). A CdS–Si heterojunction transistor. *IEEE Trans. Electron Devices* **ED–12**, 509.

1032. PAGE, D. J. (1968). A cadmium sulfide–silicon composite resonator. *Proc. IEEE* **56**, 1748.

1033. PAGE, D. J., KAYALI, A. A., and WRIGHT, G. T. (1962). Some observations of space charge limited current in CdS crystals. *Proc. Phys. Soc. (London)* **80**, 1133.

1034. PAKSWER, S., and PRATINIDHI, K. (1963). Negative resistance in thin anodic oxide films. *J. Appl. Phys.* **34**, 711.

1035. PALMER, D. R., MORRISON, S. R., and DAUENBAUGH, C. E. (1960). Electrical properties of cleaved germanium surfaces. *Phys. Chem. Solids* **14**, 27.

1036. PALMER, D. R., MORRISON, S. R., and DAUENBAUGH, C. E. (1962). Density and energy of surface states on cleaved surfaces of Ge. *Phys. Rev.* **129**, 608.

1037. PAMPLIN, P. R. (1970). II IV V_2 and I III IV_2 V_4 compounds. *Proc. Int. Conf Phys. Chem. Semicond. Heterojunctions Layer Structures, Budapest, 1970* **1**, 327. Hung. Acad. Sciences, Budapest, Hungary.

1038. PANISH, M. B. (1966). The Ga–As–Si ternary phase system. *J. Electrochem. Soc.* **113**, 1226.

1039. PANISH, M. B., and CASEY, H. C., JR. (1969). Temperature dependence of the energy gap in GaAs and GaP. *J. Appl. Phys.* **40**, 163.

1040. PANISH, M. B., and HAYASHI, I. (1970). Low threshold injection lasers utilizing a *p–n* heterojunction and a *p–p* heterojunction between GaAs and $Al_xGa_{1-x}As$. *Proc.*

Int. Conf. Phys. Chem. Semicond. Heterojunctions Layer Structures, Budapest, 1970 **2**, 419. Hung. Acad. Sciences, Budapest, Hungary.

1041. PANISH, M. B., and SUMSKI, S. (1969). Ga–Al–As: Phase, thermodynamic and optical properties. *Phys. Chem. Solids* **30**, 129.

1042. PANISH, M. B., and SUMSKI, S. (1970). Ga–As–Si: Phase studies and electrical properties of solution-grown Si-doped GaAs. *J. Appl. Phys.* **41**, 3195.

1043. PANISH, M. B., QUEISSER, H. J., DERICK, L., and SUMSKI, S. (1966). Photoluminescence and solution growth of gallium arsenide. *Solid-State Electron.* **9**, 311.

1044. PANISH, M. B., HAYASHI, I., and SUMSKI, S. (1969). A technique for the preparation of low-threshold room-temperature GaAs laser diode structures. *IEEE J. Quantum Electron.* **QE–5**, 210.

1045. PANISH, M. B., HAYASHI, I., and SUMSKI, S. (1970). Double-heterostructure injection lasers with room-temperature thresholds as low as 2300 A/cm². *Appl. Phys. Lett.* **16**, 326.

1046. PAOLA, C. R. (1970). Metallic contacts for gallium arsenide. *Solid-State Electron.* **13**, 1189.

1047. PAOLA, C. R., and KNIGHT, S. (1968). Ohmic contacts for GaAs bulk effect devices. *Electrochem. Soc. Extended Abstr., 134th Nat. Meeting, Montreal, Fall 1968*, Abstract No. 497.

1048. PAOLI, T. L., and RIPPER, J. E. (1969a). Coupled longitudinal mode pulsing in semiconductor lasers. *Phys. Rev. Lett.* **22**, 1085.

1049. PAOLI, T. L., and RIPPER, J. E. (1969b). Optical pulses from cw GaAs injection lasers. *Appl. Phys. Lett.* **15**, 105.

1050. PAOLI, T. L., and RIPPER, J. E. (1970a). Frequency stabilization and narrowing of optical pulses from cw GaAs junction lasers. *IEEE J. Quantum Electron.* **QE–6**, 335.

1051. PAOLI, T. L., and RIPPER, J. E. (1970b). Self-stabilization and narrowing of optical pulses from GaAs junction lasers by injection current feedback. *Appl. Phys. Lett.* **16**, 96.

1052. PAOLI, T. L., RIPPER, J. E., and ZACHOS, T. H. (1969). Resonant modes of GaAs junction lasers—II: High-injection level. *IEEE J. Quantum Electron.* **QE–5**, 271.

1053. PAPAZIAN, S. A., and REISMAN, A. (1968). Epitaxial deposition of germanium on semi-insulating GaAs. *J. Electrochem. Soc.* **115**, 961.

1054. PARKER, G. H., and MEAD, C. A. (1969a). The effect of trapping states on tunneling in metal–semiconductor junctions. *Appl. Phys. Lett.* **14**, 21.

1055. PARKER, G. H., and MEAD, C. A. (1969b). Tunneling in CdTe Schottky barriers. *Phys. Rev.* **184**, 780.

1056. PARKER, G. H., McGILL, T. C., and MEAD, C. A. (1968). Electric field dependence of GaAs Schottky barriers. *Solid-State Electron.* **11**, 201.

1057. PATAKI, G., NEMETH-SALLAY, M., and LORINCZY, A. (1970). A simple non-destructive method for detection of defects in semiconductor structures. *Proc. Int. Conf. Phys. Chem. Semicond. Heterojunctions Layer Structures, Budapest, 1970* **5**, 151. Hung. Acad. Sciences, Budapest, Hungary.

1058. PATEL, J. R., and CHAUDHURI, A. R. (1966). Charged impurity effects on the deformation of dislocation-free germanium. *Phys. Rev.* **143**, 601.

1059. PAULING, L. (1960). "The Nature of the Chemical Bond." Cornell Univ. Press, Ithaca, New York.

1060. PAULNACK, C. L., and HOWELLS, B. F., JR. (1968). Contact resistance measurements of metals to *p* and *n* silicon. *Electrochem. Soc. Extended Abstr., 134th Nat. Meeting, Montreal, Fall 1968*, Abstract No. 503.

1061. PAVELETS, S. YU., FEDORUS, G. A., and KONONETS, YA. F. (1970). Optimum thickness of cuprous sulfide films in *n*–*p* CdS–Cu$_{2-x}$S photocells. *Sov. Phys.–Semicond.* **4**, 282.

1062. PAYNE, R. T. (1968). Zero bias effects in GaAs $p-n$ junctions. *Bull. Amer. Phys. Soc.* **13**, 455.

1063. PEARSON, G. L. (1949). Electrical properties of crystal grain boundaries in Ge. *Phys. Rev.* **76**, 459.

1064. PECENY, T. (1964). Proposal for semiconductor laser with wide-gap emitters. *Phys. Status Solidi* **6**, 651.

1065. PELL, E. M. (1956). Influence of electric field in diffusion region upon breakdown in germanium $n-p$ junctions. *J. Appl. Phys.* **28**, 459.

1066. PENLEY, J. C. (1962). Tunneling through thin films with traps. *Phys. Rev.* **128**, 596.

1067. PERLMAN, S. S. (1963). $p-n$ heterojunctions with experiments on germanium–gallium arsenide structures. Ph.D. Thesis, Carnegie Inst. of Technol., Pittsburgh, Pennsylvania.

1068. PERLMAN, S. S. (1964). Heterojunction photovoltaic cells. *Advan. Energy Convers.* **4**, 184.

1069. PERLMAN, S. S. (1969). Barrier height diminution in Schottky diodes due to electrostatic screening. *IEEE Trans. Electron Devices* **ED–16**, 450.

1070. PERLMAN, S. S., and FEUCHT, D. L. (1964). $p-n$ heterojunctions. *Solid-State Electron.* **7**, 911.

1071. PERLMAN, S. S., and FEUCHT, D. L. (1965). Abrupt heterojunction emitters for transistors. *J. Electron. Contr.* **18**, 159.

1072. PERLMAN, S. S., and WILLIAMS, R. M. (1963). Review of experimental evidence supporting the Shockley model in epitaxial $p-n$ heterojunctions. *IEEE Trans. Electron Devices* **ED–10**, 335.

1073. PETERS, D. W. (1965). Alumina polyphase heterojunction. *J. Amer. Ceram. Soc.* **48**, 220.

1074. PETERSON, J. M., McKAIG, H. L., and DePRISCO, C. F. (1962). Ultrasonic welding in electronic devices. *IRE Int. Conv. Rec., Part* 6 **10**, 3.

1075. PFANN, W. G. (1955). Temperature gradient zone melting. *Trans. AIME* **203**, 961.

1076. PFANN, W. G., BENSON, K. E., and WERNICK, J. H. (1957). Some aspects of Peltier heating at liquid solid interfaces in germanium. *J. Electron.* **2**, 597.

1077. PHELAN, R. J. (1967). InSb–GaAsP infrared to visible light converter. *Proc. IEEE* **55**, 1501.

1078. PHELAN, R. J., and DIMMOCK, J. O. (1967). InSb MOS infrared detector. *Appl. Phys. Lett.* **10**, 55.

1079. PHILLIPS, J. C. (1970a). Elementary excitations at metal–semiconductor interfaces. *Phys. Rev.* **1**, B593.

1080. PHILLIPS, J. C. (1970b). The chemical bond and solid-state physics. *Phys. Today* **23**, 23.

1081. PICUS, G. S., DuBOIS, D. F., VAN ATTA, L. B. (1968). Influence of ionized impurity scattering on Gunn effect and impact ionization in CdTe. *Appl. Phys. Lett.* **12**, 81.

1082. PIERRON, E. D., BURD, J. W., and McNEELY, J. B. (1970). Defect studies of semiconductor materials. *Metallurgical Trans. AIME* **1**, 639.

1083. PIKHTIN, A. N., POPOV, V. A., and YAS'KOV, D. A. (1970). Uses of a laser beam in production of ohmic contacts with semiconductors. *Sov. Phys.–Semicond.* **3**, 1383.

1084. PILKUHN, M. H., and RUPPRECHT, H. (1966). Spontaneous and stimulated emission from GaAs diodes with three-layer structures. *J. Appl. Phys.* **37**, 3621.

1085. PILKUHN, M. H., and RUPPRECHT, H. (1967). Optical and electrical properties of epitaxial and diffused GaAs injection lasers. *J. Appl. Phys.* **38**, 5.

1086. PINSKER, T. N. (1968). Hot electrons in a metal–semiconductor system—diffuse reflection from the surface. *Sov. Phys.–Semicond.* **2**, 203.

1087. PIPER, W. W., and POLICK, S. J. (1961). Vapor-phase growth of single crystals of II–VI compounds. *J. Appl. Phys.* **32**, 1278.

1088. PIZZARELLO, F. A. (1962). Preparation of solid solutions of GaP and GaAs by a gas phase reaction. *J. Electrochem. Soc.* **109**, 226.

1089. PIZZARELLO, F. A. (1967). The effect of metal contacts on acoustic generation in CdS thin films. *J. Appl. Phys.* **38**, 1752.

1090. PLASKETT, T. S. (1969). The synthesis of bulk GaP from Ga solutions. *J. Electrochem. Soc.* **116**, 1722.

1091. PLASKETT, T. S., BLUM, S. E., and FOSTER, L. M. (1967). The preparation and properties of large, solution grown GaP crystals. *J. Electrochem. Soc.* **114**, 1303.

1092. PLUMMER, A. R. (1958). The effect of heat treatment on the breakdown characteristics of silicon *pn* junctions. *J. Electron. Contr.* **5**, 405.

1093. POKORNY, J., and FREMUNT, R. (1970). Doping profile measurements of epitaxial GaAs layers. *Proc. Int. Conf. Phys. Chem. Semicond. Heterojunctions Layer Structures, Budapest, 1970* **4**, 215. Hung. Acad. Sciences, Budapest, Hungary.

1094. POLGAR, P., MOUYARD, A., and SHINER, B. (1970). A high-current metal–semiconductor rectifier. *IEEE Trans. Electron Devices* **ED–17**, 725.

1095. POLLACK, S. R. (1963). Schottky field emission through insulating layers. *J. Appl. Phys.* **34**, 877.

1096. POON, H. C., GUMMEL, H. K., and SCHARFETTER, D. L. (1969). High injection in epitaxial transistors. *IEEE Trans. Electron Devices* **ED–16**, 455.

1097. PORTNOY, W. M., and LEEDY, H. M. (1968). A monolithic single-ended X-band mixer circuit. *IEEE J. Solid-State Circuits* **SC–3**, 31.

1098. POSTNIKOV, V. V., LOGINOVA, R. G., and OVSYANNIKOV, M. I. (1966). Vacuum preparation of epitaxial layers of Si. *Sov. Phys.–Crystallogr.* **10**, 495.

1099. POTTER, A. E., JR., BERRY, W. B., BRANDHORST, H. W., JR., and SCHALLA, R. L. (1968). Effect of green light on spectral response of cuprous sulfide–cadmium sulfide photovoltaic cells. NASA Tech. Note TN D4333.

1100. POWELL, C. F., OXLEY, J. H., and BLOCHER, J. M., JR. (eds.) (1966). "Vapor Deposition." Wiley, New York.

1101. POWELL, R. J. (1970). Interface barrier energy determination from voltage dependence of photo-injected currents. *J. Appl. Phys.* **41**, 2424.

1102. PRATT, G. W. (1955). Localized states at the boundary of two crystals. *Phys. Rev.* **98**, 1543.

1103. PRESTON, J. S. (1950). The constitution and mechanism of the Se rectifier photocell. *Proc. Roy. Soc. Ser. A* **202**, A449.

1104. PRICE, P. J. (1962). Transmission of Bloch waves through crystal interfaces. *Proc. Int. Conf. Phys. Semicond., Exeter, 1962*, p. 99. Inst. of Phys. and the Phys. Soc., London.

1105. PRIMAK, W., KAMPWIRTH, R., and DAYAL, Y. (1967). Peroxide etching of germanium. *J. Electrochem. Soc.* **114**, 88.

1106. PRINCE, M. B. (1953). Drift mobilities in semiconductors—1. Germanium. *Phys. Rev.* **92**, 681.

1107. PRIOR, A. C. (1959). The field dependence of carrier mobility in Si and Ge. *Phys. Chem. Solids* **12**, 175.

1108. PRITCHARD, R. L. (1967). "Electrical Characteristics of Transistors." McGraw-Hill, New York.

1109. PRITCHARD, T. I., HAMPSHIRE, M. J., and TOMLINSON, R. D. (1970). Photo-voltage properties of CdSe–Ge isotype heterojunctions. *Phys. Status Solidi* (a) **3**, 411.

1110. PUROHIT, R. K. (1967). GaP–GaAs *n–p* heterojunctions. *Phys. Status Solidi* **24**, K57.

1111. RAI-CHOUDHURY, P. (1969). Epitaxial gallium arsenide from trimethyl gallium and arsine. *J. Electrochem. Soc.* **116**, 1745.

1112. RAMACHANDRAN, T. B., and MORONEY, W. J. (1964). Special characteristic, efficiency and speed of GaAs$_x$P$_{1-x}$GaAs photodiode. *Proc. IEEE* **52**, 1358.

1113. RAMACHANDRAN, T. B., and SANTOSUOSSO, R. P. (1966). Contacting *n*-type high resistivity GaAs for Gunn oscillators. *Solid-State Electron.* **9**, 733.

1114. RAMACHANDRAN, T. B., CHOW, K. K., MORONEY, W. J., and OLENDZENSKY, P. (1965). Photomixing in a GaAs$_{(x)}$P$_{(1-x)}$GaAs heterodiode. *J. Appl. Phys.* **36**, 2594.

1115. RAPPAPORT, P., and WYSOCKI, J. J. (1965). The photovoltaic effect. *In* "Photoelectronic Materials and Devices" (S. Larach, ed.), Chap. 6. Van Nostrand–Reinhold, Princeton, New Jersey.

1116. RAYBOLD, R. L. (1960). Electroless plated contact to silicon carbide. *Rev. Sci. Instrum.* **31**, 781.

1117. READ, W. T., JR. (1954). Theory of dislocations in germanium. *Phil. Mag.* **45**, 775.

1118. REBANE, K. S. K., TAMMIK, A. A., and TIGANE, I. F. (1970). Early stages of the epitaxial growth of ZnS and ZnSe on NaCl substrate. *Proc. Int. Conf. Phys. Chem. Semicond. Heterojunctions Layer Structures, Budapest, 1970* **3**, 229. Hung. Acad. Sciences, Budapest, Hungary.

1119. REDIKER, R. H. (1965). Semiconductor lasers. *Phys. Today* **18**, 42.

1120. REDIKER, R. H., and QUIST, T. M. (1963). Properties of GaAs alloy diodes. *Solid-State Electron.* **6**, 657.

1121. REDIKER, R. H., QUIST, T. M., and LAX, B. (1963). High speed heterojunction photo diodes and beam of light transistors. *Proc. IEEE* **51**, 218.

1122. REDIKER, R. H., STOPEK, S., and WARD, J. H. R. (1964). Interface–alloy epitaxial heterojunctions. *Solid-State Electron.* **7**, 621.

1123. REDIKER, R. H., STOPEK, S., and HINKLEY, E. D. (1965). Electrical and electro-optical properties of interface–alloy heterojunctions. *Trans. AIME* **233**, 463.

1124. REICHENBAUM, G. (1959). Improved semiconductor-to-copper soldered contact. *Brit. J. Appl. Phys.* **10**, 469(L).

1125. REISMAN, A., and ALYANAKYAN, S. A. (1964). Thermodynamic analyses of open tube germanium disproportionation reactions. *J. Electrochem. Soc.* **111**, 1154.

1126. REISMAN, A., and BERKENBLIT, M. (1965). Substrate orientation effects and germanium epitaxy in an open tube HI transport system. *J. Electrochem. Soc.* **112**, 315.

1127. REISMAN, A., and BERKENBLIT, M. (1966). Kinetics of the reaction HI–⟨111⟩ Ge. *J. Electrochem. Soc.* **113**, 146.

1128. REISMAN, A., and ROHR, R. (1964). Room temperature chemical polishing of Ge and GaAs. *J. Electrochem. Soc.* **111**, 1425.

1129. REISMAN, A., BERKENBLIT, M., and ALYANAKYAN, S. A. (1965). Transpiration studies of the Ge–I$_2$–inert gas system. *J. Electrochem. Soc.* **112**, 241.

1130. REISS, H. (1968). Rotation and translation of islands in the growth of heteroepitaxial films. *J. Appl. Phys.* **39**, 5045.

1131. REVESZ, A. G., and ZAININGER, K. H. (1968). The Si–SiO$_2$ solid-state interface system. *RCA Rev.* **29**, 22.

1132. REYNOLDS, J. H. (1967). A semiconductor–metal–semiconductor light detector. *Trans. AIME* **239**, 326.

1133. RIBEN, A. R. (1965). *n*Ge–*p*GaAs Heterojunctions. Ph.D. Thesis, Carnegie Inst. of Technol., Pittsburgh, Pennsylvania.

1134. RIBEN, A. R., and FEUCHT, D. L. (1964). Evidence of tunneling in non-degenerate Ge–GaAs heterojunctions. *IEEE Trans. Electron Devices* **ED–11**, 534.

1135. RIBEN, A. R., and FEUCHT, D. L. (1966a). Electrical transport in *n*Ge–*p*GaAs heterojunctions. *Int. J. Electron.* **20**, 583.

1136. RIBEN, A. R., and FEUCHT, D. L. (1966b). *n*Ge–*p*GaAs heterojunctions. *Solid-State Electron.* **9**, 1055.

1137. RIBEN, A. R., OLDHAM, W. G., and FEUCHT, D. L. (1965). Resistivity control of Ge grown by GeI$_2$ disproportionation. *J. Appl. Phys.* **36**, 3685.

1138. RIBEN, A. R., FEUCHT, D. L., and OLDHAM, W. G. (1966). Preparation of Ge–Si and Ge–GaAs heterojunctions. *J. Electrochem. Soc.* **113**, 245.

1139. RIBENYI, A. (1970). Schottky diode in TTL circuits. *Proc. Int. Conf. Phys. Chem. Semicond. Heterojunctions Layer Structures, Budapest, 1970* **5**, 349. Hung. Acad. Sciences, Budapest, Hungary.

1140. RICHARDS, J. L., HART, P. B., and GALLONE, L. M. (1963). Epitaxy of compound semiconductors by flash evaporation. *J. Appl. Phys.* **34**, 3418.

1141. RICHARDSON, J. R., and BAERTSCH, R. D. (1969). Zinc sulfide Schottky barrier ultraviolet detectors. *Solid-State Electron.* **12**, 393.

1142. RICHMAN, D. (1968). Vapor phase growth and properties of AlP. *J. Electrochem. Soc.* **115**, 945.

1143. RICHMAN, D., and ARLETT, R. H. (1969a). Preparation and properties of homoepitaxial silicon grown at low temperatures from silane. *In* "Semiconductor Silicon" (R. R. Habrecht and E. L. Kern, eds.) p. 200. Electrochem. Soc., New York.

1144. RICHMAN, D., and ARLETT, R. H. (1969b). Low-temperature epitaxial growth of single crystalline silicon from silane. *J. Electrochem. Soc.* **116**, 872.

1145. RICHMAN, D., and NUESE, C. J. (1970). Preparation and properties of In$_x$Ga$_{1-x}$P alloys. *Tech. Conf. Electron. Magnetic Mater. Computers, August 1970.* Metallurgical Soc., AIMMPE.

1146. RICHMAN, D., CHIANG, Y. S., and ROBINSON, P. H. (1970). Low-temperature vapor growth of homoepitaxial silicon. *RCA Rev.* **31**, 613.

1147. RIDEOUT, V. L., and CROWELL, C. R. (1967). Pressure sensitivity of gold–potassium tantalate Schottky barrier diodes. *Appl. Phys. Lett.* **10**, 329.

1148. RILEY, T. J. (1966). The heterojunction triode. Ph.D. Thesis, Univ. of Birmingham, United Kingdom.

1149. RINDNER, W., and LAVINE, J. M. (1962). Trace-plating—A new semiconductor device fabrication technique. *Solid-State Electron.* **5**, 85.

1150. RIPPER, J. E. (1969). Time delays and Q switching in junction lasers: I. Theory. *IEEE J. Quantum Electron.* **QE–5**, 391.

1151. RIPPER, J. E. (1970a). Analysis of frequency modulation of junction lasers by ultrasonic waves. *IEEE J. Quantum Electron.* **QE–6**, 129.

1152. RIPPER, J. E. (1970b). Comment on the reliability of GaAs stripe-geometry junction lasers. *IEEE J. Quantum Electron.* **QE–6**, 372.

1153. RIPPER, J. E. (1970c). Semiconductor lasers—Two years later. *IEEE J. Quantum Electron.* **QE–6**, 275.

1154. RIPPER, J. E., and DYMENT, J. C. (1968). Internal Q switching in GaAs junction lasers. *Appl. Phys. Lett.* **12**, 365.

1155. RIPPER, J. E., and DYMENT, J. C. (1969). Time delays and Q switching in junction lasers: II. Computer calculations and comparison with experiments. *IEEE J. Quantum Electron.* **QE–5**, 396.

1156. RIPPER, J. E., and PAOLI, T. L. (1969). Frequency pulling and pulse position modulation of pulsing cw GaAs injection lasers. *Appl. Phys. Lett.* **15**, 203.

1157. RIPPER, J. E., and PAOLI, T. L. (1970a). Bistable operation of cw junction lasers due to saturable absorbing centers. *Proc. IEEE* **58**, 178.

1158. RIPPER, J. E., and PAOLI, T. L. (1970b). Locking of spontaneously pulsing cw GaAs injection lasers by fractional-harmonic current modulation. *IEEE J. Quantum Electron.* **QE–6**, 326.

1159. RIPPER, J. E., and PAOLI, T. L. (1970c). Optical coupling of adjacent stripe-geometry junction lasers. *Appl. Phys. Lett.* **17**, 371.

1160. RIPPER, J. E., PAOLI, T. L., and DYMENT, J. C. (1970). Characteristics of bistable cw GaAs junction lasers operating above the delay-transition temperature. *IEEE J. Quantum Electron.* **QE–6**, 300.

1161. RIVIERE, J. C. (1966). The work function of gold. *Appl. Phys. Lett.* **8**, 172.

1162. ROBERTS, G. I., and CROWELL, C. R. (1968). Surface state and interface effects on the capacitance–voltage relationship in Schottky barriers. *Bull. Amer. Phys. Soc.* **13**, 455; *J. Appl. Phys.* **40**, 3726.

1163. ROBERTS, G. I., and CROWELL, C. R. (1970). Capacitive energy level spectroscopy of deep-lying semiconductor impurities using Schottky barriers. *J. Appl. Phys.* **41**, 1767.

1164. ROBINSON, P. H. (1963). Transport of gallium arsenide by a close-spaced technique. *RCA Rev.* **24**, 574.

1165. ROBINSON, R. J., and KUN, Z. K. (1969). Visible-light emitting diodes using (II–VI)–(III–V) systems. *Appl. Phys. Lett.* **15**, 371.

1166. RODRIGUEZ, V., and NICOLET, M. A. (1969). Drift velocity of electrons in silicon at high electric fields from $4.2°$ to $300°K$. *J. Appl. Phys.* **40**, 496.

1167. ROLLETT, J. M. (1959). The characteristic frequencies of a drift transistor. *J. Electron. Contr.* **7**, 193.

1168. ROSE, A. (1955). Space charge limited currents in solids. *Phys. Rev.* **97**, 1538.

1169. ROSS, B. (1970). Properties of $InAs_{1-x}P_x$ lasers. *Electrochem. Soc. Extended Abstr., 137th Nat. Meeting, Los Angeles, Spring 1970*, p. 215.

1170. ROSZTOCZY, F. E., and STEIN, W. W. (1970). Preparation of semiconductor heterojunctions by liquid phase epitaxy. *Proc. Int. Conf. Phys. Chem. Semicond. Heterojunctions Layer Structures, Budapest, 1970* **1**, 333. Hung. Acad. Sciences, Budapest, Hungary.

1171. ROSZTOCZY, F. E., ERMANIS, F., HAYASHI, I., and SCHWARTZ, B. (1968). Germanium-doped gallium arsenide. *Bull. Amer. Phys. Soc.* **13**, 375.

1172. ROTH, H., BERNARD, W., ZELDES, P., and SCHMID, A. P. (1963). Voltage-annealing of radiation damage in tunnel diodes. *J. Appl. Phys.* **34**, 669.

1173. RUBENSTEIN, M. (1962). Solubilities of gallium phosphide in metallic solvents. *Electrochem. Soc. Extended Abstr., 137th Nat. Meeting, Los Angeles, Spring 1970*, p. 129.

1174. RUBENSTEIN, M. (1966a). Solubilities of GaAs in metallic solvents. *J. Electrochem. Soc.* **113**, 752.

1175. RUBENSTEIN, M. (1966b). Solubilities of some II–VI compounds in bismuth. *J. Electrochem. Soc.* **113**, 623.

1176. RUBENSTEIN, M. (1968). Solution growth of some II–VI compounds using tin as a solvent. *J. Cryst. Growth* **3,4**, 309.

1177. RUBENSTEIN, M. (1969a). Liquidus solubilities of CdS in a metals solvent. *Trans. AIME* **245**, 457.

1178. RUBENSTEIN, M. (1969b). A portion of the Zn–S liquidus phase diagram. *Amer. Comm. Cryst. Growth, Conf. Cryst. Growth, Gaithersburg, August 1969.*

1179. RUBENSTEIN, M., and RYAN, F. M. (1967). Growth of CdS from metallic solution. *In* "II–VI Semiconducting Compounds" (D. G. Thomas, ed.), p. 402. Benjamin, New York.

1180. RUBINOVA, É. É., NOVIKOV, S. R., and KONOPLEVA, R. S. (1970). Oscillations of the photocurrent in Ge with radiation defects. *Sov. Phys.–Semicond.* **3**, 1294.

1181. RUEHRWEIN, R. A. (1965). Use of Hydrogen Halide and Hydrogen in Separate Streams as Carrier Gases in Vapor Deposition of II–VI Compounds. U.S. Patent 3,224,912.

1182. RUNYAN, W. R. (1969). The status of silicon epitaxy. *In* "Semiconductor Silicon" (R. R. Habrecht and E. L. Kern, eds.). Electrochem. Soc., New York.

1183. RUNYAN, W. R., and ALEXANDER, E. G. (1967). An experimental study of drift-field silicon solar cells. *IEEE Trans. Electron Devices* **ED–14**, 3.

1184. RUPPRECHT, H., WOODALL, J. M., and PETTIT, G. D. (1967). Efficient visible electroluminescence at 300°K from $Ga_{1-x}Al_xAs$ $p-n$ junctions grown by liquid-phase epitaxy. *Appl. Phys. Lett.* **11**, 81.

1185. RUSCH, W. V. T., and BURRUS, C. A. (1968). Planar millimeter wave epitaxial silicon Schottky-barrier converter diodes. *Solid-State Electron.* **11**, 517.

1186. RUSSELL, G. J., IP, H. K., and HANEMAN, D. (1966). Vacuum thermal decomposition of III–V compound surfaces. *J. Appl. Phys.* **37**, 3328.

1187. RUTH, R. P., MARINACE, J. C., and DUNLAP, W. C., JR. (1960). Vapor deposited single-crystal germanium. *J. Appl. Phys.* **31**, 995.

1188. RUTZ, R. F. (1963). Transistor-like device using optical coupling between diffused $p-n$ junctions in GaAs. *Proc. IEEE* **51**, 470.

1189. RUTZ, R. F. (1966). High-Gain Photon-Coupled Semiconductor Device. U.S. Patent 3,278,814.

1190. RYBKA, V., KREJCI, P., DUDROVA, E., and SEVCIK, Z. (1970). The properties of Ge films prepared by plasma sputtering on semi-insulating substrates *Proc. Int. Conf. Phys. Chem. Semicond. Heterojunctions Layer Structures, Budapest, 1970* **3**, 239. Hung. Acad. Sciences, Budapest, Hungary.

1191. RYDER, E. J. (1953). Mobility of holes and electrons in high electric fields. *Phys. Rev.* **90**, 766.

1192. RYU, I., and TAKAHASHI, K. (1965). Preparation of Ge–GaAs heterojunctions by vacuum evaporation. *Jap. J. Appl. Phys.* **4**, 850.

1193. SAGAR, A., POLLAK, M., and LEHMANN, W. (1968a). Piezoresistance and piezo-Hall effects in n-ZnSe. *Phys. Rev.* **174**, 859.

1194. SAGAR, A., LEHMANN, W., and FAUST, J. W., JR. (1968b). Etchants for ZnSe. *J. Appl. Phys.* **39**, 5336.

1195. SAH, C. T. (1961). Electronic processes and excess current in gold-doped narrow silicon junctions. *Phys. Rev.* **123**, 1594.

1196. SAH, C. T., and REDDI, V. G. K. (1964). Frequency dependence of the reverse biased capacitance of gold doped silicon p^+n step junctions. *IEEE Trans. Electron Devices* **ED–11**, 345.

1197. SAH, C. T., NOYCE, R. N., and SHOCKLEY, W. (1957). Carrier generation and recombination in $p-n$ junctions and $p-n$ junction characteristics. *Proc. IRE* **45**, 1228.

1198. SAHAI, R., and MILNES, A. G. (1970). Heterojunction solar cell calculations. *Solid-State Electron.* **13**, 1289.

1199. SAKAI, Y., and TAKAHASHI, K. (1963). Preparation and properties of vacuum-deposited germanium thin films. *Jap. J. Appl. Phys.* **2**, 629.

1200. SALKOV, E. A. (1965). Some properties of a p(SiC)–n(CdS) junction. *Sov. Phys.–Solid State* **7**, 227.

1201. SALOW, H., and GROBE, E. (1968). Ohmic contact for Gunn elements. *Z. Angew. Phys.* **25**, 137.

1202. SALTICH, J. (1968). Study of Schottky barriers on n-type silicon as a function of doping density. *J. Electrochem. Soc.* **115**, 323c.

1203. SALTICH, J. (1969). Investigation of Schottky barrier heights on n-type silicon as a function of doping density. *In* "Ohmic Contacts to Semiconductors" (B. Schwartz. ed.). Electrochem. Soc., New York.

1204. SALTICH, J. L., and TERRY, L. E. (1970). Effects of pre- and post-annealing treatments on silicon Schottky barrier diodes. *Proc. IEEE* **58**, 492.

1205. SANDERS, T. J. (1969). The Photovoltaic Spectral Response of the Epitaxially Grown Graded Band Gap Silicon–Germanium Heterojunction. Ph.D. Thesis, Purdue Univ., Lafayette, Indiana. (Also University Microfilms, Ann Arbor, Michigan, Order No. 69-17,251.)

1206. SAN-MEI, K. (1963). The preparation and properties of vapour grown GaAs–GaP alloys. *J. Electrochem. Soc.* **110**, 991.

1207. SARANYEVICH, Y. R., STRICHA, V. I., and SHEKA, D. I. (1970). The influence of differences in effective masses on the transferring of charge carrier through hetero-junctions. *Abstr. Int. Conf. Phys. Chem. Semicond. Heterojunctions Layer Structures, Budapest, 1970*, p. 70. Hung. Acad. Sciences, Budapest, Hungary.

1208. SAUL, R. H. (1968). The defect structure of GaP crystals grown from gallium solutions, vapor phase and liquid phase epitaxial depositions. *J. Electrochem. Soc.* **115**, 1184.

1209. SAUL, R. H. (1969). Effect of a $GaAs_xP_{1-x}$ transition zone on the perfection of GaP crystals grown by deposition onto GaAs substrates. *J. Appl. Phys.* **40**, 3273.

1210. SAUL, R. H., and HACKETT, W. H., JR. (1970). Distribution of impurities in Zn, O-doped GaP liquid-phase epitaxial layers. *Electrochem. Soc. Extended Abstr., 137th Nat. Meeting, Los Angeles, Spring 1970*, p. 220.

1211. SAUL, R. H. ARMSTRONG, J., and HACKETT, W. H., JR. (1969). Fabrication of red electroluminescent GaP diodes with external quantum efficiency of 7%. *Appl. Phys. Lett.*, **15**, 229.

1212. SAWYER, D. E. (1968). Photosensitive barium titanate Schottky diodes. *Appl. Phys. Lett.* **13**, 392.

1213. SAXENA, A. N. (1969a). Forward current–voltage characteristics of Schottky barriers on *n*-type silicon. *Surface Sci.* **13**, 151.

1214. SAXENA, A. N. (1969b). Forward current–voltage characteristics and differential resistance peak of a Schottky barrier diode on heavily doped silicon. *Appl. Phys. Lett.* **14**, 11.

1215. SCHAFER, H. (1953). Formation of silicon chlorides of higher molecular weight in the hot–cold tube. *Z. Anorg. Allgem. Chem.* **274**, 265.

1216. SCHAFER, H., and NICKL, J. (1953). The equilibrium: $Si + SiCl_4 \leftrightarrows 2SiCl_2$ and the thermochemical properties of $SiCl_2$ gas. *Z. Anorg. Allgem. Chem.* **274**, 250.

1217. SCHAFER, H., JACOB, H., and ETZEL, K. (1956). I. The transport of solids in a temperature gradient with the help of heterogeneous equilibriums. *Z. Anorg. Allgem. Chem.* **286**, 27.

1218. SCHAFFT, H. A. (1967). Second breakdown—A comprehensive review. *Proc. IEEE* **55**, 1272.

1219. SCHARFETTER, D. L. (1965). Minority carrier injection and charge storage in epitaxial Schottky barrier diodes. *Solid-State Electron.* **8**, 299.

1220. SCHEER, J. J. (1960). Some preliminary experiments concerning the influence of band bending on photo-electric emission. *Philips Res. Rep.* **15**, 584.

1221. SCHEER, J. J., and VAN LAAR, J. (1963). Photo-emission from semiconductor surfaces. *Phys. Lett.* **3**, 246.

1222. SCHEER, J. J., and VAN LAAR, J. (1965). GaAs–Cs: A new type of photoemitter. *Solid State Commun.* **3**, 189.

1223. SCHIBLI, E. (1967). Deep Impurities in Silicon. Ph.D. Thesis, Carnegie Inst. of Technol., Pittsburgh, Pennsylvania.

1224. SCHIBLI, E., and MILNES, A. G. (1967/1968). Lifetime and capture cross-section studies of deep impurities in silicon. *Mater. Sci. Eng.* **2**, 229.

1225. SCHIBLI, E., and MILNES, A. G. (1968). Effects of deep impurities on n^+p junction reverse biased small signal capacitance. *Solid-State Electron.* **11**, 323.

1226. SCHLEGEL, E. S. (1967). A bibliography of metal–insulator–semiconductor studies. *IEEE Trans. Electron Devices* **ED–14**, 728.

1227. SCHLEGEL, E. S. (1968). Additional bibliography of metal–insulator–semiconductor studies. *IEEE Trans. Electron Devices* **ED–15**, 951.

1228. SCHMIDLIN, F. W., ROBERTS, G. G., and LAKATOS, A. I. (1968). Resistance limited currents in solids with blocking contacts. *Appl. Phys. Lett.* **13**, 353.

1229. SCHMIDT, R. (1962a). Electrodes to Semiconductor Wafers. U.S. Patent 3,239,376.

1230. SCHMIDT, R. (1962b). Semiconductor Contact. U.S. Patent 3,231,421.

1231. SCHMIDT, R., and WERNICK, J. H. (1966). Aluminum–Gold Contact to Silicon and Germanium. U.S. Patent 3,403,308.

1232. SCHMIDT, W. A. (1966). Evaporated ohmic contacts on GaAs. *J. Electrochem. Soc.* **113**, 860.

1233. SCHNEIDER, M. V. (1966). Schottky barrier photodiodes with antireflection coating. *Bell Syst. Tech. J.* **45**, 1611.

1234. SCHOLTEN, P. C. (1966). Indium contacts on CdS. *Solid-State Electron.* **9**, 1142.

1235. SCHULZE, R. G. (1966). Some characteristics of GaAs–Ge epitaxy. *J. Appl. Phys.* **37**, 4295.

1236. SCHWARTZ, B., and SARACE, J. C. (1966). Low-temperature alloy contacts to gallium arsenide using metal halide fluxes. *Solid-State Electron.* **9**, 859.

1237. SCHWARTZ, R. F., and SPRATT, J. P. (1962). A tunnel emission device. *Proc. IRE* **50**, 467.

1238. SCOTT-MONCK, J. A., and LEARN, A. J. (1968). Rectifying contacts under evaporated CdS. *Proc. IEEE* **56**, 68.

1239. SEARLE, C. L., BOOTHROYD, A. R., ANGELO, E. J., GRAY, P. E., and PEDERSON, D. O. (1964). "Elementary Circuit Properties of Transistors" (Semiconductor Electronic Education Committee), Vol. 3. Wiley, New York.

1240. SEDGWICK, T. O., and AGULE, B. J. (1966). Bourdon gauge determination of equilibrium in the ZnSe–I$_2$ system. *J. Electrochem. Soc.* **113**, 54.

1241. SEIDEL, T., and SCHARFETTER, D. (1967). Dependence of hole velocity upon electric field and hole density for p type Si. *Phys. Chem. Solids* **28**, 2563.

1242. SEKI, H., and KINOSHITA, M. (1968). Epitaxial growth of InP on GaAs in an open flow system. *Jap. J. Appl. Phys.* **7**, 1142.

1243. SEKI, H., MORIYAMA, K., MATUMOTO, S., and URAMOTO, M. (1967). Thermodynamic study of the growth rate of epitaxial GaAs by GaAs/AsCl$_3$/H$_2$ system. *Jap. J. Appl. Phys.* **6**, 785.

1244. SEKI, H., MORIYAMA, K., ASAKAWA, I., and HORIE, S. (1968). Thermodynamic study of the transport and epitaxial growth of GaAs in an open tube. *Jap. J. Appl. Phys.* **7**, 1324.

1245. SELLE, B., KRISPIN, P., and MAEGE, J. (1970). Electrical and photoelectrical properties of p-Cu$_x$S-n-CdS-heterojunctions. *Proc. Int. Conf. Phys. Chem. Semicond. Heterojunctions Layer Structures, Budapest, 1970* **2**, 247. Hung. Acad. Sciences, Budapest, Hungary.

1246. SELLO, H. (1968). Ohmic contacts and integrated circuits. *Electrochem. Soc. Extended Abstr., 134th Nat. Meeting, Montreal, Fall 1968*, Abstract No. 509.

1247. SELLO, H. (1969). Ohmic contacts and integrated circuits. *In* "Ohmic Contacts to Semiconductors" (B. Schwartz, ed.). Electrochem. Soc., New York.

1248. SEMILETOV, S. A., and AGALARZADE, P. S. (1963). Production of thin films of InSb by vacuum evaporation. *Sov. Phys.–Crystallogr.* **8**, 231.

1249. SENECHAL, R. R., and BASINSKI, J. (1968). Capacitance measurements on Au–GaAs Schottky barriers. *J. Appl. Phys.* **39**, 4581.

1250. SERAPHIN, B. O., and BENNETT, H. E. (1967). Optical constants. *In* "Semiconductors and Semimetals" (R. K. Willardson and A. C. Beer, eds.), Vol. 3, Chap. 12. Academic Press, New York.

1251. SERDYUK, V., and BUBE, R. H. (1967). Contact mechanism for dark conductivity maximum in CdS. *J. Appl. Phys.* **38**, 2399.

1252. SERIZAWA, H., EGUCHI, O., TSUJIMOTO, Y., and FUKAI, M. (1970). Interface observation of ZnSe–ZnTe heterojunctions. *J. Appl. Phys.* **41**, 5032.

1253. SERREZE, H., FISCHLER, S., and SAWYER, D. (1968). GaSb–ZnTe heterojunction. *J. Appl. Phys.* **39**, 5330.

1254. SHAO, J., and WRIGHT, G. T. (1961). Characteristics of the space-charge-limited dielectric diode at high frequencies. *Solid-State Electron.* **3**, 291.

1255. SHARAPOV, B. N. (1970). Threshold currents of injection lasers with one and two heterojunctions. *Sov. Phys.–Semicond.* **4**, 948.

1256. SHARPLESS, W. M. (1958). Fabrication of Semiconductor Devices. U.S. Patent 2,995,475.

1257. SHARPLESS, W. M. (1959). High frequency gallium arsenide point-contact rectifiers. *Bell Syst. Tech. J.* **38**, 259.

1258. SHASKOV, Y. M. (1960). "Metallurgy of Semiconductors." Pittman, New York.

1259. SHAW, D. W. (1968). Influence of substrate temperature on GaAs epitaxial deposition rates. *J. Electrochem. Soc.* **115**, 405.

1260. SHEFTAL, N. N. (ed.) (1969). "Growth of Crystals," Vol. 8. Plenum Press, New York.

1261. SHEPHERD, F. D., JR. (1967). Tunneling enhanced transistor. U.S. Patent 3,317,801.

1262. SHEWCHUN, J. (1966). Phonon spectroscopy of Ge–Si tunnel heterojunctions. *Phys. Rev.* **141**, 775.

1263. SHEWCHUN, J., and WEI, L. Y. (1964). Germanium–silicon alloy heterojunctions. *J. Electrochem. Soc.* **111**, 1145.

1264. SHIER, J. S. (1970). Ohmic contacts to silicon carbide. *J. Appl. Phys.* **41**, 771.

1265. SHIH, K. K. (1970). Epitaxial growth of GaAs$_x$P$_{1-x}$ from the liquid phase. *J. Electrochem. Soc.* **117**, 387.

1266. SHIH, K. K., PETTIT, G. D., and LORENZ, M. R. (1968). Electroluminescence from Ge-doped GaP *p–n* junctions. *J. Appl. Phys.* **39**, 1557.

1267. SHINODA, D. (1968). Ohmic contacts to silicon using evaporated metal silicides. *Electrochem. Soc. Extended Abstr., 134th Nat. Meeting, Montreal, Fall 1968*, Abstract 502.

1268. SHINODA, D. (1969). Ohmic contacts to silicon using evaporated metal silicides. *In* "Ohmic Contacts to Semiconductors" (B. Schwartz, ed.). Electrochem. Soc., New York.

1269. SHIRAFUJI, J., and NAKAYAMA, M. (1964). GaAs–Ge alloyed junction. *Jap. J. Appl. Phys.* **3**, 801.

1270. SHOCKLEY, W. (1939). On the surface states associated with a periodic potential. *Phys. Rev.* **56**, 317.

1271. SHOCKLEY, W. (1950). "Electrons and Holes in Semiconductors." Van Nostrand–Reinhold, Princeton, New Jersey.

1272. SHOCKLEY, W. (1951). Circuit Element Using Semiconductor Material. U.S. Patent 2,569,347.

1273. SHOCKLEY, W. (1953). Dislocations and edge states in the diamond crystal structure. *Phys. Rev.* **91**, 228.

1274. SHOCKLEY, W., and PRIM, R. C. (1953). Space-charge limited emission in semiconductors. *Phys. Rev.* **90**, 753.

1275. SHOCKLEY, W., and READ, W. T., JR. (1952). Statistics of the recombination of holes and electrons. *Phys. Rev.* **87**, 835.

1276. SHULMAN, R. G., and McMAHON, M. E. (1953). Recovery current in germanium *p–n* junction diodes. *J. Appl. Phys.* **24**, 1257.

1277. SHUMKA, A. (1969). A germanium solid-state triode. *J. Appl. Phys.* **40**, 438.

1278. SIGMUND, H. (1970). Photon-assisted tunneling in Ge–GaAs heterojunctions. *Proc. Int. Conf. Phys. Chem. Semicond. Heterojunctions Layer Structures, Budapest, 1970* **2**, 435. Hung. Acad. Sciences, Budapest, Hungary.

1279. SIHVONEN, Y. T., and BOYD, D. R. (1958). Ohmic probe contacts to CdS crystals. *J. Appl. Phys.* **29**, 1143.

1280. SIHVONEN, Y. T., and BOYD, D. R. (1960). Transparent indium contacts to CdS. *Rev. Sci. Instrum.* **31**, 992.

1281. SIMMONS, J. G. (1963a). Electric tunnel effect between dissimilar electrodes separated by a thin insulating film. *J. Appl. Phys.* **34**, 2581.

1282. SIMMONS, J. G. (1963b). Generalized formula for the electric tunnel effect between similar electrodes separated by a thin insulating film. *J. Appl. Phys.* **34**, 1793.

1283. SIMMONS, J. G. (1965). The electric-tunnel effect and its use in determining properties of surface oxides. *Trans. AIME* **233**, 485.

1284. SIMMONS, J. G. (1967). Poole–Frenkel effect and Schottky effect in metal–insulator–metal systems. *Phys. Rev.* **155**, 657.

1285. SIMON, R. E., and SPICER, W. E. (1960). Photoemission from Si induced by an internal electric field. *Phys. Rev.* **119**, 621.

1286. SIMON, R. E., and WILLIAMS, B. F. (1968). Secondary-electron emission. *IEEE Trans. Nucl. Sci.* **NS–15**, 167.

1287. SIMON, R. E., SOMMER, A. H., TIETJEN, J. J., and WILLIAMS, B. F. (1968). New high-gain dynode for photomultipliers. *Appl. Phys. Lett.* **13**, 355.

1288. SIMON, R. E., SOMMER, A. H., TIETJEN, J. J., and WILLIAMS, B. F. (1969). $GaAs_{1-x}P_x$ as a new high quantum yield photoemissive material for the visible spectrum. *Appl. Phys. Lett.* **15**, 43.

1289. SIMPSON, J., and ARMSTRONG, H. L. (1953). High inverse voltage germanium rectifiers. *J. Appl. Phys.* **24**, 25.

1290. SINGH, H. P., and DAYAL, B. (1967). X-ray determination of the thermal expansion of ZnSe. *Phys. Status Solidi* **23**, K93.

1291. SIRTL, E. (1969). Thermodynamics of silicon vapor deposition. *J. Electrochem. Soc.* **116**, 148c.

1292. SLEGER, K. (1971). ZnSe–GaAs Heterojunction Transistors. Ph.D. Thesis, Carnegie-Mellon Univ.,

1293. SLEGER, K., MILNES, A. G., and FEUCHT, D. L. (1970). ZnSe–GaAs and ZnSe–Ge heterojunction transistors. *Proc. Int. Conf. Phys. Chem. Semicond. Heterojunctions Layer Structures, Budapest, 1970* **1**, 73. Hung. Acad. Sciences, Budapest, Hungary.

1294. SMITH, B. L. (1969). Near ideal Au–GaP Schottky diodes. *J. Appl. Phys.* **40**, 4675.

1295. SMITH, R. A. (1959). "Semiconductors." Cambridge Univ. Press, London and New York.

1296. SMITH, R. A. (1961). "Wave Mechanics of Crystalline Solids." Chapman & Hall, London.

1297. SMITH, R. T. J. (1969). Evidence for a native donor in ZnSe from high temperature electrical measurements. *Solid State Commun.* **7**, 1757.

1298. SMITH, R. W. (1955a). Properties of ohmic contacts to cadmium sulfide single crystals. *Phys. Rev.* **97**, 1525.

1299. SMITH, R. W. (1955b). A mathematical analysis of solute redistribution during solidification. *Can. J. Phys.* **33**, 723.

1300. SOKOLOVA, E. B. (1970). Cascade capture of carriers by a linear dislocation. *Sov. Phys.–Semicond.* **3**, 1266.

1301. SOMMER, A. H. (1968). "Photoemissive Materials." Wiley, New York.

1302. SOMMER, A. H., and SPICER, W. E. (1965). Photoelectric emission. *In* "Photoelectronic Materials and Devices" (S. Larach, ed.), Chap. 4. Van Nostrand–Reinhold, Princeton, New Jersey.

1303. SOMMER, A. H., WHITAKER, H. H., and WILLIAMS, B. F. (1970). Thickness of Cs and Cs–O films on GaAs(Cs) and GaAs(Cs–O) photocathodes. *Appl. Phys. Lett.* **17**, 273.

1304. SOMMERS, H. S., JR. (1960). Solid-State Charge Carrier Valve. U.S. Patent 2,961,475.

1305. SONNENBERG, H. (1969a). Low-work-function surfaces for negative-electron-affinity photoemitters. *Appl. Phys. Lett.* **14**, 289.

1306. SONNENBERG, H. (1969b). Effect of acceptor density on photoemission from p–GaAs–Cs and –Cs_2O. *J. Appl. Phys.* **40**, 3414.

1307. SONNENBERG, H. (1970). InAsP–Cs_2O, a high efficiency infrared-photocathode. *Appl. Phys. Lett.* **16**, 245.

1308. SORENSEN, H. O. (1966). Quantitative comparison of microwave solid state detectors. *IEEE Trans. Microwave Theory Tech.* **14**, 588.

1309. SOSNOWSKI, L. (1959). Electronic processes at grain boundaries. *Phys. Chem. Solids* **8**, 142.

1310. SOSNOWSKI, L. (1970). Theory of high photovoltages in semiconducting films. *Proc. Int. Conf. Phys. Chem. Semicond. Heterojunctions Layer Structures, Budapest, 1970* **4**, 29. Hung. Acad. Sciences, Budapest, Hungary.

1311. SPENCE, W., GUITERREZ, W. A., and WILSON, H. L. (1966). A CdS–ZnS thin film rectifier for operation at temperatures up to 500°C. *Proc. IEEE* **54**, 294.

1312. SPENKE, E. (1958). "Electronic Semiconductors." McGraw-Hill, New York.

1313. SPITZER, W. G., and FAN, H. Y. (1957). Determination of optical constants and carrier effective mass of semiconductors. *Phys. Rev.* **106**, 882.

1314. SPITZER, W. G., and MEAD, C. A. (1963). Barrier height studies on metal–semiconductor systems. *J. Appl. Phys.* **34**, 3061.

1315. SPITZER, W. G., and MEAD, C. A. (1964a). Conduction band minima of $Ga(As_{1-x}P_x)$. *Phys. Rev.* **133**, A872.

1316. SPITZER, W. G., and MEAD, C. A. (1964b). Fermi level position at metal–semiconductor interfaces. *Phys. Rev.* **134**, A713.

1317. SPITZER, W. G., and WHELAN, J. M. (1959). Infrared absorption and electron effective mass in n-type gallium arsenide. *Phys. Rev.* **114**, 59.

1318. SPITZER, W. G., CROWELL, C. R., and ATALLA, M. M. (1962). Mean free path of photo excited electrons in Au. *Phys. Rev. Lett.* **8**, 57.

1319. SPRATT, J. P. (1962). A tunnel emission device. *Proc. IRE* **50**, 467.

1320. SPRATT, J. P., SCHWARTZ, R. F., and KANE, W. M. (1961). Hot electrons in metal films: Injection and collection. *Phys. Rev. Lett.* **6**, 341.

1321. SPRINGGATE, W. F. (1969). Investigation into the mechanism of degradation of solar cells with silver–titanium contacts. *In* "Ohmic Contacts to Semiconductors" (B. Schwartz, ed.). Electrochem. Soc., New York.

1322. SREEDHAR, A. K. (1969). Efficiency calculations of heterojunction solar energy converters. *IEEE Trans. Electron Devices* **ED–16**, 309.

1323. STEINBERG, R. F., and SCRUGGS, D. M. (1966). Preparation of epitaxial GaAs films by vacuum evaporation of the elements. *J. Appl. Phys.* **37**, 4586.

1324. STEINRISSER, F., DAVIS, L. C., and DUKE, C. B. (1968). Electron and phonon tunneling spectroscopy in metal–germanium contacts. *Phys. Rev.* **176**, 912.

1325. STERN, R. (1966). Stimulated emission in semiconductors. *In* "Semiconductors and Semimetals" (R. K. Willardson and A. C. Beer, eds.), Vol. 2. Academic Press, New York.

1326. STEVENSON, J. R., and HENSLEY, E. B. (1961). Thermionic and photoelectric emission from magnesium oxide. *J. Appl. Phys.* **32**, 166.

1327. STRAIN, R. J. (1968). Designing transparent contacts for M-I-S lamps. *Electrochem. Soc. Extended Abstr., 134th Nat. Meeting, Montreal, Fall 1968*, Abstract No. 500.

1328. STRATTON, R. (1962a). Volt–current characteristics for tunneling through insulating films. *Phys. Chem. Solids* **23**, 1177.

1329. STRATTON, R. (1962b). Theory of field emission from semiconductors. *Phys. Rev.* **125**, 67.

1330. STRATTON, R. (1962c). Diffusion of hot and cold electrons in semiconductor barriers. *Phys. Rev.* **126**, 2002.

1331. STRATTON, R. (1964). Energy distribution of field emitted electrons. *Phys. Rev.* **135**, A794.

1332. STRATTON, R. (1965). On approximate thermionic and field emission equations. *Solid-State Electron.* **8**, 175.

1333. STRATTON, R., and PADOVANI, F. A. (1967). Differential resistance peaks of Schottky barrier diodes. *Solid-State Electron.* **10**, 813.

1334. STRATTON, R., and PADOVANI, F. A. (1968). Influence of ellipsoidal energy surfaces on the differential resistance of Schottky barriers. *Phys. Rev.* **175**, 1072.

1335. STRAUMANIS, M. E., and KIM, C. D. (1965). Phase extent of gallium arsenide determined by the lattice constant and density method. *Acta Crystallogr.* **19**, 256.

1336. STREHLOW, W. H. (1969). Chemical polishing of II–VI compounds. *J. Appl. Phys.* **40**, 2928.

1337. STREHLOW, W. H. (1970). Study of the morphology of epitaxial CdS films. *J. Appl. Phys.* **41**, 1810.

1338. STREHLOW, W. H., and COOK, E. L. (1969). Epitaxy of CdS on SrF_2. *Phys. Rev.* **188**, 1256.

1339. STRINGFELLOW, G. B., and BUBE, R. H. (1967). Photoelectric properties of p type ZnSe: Cu crystals. *In* "II–VI Semiconducting Compounds" (D. G. Thomas, ed.). Benjamin, New York.

1340. STRINGFELLOW, G. B., and GREENE, P. E. (1969). Dislocations in $GaAs_{1-x}P_x$. *J. Appl. Phys.* **40**, 502.

1341. STRINGFELLOW, G. B., and GREENE, P. E. (1970). Steady-state solution growth of InAsSb alloys. *J. Electrochem. Soc.* **117**, 95c.

1342. SUBASHIEV, V. K. (1970). The quasielectrical field in a graded heterojunction. *Abstr. Int. Conf. Phys. Chem. Semicond. Heterojunctions Layer Structures, Budapest, 1970*, p. 73. Hung. Acad. Sciences, Budapest, Hungary.

1343. SUGANO, T., MORINO, A., KOSHIGA, F., MISHIMA, K., NISHI, T., and MATSUDA, S. (1967). Characteristics of contacts between silicon and a few kinds of metals. *J. Inst. Electron. Comm. Eng. Jap.* **50**, 1045.

1344. SUGANO, T., CHOU, H. K., YOSHIDA, M., and NISHI, T. (1968). Chemical deposition of Mo on Si. *Jap. J. Appl. Phys.* **7**, 1028.

1345. SULLIVAN, M. V., and EIGLER, J. H. (1956). Five metal hydrides as alloying agents on silicon. *J. Electrochem. Soc.* **103**, 218.

1346. SULLIVAN, M. V., and EIGLER, J. H. (1957). Electroless nickel plating for making ohmic contacts to silicon. *J. Electrochem. Soc.* **104**, 226.

1347. SULLIVAN, M. V., and KOLB, G. A. (1963). Chemical polishing of gallium arsenide in bromine-methanol. *J. Electrochem. Soc.* **110**, 585.

1348. Suzuki, H. (1966). Electron affinity of semiconducting compound CdSe. *Jap. J. Appl. Phys.* **5**, 1253.

1349. Swank, R. K. (1967). Surface properties of II–VI compounds. *Phys. Rev.* **153**,844.

1350. Swank, R. K., Aven, M., and Devine, J. Z. (1969). Barrier heights and contact properties of *n*-type ZnSe crystals. *J. Appl. Phys.* **40**, 89.

1351. Swanson, J. A. (1954). Diode theory in the light of hole injection. *J. Appl. Phys.* **25**, 314.

1352. Sze, S. M. (1969). "Physics of Semiconductor Devices," Chap. 8. Wiley, New York.

1353. Sze, S. M., and Gibbons, G. (1966). Avalanche breakdown voltages of abrupt and linearly graded *p–n* junctions in Ge, Si, GaAs and GaP. *Appl. Phys. Lett.* **8**, 111.

1354. Sze, S. M., and Gummel, H. K. (1966). Appraisal of semiconductor–metal–semi-conductor transistor. *Solid-State Electron.* **9**, 751.

1355. Sze, S. M., Crowell, C. R., and Kahng, D. (1964). Photoelectric determination of the image force dielectric constant for hot electrons in Schottky barriers. *J. Appl. Phys.* **35**, 2534.

1356. Sze, S. M. Crowell, C. R., Carey, G. P., and LaBate, E. E. (1966). Hot-electron transport in semiconductor–metal–semiconductor structures. *J. Appl. Phys.* **37**, 2690.

1357. Szekely, C., Bertoti, I., Stubnya, G., Varga, L., and Gutai, L. (1970). Preparation, structure and some properties of Si–Ge and Si–Si$_{1-x}$Ge$_x$ heterojunctions. *Proc. Int. Conf. Phys. Chem. Semicond. Heterojunctions Layer Structures, Budapest, 1970* **1**, 341. Hung. Acad. Sciences, Budapest, Hungary.

1358. Szentpali, B., Varga, L., Bertoti, I., and Szekely, C. (1970). Defect structure in heterojunctions: GaAs–Ge. *Proc. Int. Conf. Phys. Chem. Semicond. Heterojunctions Layer Structures, Budapest, 1970* **1**, 349. Hung. Acad. Sciences, Budapest, Hungary.

1359. Szydlo, N., Poirier, R., and Kleefstra, M. (1970). Reverse I–V characteristics of the Na–Si Schottky barrier. *Appl. Phys. Lett.* **17**, 477.

1360. Tada, K., and Laraya, J. J. R. (1967). Reduction of the storage time of a transistor using a Schottky-barrier diode. *Proc. IEEE* **55**, 2064.

1361. Takabayashi, M. (1962). Epitaxial vapor growth of single crystal Ge. *Jap. J. Appl. Phys.* **1**, 22.

1362. Takahashi, K. (1969). General solutions of photovoltaic effects in semiconductor junctions. *Int. J. Electron.* **26**, 253.

1363. Takahashi, K. (1970). General solutions of photovoltaic effects in semiconductor junctions. *Proc. Int. Conf. Phys. Chem. Semicond. Heterojunctions Layer Structures, Budapest, 1970* **2**, 455. Hung. Acad. Sciences, Budapest, Hungary.

1364. Takahashi, K., Baker, W. D., and Milnes, A. G. (1969). ZnTe–InAs heterojunctions. *Int. J. Electron.* **27**, No. 4, 383.

1365. Takeda, Y., Hirai, T., and Hirao, M. (1965). Phase diagram for the pseudo-binary system germanium and gallium arsenide. *J. Electrochem. Soc.* **112**, 363.

1366. Tallman, R. L., Chu, T. L., Gruber, G. A., Oberly, J. J., and Wolley, E. D. (1966). Epitaxial growth of Si on hexagonal SiC. *J. Appl. Phys.* **37**, 1588.

1367. Taneja, N. D. (1971). Anisotype heterojunctions involving wide band gap II–VI semiconductors. Ph.D. Thesis, Univ. British Columbia, Canada.

1368. Tansley, T. L. (1966a). Heterojunction boundary conditions. *J. Appl. Phys.* **37**, 3908.

1369. Tansley, T. L. (1966b). Heterojunctions applied. *New Sci.* **31**, 316.

1370. Tansley, T. L. (1966c). Forward bias current–voltage characteristics for a hetero-junction in which tunneling dominates. *Phys. Status Solidi* **18**, 105.

1371. Tansley, T. L. (1970). Forward bias conduction properties of anisotype hetero-junctions. *Proc. Int. Conf. Phys. Chem. Semicond. Heterojunctions Layer Structures, Budapest, 1970* **2**, 123. Hung. Acad. Sciences, Budapest, Hungary.

1372. TANSLEY, T. L., and NEWMAN, P. C. (1967). Measurements of heterojunctions alloyed on to GaAs. *Solid-State Electron.* **10**, 497.

1373. TAO, T. F., and HSAI, Y. (1968). Dopant dependency of fermi level location in heavily doped silicon. *Appl. Phys. Lett.* **13**, 291.

1374. TAUC, J. (1957). Generation of emf in semiconductors. *Rev. Mod. Phys.* **28**, 307.

1375. TAYLOR, N. J. (1969). The role of Auger electron spectroscopy in surface elemental analysis. *Vacuum* **19**, 575–578.

1376. TAYLOR, R. C. (1969). The effect of substrate orientation on the incorporation of zinc and selenium in vapor-grown gallium phosphide. *J. Electrochem. Soc.* **116**, 383.

1377. TAYLOR, R. C., WOODS, J. F., and LORENZ, M. R. (1968). Electrical and optical properties of vapor-grown GaP. *J. Appl. Phys.* **39**, 5404.

1378. TAYLOR, T. C. (1965). Device Composed of Different Semiconductive Materials. U.S. Patent 3,176,204.

1379. TAYLOR, W. E., ODELL, N. H., and FAN, H. Y. (1952). Grain boundaries in germanium. *Phys. Rev.* **88**, 867.

1380. TERAMOTO, I., IWASA, H., and TAI, H. (1968). Contact resistance of electroless nickel on silicon. *J. Electrochem. Soc.* **115**, 912.

1381. TERMAN, L. M. (1962). An investigation of surface states at a silicon/silicon oxide interface employing metal-oxide–silicon diodes. *Solid-State Electron.* **5**, 285.

1382. TERRY, L. E. (1968). Metal–silicon contacts and contact resistance. *Trans. Electrochem. Soc. Meeting, Montreal, October 1968,* Abstract No. 504.

1383. TEWINKEL, J. (1959). Drift transistor—simplified electrical characterization. *Electron. Radio Eng.* **36**, 280.

1384. THEURERER, H. C. (1961). Epitaxial silicon films by the hydrogen reduction of SiCl₄. *J. Electrochem. Soc.* **109**, 301c.

1385. THEURERER, H. C., and CHRISTENSEN, H. (1960). Epitaxial films of silicon and germanium by halide reduction. *J. Electrochem. Soc.* **107**, 268c.

1386. THOMAS, D. E., and MOLL, J. L. (1958). Junction transistor short circuit current gain and phase determination. *Proc. IRE* **46**, 1177.

1387. THOMAS, P., and QUEISSER, H. J. (1968). Electron–phonon coupling in the barriers of GaAs Schottky diodes. *Phys. Rev.* **175**, 983.

1388. THOMAS, R. N., and FRANCOMBE, M. H. (1968). Low temperature epitaxial growth of *PN* junctions by UHV sublimation. *Appl. Phys. Lett.* **13**, 270.

1389. THOMAS, R. W. (1969). Growth of single crystal GaP from organometallic sources. *J. Electrochem. Soc.* **116**, 1449.

1390. THOMPSON, H. W., JR., and REENSTRA, A. L. (1967). Electrical properties of the interface region of the Ge–Si alloyed heterojunctions. *J. Appl. Phys.* **38**, 4739.

1391. THURMOND, C. D. (1965). Phase equilibria in the GaAs and the GaP systems. *Phys. Chem. Solids* **26**, 785.

1392. THURMOND, C. D., and FROSCH, C. J. (1962). Thermodynamic properties of GaP and GaAs. *J. Electrochem. Soc.* **109**, 301c.

1393. THURMOND, C. D., and KOWALCHIK, M. (1960). Germanium and silicon liquidus curves. *Bell Syst. Tech. J.* **39**, 169.

1394. TIETJEN, J. J., and AMICK, J. A. (1966). The preparation and properties of vapor-deposited epitaxial $GaAs_{1-x}P_x$ using arsine and phosphine. *J. Electrochem. Soc.* **113**, 724.

1395. TIETJEN, J. J., MARUSKA, H. P., and CLOUGH, R. B. (1969). The preparation and properties of vapor-deposited epitaxial $InAs_{1-x}P_x$ using arsine and phosphine. *J. Electrochem. Soc.* **116**, 492.

1396. TIETJEN, J. J., ENSTROM, R. E., and RICHMAN, D. (1970). Vapor-phase growth of several III–V compound semiconductors. *RCA Rev.* **31**, 635.

1397. TIHANYI, J. (1970). Investigations on Al–Mo–SiO$_2$–Si systems. *Abstr. Int. Conf. Phys. Chem. Semicond. Heterojunctions Layer Structures, Budapest, 1970,* p. 160. Hung. Acad. Sciences, Budapest, Hungary.

1398. TILLER, W. A. (1958). Production of dislocations during growth from the melt. *J. Appl. Phys.* **29**, 611.

1399. TOTTA, P. A., and SOPHER, R. P. (1969). SLT device metallurgy and its monolithic extension. *IBM J. Res. Develop.* **13**, 226.

1400. TOUSEK, JANA, and TOUSEK, J. (1970). On the rectifying contacts of Au–CdTe structures. *Proc. Int. Conf. Phys. Chem. Semicond. Heterojunctions Layer Structures, Budapest, 1970* **5**, 363. Hung. Acad. Sciences, Budapest, Hungary.

1401. TRAMPOSCH, R. F. (1969). Epitaxial films of germanium deposited on sapphire via chemical vapor transport. *J. Electrochem. Soc.* **116**, 654.

1402. TRUMBORE, F. A., and TARTAGLIA, A. A. (1958). Resistivities and hole mobilities in very heavily doped Ge. *J. Appl. Phys.* **29**, 1511.

1403. TRUMBORE, F. A., KOWALCHIK, M., and WHITE, H. G. (1967). Efficient electroluminescence in GaP p–n junctions grown by liquid-phase epitaxy on vapor-grown substrates. *J. Appl. Phys.* **38**, 1987.

1404. TSUI, D. C. (1969). Surface plasmon excitation by tunneling electrons in GaAs–Pb tunnel junctions. *Bull. Amer. Phys. Soc.* **14**, 414.

1405. TSUJIMOTO, Y., NAKAJIMA, T., ONODERA, Y., and TUKAI, M. (1967). Preparation of ZnSe$_x$Te$_{1-x}$ crystal by vapor transport. *Jap. J. Appl. Phys.* **6**, 1014.

1406. TUBOTA, H. (1963). Electrical properties of A^{II}B^{VI} compounds, CdSe and ZnTe. *Jap. J. Appl. Phys.* **2**, 259.

1407. TURNBULL, A. A., and EVANS, G. B. (1968). Photoemission from GaAs–Cs–O. *Brit. J. Appl. Phys. (J. Phys. D)* **1**, 155.

1408. TURNER, D. R. (1959). Electroplating metal contacts on germanium and silicon. *J. Electrochem. Soc.* **106**, 786.

1409. TURNER, D. R., and SAUER, H. A. (1960a). Ohmic contacts to semiconductor ceramics. *J. Electrochem. Soc.* **107**, 250.

1410. TURNER, D. R., and SAUER, H. A. (1960b). Ohmic contacts to semiconductors. *J. Electrochem. Soc.* **107**, 251.

1411. TURNER, M. J., and RHODERICK, E. H. (1968). Metal–silicon Schottky barriers. *Solid-State Electron.* **11**, 291.

1412. UEBBING, J. J. (1970). Use of Auger electron spectroscopy in determining the effect of carbon and other surface contaminants on GaAs–Cs–O photocathodes. *J. Appl. Phys.* **41**, 802.

1413. UEBBING, J. J., and BELL, R. L. (1967). Cesium–GaAs Schottky barrier height. *Appl. Phys. Lett.* **11**, 357.

1414. UEBBING, J. J., and BELL, R. L. (1968). Improved photoemitters using GaAs and InGaAs. *Proc. IEEE* **56**, 1624.

1415. UNTERKOFLER, G. (1963). Theoretical curves of tunnel resistivity vs. voltage. *J. Appl. Phys.* **34**, 3143.

1416. UNVALA, B. A. (1962). Epitaxial growth of Si by vacuum evaporation. *Nature (London)* **194**, 966.

1417. UNVALA, B. A. (1963). Single crystal growth of silicon by vacuum deposition. *Vide* **104**, 109.

1418. UNVALA, B. A. (1964). Growth of epitaxial silicon layers by vacuum evaporation. *Phil. Mag.* **9**, 691.

1419. VALDES, L. B. (1961). "The Physical Theory of Transistors." McGraw-Hill, New York.

1420. VALENTA, M. W., and RAMASASATRY, C. (1957). Effect of heavy doping on the self-diffusion of germanium. *Phys. Rev.* **106**, 73.

1421. VALLESE, L. M. (1970). Get laser data on coherence and radiance. *Microwaves*, **9**, No. 11, 45.

1422. VALOV, YU. A., GORYUNOVA, N. A., LEONOV, E. I., and ORLOV, V. M. (1970). Preparation and properties of some ternary compound $A^{II}B^{IV}C^V$ layers and of heterojunctions based on them. *Abstr. Int. Conf. Phys. Chem. Semicond. Heterojunctions Layer Structures, Budapest, 1970*, p. 135. Hung. Acad. Sciences, Budapest, Hungary.

1423. VAN DER MERWE, J. H. (1963). Crystal interfaces, part II, finite overgrowths. *J. Appl. Phys.* **34**, 123.

1424. VAN DER MERWE, J. H. (1964). Interfacial misfit and bonding between oriented films and their substrates. *In* "Single Crystal Films" (M. H. Francombe and H. Sato, eds.) p. 139. Macmillan, New York.

1425. VAN DER PAUW, L. J. (1958a). A method of measuring the resistivity and Hall coefficient on lamellae of arbitrary shape. *Philips Tech. Rev.* **20**, 220.

1426. VAN DER PAUW, L. J. (1958b). A method of measuring specific resistivity and Hall effect of discs of arbitrary shape. *Philips Res. Rep.* **13**, 1.

1427. VAN DER ZIEL, A. (1968). "Solid State Electronics," 2nd ed. Prentice-Hall, Englewood Cliffs, New Jersey.

1428. VAN DE WIELE, F., and VAN OVERSTRAETEN, R. (1967). Influence of interface states on the capacitance of homo- and heterojunctions. *Electron. Lett.* **3**, 218.

1429. VAN HOOK, H. J., and LENKER, E. S. (1963). The system InAs–GaAs. *Trans. AIME* **227**, 220.

1430. VAN LAAR, J., and SCHEER, J. J. (1962a). Influence of band bending on photoelectric emission from silicon single crystals. *Philips Res. Rep.* **17**, 101.

1431. VAN LAAR, J., and SCHEER, J. J. (1962b). The influence of band bending on photoelectric emission of silicon. *Proc. Int. Conf. Phys. Semicond., Exeter, July 1962*. Inst. of Phys. and the Phys. Soc. London.

1432. VAN OPDORP, C. (1969). Si–Ge Iso-type Heterojunctions. Thesis, Technical Univ., Eindhoven, October 1969. (Also *Philips Res. Rep. Suppl.* No. 10.)

1433. VAN OPDORP, C. (1970). The present state of the theory of photoeffects in heterojunctions. *Proc. Int. Conf. Phys. Chem. Semicond. Heterojunctions Layer Structures, Budapest, 1970* **2**, 191. Hung. Acad. Science, Budapest, Hungary.

1434. VAN OPDORP, C., and KANERVA, H. K. J. (1967). Current–voltage characteristics and capacitance of isotype heterojunctions. *Solid-State Electron.* **10**, 401.

1435. VAN OPDORP, C., and VRAKKING, J. (1967). Photo-effects in isotype heterojunctions. *Solid-State Electron.* **10**, 955.

1436. VAN OVERSTRAETEN, R. J., and NUYTS, W. (1969). Theoretical investigation of the efficiency of drift-field solar cells. *IEEE Trans. Electron Devices* **ED–16**, 632.

1437. VAN OVERSTRAETEN, R. J., and VAN DE WIELE, F. (1968). Interpretation for homojunctions of the difference between the theoretical and the experimental built-in potential determined from capacitance measurements, using interface states. *IEEE Trans. Electron Devices* **ED–15**, 164.

1438. VAN RUYVEN, L. J. (1964). The position of the Fermi level at a heterojunction interface. *Phys. Status Solidi* **5**, K109.

1439. VAN RUYVEN, L. J., and DEKKER, W. (1962). Epitaxial growth of Ge on GaP. *Physica* **28**, 307.

1440. VAN RUYVEN, L. J., and DEV, I. (1966). Position dependent edge emission from zinc–cadmium sulphide graded heterojunctions. *J. Appl. Phys.* **37**, 3324.

1441. VAN RUYVEN, L. J., and WILLIAMS, F. E. (1967a). Anti-Stokes' light converter based on graded-band-gap semiconductors. *Solid-State Electron.* **10**, 1159.

1442. VAN RUYVEN, L. J., and WILLIAMS, F. E. (1967b). Electronic transport in graded-band-gap semiconductors. *Amer. J. Phys.* **35**, 705.

1443. VAN RUYVEN, L. J., PAPENHUIJZEN, J. M. P., and VERHOEVEN, A. C. J. (1965). Optical phenomena in Ge–GaP heterojunctions. *Solid-State Electron.* **8**, 631.

1444. VANYUKOV, A. V., KIREEV, P. S., FIGUROVSKII, E. N., KOROVIN, A. P., and VISHNYAKOVA, Z. P. (1970). Properties of *n–p* CdSe–ZnTe heterojunctions. *Sov. Phys.–Semicond.* **3**, 1297.

1445. VARSHNI, Y. P. (1967). Band-to-band recombination in groups IV, VI, and III–V semiconductors. *Phys. Status Solidi* **20**, 9.

1446. VASICEK, A. (1960). "Optics of Thin Films." North-Holland Publ., Amsterdam.

1447. VASIL'EV, V. D., and TIKHONOVA, A. A. (1967). Formation temperatures for evaporated epitaxial germanium films. *Sov. Phys.–Crystallogr.* **11**, 575.

1448. VIVILOV, V. S., PLOTNIKOV, A. F., and SHUBIN, V. E. (1970). Possibility of using InSb-based MOS structures as image convertors. *Sov. Phys.–Semicond.* **4**, 501.

1449. VEHSE, R. C. (1970). A floating seed technique for large area liquid phase epitaxy of GaP. *Electrochem. Soc. Extended Abstr., 137th Nat. Meeting, Los Angeles, Spring 1970,* p. 196.

1450. VEINGER, A. I., IVANOV, V. G., PARITSKII, L. G., and RYVKIN, S. M. (1969). Exclusion of nonequilibrium hot carriers in germanium and silicon. *Sov. Phys.–Semicond.* **3**, 476.

1451. VERMILYEA, D. A. (1963). Flaws in anodic Ta$_2$O$_5$ films. *J. Electrochem. Soc.* **110**, 250.

1452. VIELAND, L. J., and SEIDEL, T. (1962). Behavior of germanium in gallium arsenide. *J. Appl. Phys.* **33**, 2414.

1453. VIJH, A. K. (1970). Chemical approaches to the approximate prediction of band gaps of binary semiconductors and insulators. *J. Electrochem. Soc.* **117**, 173c.

1454. VILMS, J., and WANDINGER, L. (1968). Theory of contact resistance for one type of ohmic contact. *Trans. Electrochem. Soc. Meeting, October* 1968, Abstract No. 495.

1455. VILMS, J., and WANDINGER, L. (1969). Theory of contact resistance for one type of ohmic contact. *In* "Ohmic Contacts to Semiconductors" (B. Schwartz, ed.). Electrochem. Soc., New York.

1456. VOHL, P., PERKINS, D. M., ELLIS, S. G., ADDISS, R. R., HUI, W., and NOEL, G. (1967). GaAs thin-film solar cells. *IEEE Trans. Electron Devices* **ED–14**, 26.

1457. VOLZHENSKII, D. S., and PASHKOVSKII, M. V. (1966). Fusing point ohmic contacts in an inert atmosphere. *Instrum. Exp. Tech. (USSR)* **2**, 495.

1458. VORONKOV, V. V., VORONKOVA, G. I., and IGLITSYN, M. (1970). Determination, using the Hall effect, of the parameters of energy levels at low level density. *Sov. Phys.–Semicond.* **3**, 1449.

1459. VORONKOVA, N. M., and NASLEDOV, D. N. (1965). Photomagnetic effect and photoconductivity of *n*-type GaAs. *Sov. Phys.–Solid State* **6**, 1736.

1460. VUILLOD, J., and CHAKRAVERTY, B. K. (1970). Electronic properties of thin films of ZnO. *Proc. Int. Conf. Phys. Chem. Semicond. Heterojunctions Layer Structures, Budapest, 1970* **4**, 243. Hung. Acad. Sciences, Budapest, Hungary.

1461. WAGNER, J. W., and THOMPSON, A. G. (1970). Preparation and properties of InAs$_{1-x}$P$_x$ alloys. *Electrochem. Soc. Extended Abstr., 137th Nat. Meeting, Los Angeles, Spring 1970,* p. 213.

1462. WAGNER, R. S., and ELLIS, W. C. (1964). Vapor–liquid–solid mechanism of single crystal growth. *Appl. Phys. Lett.* **4**, 89.

1463. WAIT, J. T., and HAUSER, J. R. (1968). Beta falloff in transistors at high collector current. *Proc. IEEE* **56**, 2087.

1464. WALKER, W. C., and LAMBERT, E. Y. (1957). Ohmic and rectifying contacts to semi-conducting CdS crystals. *J. Appl. Phys.* **28**, 635.

1465. WALTER, J. P., and COHEN, M. L. (1970). Calculated and measured reflectivity of ZnTe and ZnSe. *Phys. Rev. B* **1**, 2661.

1466. WALTZ, M. C. (1955). Electrical contact for transistors and diodes. *Bell Lab. Rec.* **33**, 260.

1467. WANG, S. (1963). Proposal for a two-stage semiconductor laser through tunneling and injection. *J. Appl. Phys.* **34**, 3443.

1468. WANG, S., and TSENG, C. C. (1964). Considerations regarding the use of semiconductor heterojunctions for laser operation. *Proc. IEEE* **52**, 426.

1469. WATANABE, N. (1965). Reverse biased electroluminescence in alloyed ZnTe diodes. *Jap. J. Appl. Phys.* **4**, 343.

1470. WATANABE, N. (1966). Forward and reverse biased electroluminescence in alloyed ZnTe diodes. *Jap. J. Appl. Phys.* **5**, 12.

1471. WATANABE, N., USUI, S., and KANAI, Y. (1964). Injection luminescence in ZnTe diodes. *Jap. J. Appl. Phys.* **3**, 427.

1472. WATANABE, S., and MITA, Y. (1969). CdS–PbS heterojunctions. *J. Electrochem. Soc.* **116**, 989.

1473. WATSON, H. A. (1969). *In* "Microwave Semiconductor Devices and their Circuit Applications," (L. E. Miller, ed.), Chap. 17. McGraw-Hill, New York.

1474. WEBER, E. H., and NIENEROWSKI, K. (1970). Influence of volume traps on the capacity at the CdSe–SiO$_x$ interface. *Proc. Int. Conf. Phys. Chem. Semicond. Heterojunctions Layer Structures, Budapest, 1970* **5**, 213. Hung. Acad. Sciences, Budapest, Hungary.

1475. WEI, I. Y., and SCHEWCHUN, J. (1963). Negative resistance and hysteresis in a Ge–Si heterojunction. *Proc. IEEE* **51**, 946.

1476. WEIMER, P. K. (1966). Tellurium Thin Film Field Effect Solid State Electrical Devices. U.S. Patent 3,290,569.

1477. WEINREICH, O., MATARE, H., and REED, B. (1959). The grain boundary amplifier. *Proc. Phys. Soc. (London)* **73**, 969.

1478. WEINREICH, O., DERMIT, G., and TUFTS, C. (1961). Germanium films on germanium obtained by thermal evaporation in vacuum. *J. Appl. Phys.* **32**, 1170.

1479. WEINSTEIN, M., BELL, R. O., and MENNA, A. A. (1964). Preparation and properties of GaAs–GaP, GaAs–Ge, and GaP–Ge heterojunctions. *J. Electrochem. Soc.* **111**, 674.

1480. WEISBERG, L. R. (1967). Low-temperature vacuum deposition of homoepitaxial silicon. *J. Appl. Phys.* **38**, 4537.

1481. WEMPLE, S. H. (1968). Electrical contact to n and p type ferroelectric oxides. *Trans. Electrochem. Soc. Meeting, October 1968*, Abstract 498.

1482. WEMPLE, S. H., KAHNG, D., and BRAUN, H. J. (1967a). Surface barrier diodes on semiconducting KTaO$_3$. *J. Appl. Phys.* **38**, 353.

1483. WEMPLE, S. H., KAHNG, D., BERGLUND, C. N., and VAN UITERT, L. G. (1967b). Surface barrier junctions on semiconducting ferroelectrics. *J. Appl. Phys.* **38**, 799.

1484. WERNICK, J. H. (1957). Determination of diffusivities in liquid metals by means of temperature-gradient zone melting. *J. Chem. Phys.* **25**, 47.

1485. WERTHEIM, G. K. (1958). Transient recombination of excess carriers in semiconductors. *Phys. Rev.* **109**, 1086.

1486. WHELAN, M. V. (1970). Recombination–generation currents at a thermally oxidized silicon interface. *Proc. Int. Conf. Phys. Chem. Semicond. Heterojunctions Layer Structures, Budapest, 1970* **5**, 221. Hung. Acad. Sciences, Budapest, Hungary.

1487. WHITAKER, J. (1965). Electrical properties of n-type AlAs. *Solid-State Electron.* **8**, 649.

1488. WHITCOMB, D. L. (1968). Refractory metallization systems for high current density and high temperature use in integrated circuits. *Trans. Electrochem. Soc. Meeting, October 1968*, Abstract No. 512.

1489. WHITE, H. G., and LOGAN, R. A. (1963). GaP surface-barrier diodes. *J. Appl. Phys.* **34**, 1990.

1490. WIDMER, H. (1964). Epitaxial growth of Si on Si in ultra high vacuum. *Appl. Phys. Lett.* **5**, 108.

1491. WILLARDSON, R. K., and GOERING, H. L. (eds.) (1962). "Compound Semiconductors," Vol. 1, Preparation of III–V compounds. Van Nostrand–Reinhold, Princeton, New Jersey.

1492. WILLIAMS, B. F. (1969). InGaAs–CsO, a low work function (less than 1.0 eV) photoemitter. *Appl. Phys. Lett.* **14**, 273.

1493. WILLIAMS, B. F., and SIMON, R. E. (1967). Direct measurement of hot electron–phonon interactions in GaP. *Phys. Rev. Lett.* **18**, 485.

1494. WILLIAMS, E. W., and BLACKNALL, D. M. (1967). The observation of defects in GaAs using photoluminescence at 20°K. *Trans. AIME* **239**, 387.

1495. WILLIAMS, J. D., and TERRY, L. E. (1967). Textural and electrical properties of vacuum-deposited germanium films. *J. Electrochem. Soc.* **114**, 158.

1496. WILLIAMS, R. (1962). Photoemission of holes from tin into gallium arsenide. *Phys. Rev. Lett.* **8**, 402.

1497. WILLIAMS, R. (1966). Photoemission of electrons from silicon into silicon dioxide. Effects of ion migration in the oxide. *J. Appl. Phys.* **37**, 1491.

1498. WILLIAMS, R., and BUBE, R. H. (1960). Photoemission in the photovoltaic effect in cadmium sulfide crystals. *J. Appl. Phys.* **31**, 968.

1499. WILLIAMS, R., and WRONSKI, C. R. (1968). Electron emission from the Schottky barrier structure ZnS:Pt:Cs. *Appl. Phys. Lett.* **13**, 231.

1500. WILLIAMS, V. A. (1966). High conductivity transparent contacts to ZnS. *J. Electrochem. Soc.* **113**, No. 3, 234.

1501. WILLIAMS, V. A. (1970). The heterojunction between zinc sulphide and silicon. *Proc. Int. Conf. Phys. Chem. Semicond. Heterojunctions Layer Structures, Budapest, 1970* **1**, 357. Hung. Acad. Sciences, Budapest, Hungary.

1502. WINOGRADOFF, N., and KESSLER, H. K. (1966). Radiative recombination lifetimes in laser-excited silicon. *Appl. Phys. Lett.* **8**, 99.

1503. WOLF, M. (1960). Limitations and possibilities for improvement of photovoltaic solar energy converters. *Proc. IRE* **48**, 1246.

1504. WOLF, M. (1963). Drift fields in photovoltaic solar energy converter cells. *Proc. IEEE* **51**, 674.

1505. WOLF, P. (1970). Microwave properties of Schottky-barrier field-effect transistors *IBM J. Res. Develop.* **14**, 125.

1506. WOLFE, C. M., and LINDLEY, W. T. (1969). Epitaxially grown guard rings for GaAs diodes. *J. Electrochem. Soc.* **116**, 276.

1507. WOLSKY, S. P. (1959). Ohmic Semiconductor Contacts. U.S. Patent 2,879,457.

1508. WOLTER, A. R. (1969). Metal–silicon junctions formed with metal ions. *In* "Ohmic Contacts to Semiconductors" (B. Schwartz, ed.). Electrochem. Soc., New York.

1509. WONG, J. Y., LOEBNER, E. E., and GARLIEPP, K. L. (1970). Electroluminescent properties of InAs$_{1-x}$Sb$_x$ alloys. *Electrochem. Soc. Extended Abstr., 137th Nat. Meeting, Los Angeles, Spring 1970*, p. 232.

1510. WOODALL, J. M. (1971). Isothermal solution mixing growth of thin Ga$_{1-x}$Al$_x$As layers. *J. Electrochem. Soc.* **118**, 150.

1511. WOODALL, J. M., RUPPRECHT, H., and REUTER, W. (1969). Liquid phase epitaxial growth of Ga$_{1-x}$Al$_x$As. *J. Electrochem. Soc.* **116**, 899.

1512. Woodall, J. M., Lynch, R., and Shang, D. (1970). The electroluminescent and photodetector properties of $Ga_{1-x}Al_xAs$ devices with graded compositions. *Electrochem. Soc. Extended Abstr., 137th Nat. Meeting, Los Angeles, Spring 1970,* p. 208. (Also *J. Electrochem. Soc.* **117**, 94c.)

1513. Woodbury, H. H., and Aven, M. (1968). Some diffusion and solubility measurements of Cu in CdTe. *J. Appl. Phys.* **39**, 5485.

1514. Wright, G. T. (1960). A proposed space-charge-limited dielectric triode. *J. Brit. IRE* **20**, 337.

1515. Wright, G. T. (1961). Mechanisms of space charge limited current in solids. *Solid-State Electron.* **2**, 165.

1516. Wright, G. T. (1962). The space-charge-limited dielectric triode. *Solid-State Electron.* **5**, 117.

1517. Wright, G. T. (1963). Space-charge-limited solid state devices. *Proc. IEEE* **51**, 1642.

1518. Wrigley, C. (1969). Properties of the silicon-sapphire interface in heteroepitaxy. *J. Electrochem. Soc.* **116**, 146c.

1519. Wurst, E. C., and Borneman, E. H. (1957). Rectification properties of metal–silicon contacts. *J. Appl. Phys.* **28**, 235.

1520. Wustenhagen, J. (1964). Contacts to GaAs planar transistor. *Z. Naturforsch.* **19a**, No. 12, 1433.

1521. Wysocki, J. J. (1961). The effect of series resistance on photovoltaic solar energy conversion. *RCA Rev.* **22**, 57.

1522. Wysocki, J. J., and Rappaport, P. (1960). Effect of temperature on photovoltaic solar energy conversion. *J. Appl. Phys.* **31**, 571.

1523. Wysocki, J. J., Rappaport, P., Davison, E., and Loferski, J. J. (1966). Low-energy proton bombardment of GaAs and Si solar cells. *IEEE Trans. Electron Devices* **ED–13**, 420.

1524. Yamamoto, T., and Ota, Y. (1966). The capacitance variation of GaAs surface barrier diode by photoeffect. *Proc. IEEE* **54**, 691.

1525. Yamamoto, T., and Ota, Y. (1968). The gold *n*-type GaAs surface barrier diode and its application to photocapacitors. *Solid-State Electron.* **11**, 219.

1526. Yamato, Y. (1965). Ge–ZnSe heterojunctions. *Jap. J. Appl. Phys.* **4**, 541.

1527. Yasuda, Y. (1970). Epitaxial growth of silicon films evaporated on sapphire and spinel substrates. *Proc. Int. Conf. Phys. Chem. Semicond. Heterojunctions Layer Structures, Budapest, 1970* **3**, 265. Hung. Acad. Sciences, Budapest, Hungary.

1528. Yawata, K. (1965). Investigation of Ge–GaAs tunnel heterodiodes. *NEC Res. Develop.* (*Japan*) **7**, 26.

1529. Yawata, S., and Anderson, R. L. (1965). Optical modulation of current in Ge–Si *n–n* heterojunctions. *Phys. Status Solidi* **12**, 297.

1530. Yeargan, J. R., and Taylor, H. L. (1968). Conduction properties of pyrolytic silicon nitride films. *J. Electrochem. Soc.* **115**, 273.

1531. Yeh, C., and Liu, S. G. (1969). Breakdown characteristics of $Al_xGa_{1-x}As$ avalanche diodes. *Appl. Phys. Lett.* **15**, 391.

1532. Yep, T. O., and Archer, R. J. (1967). Investigation and development of semiconductor metal cathodes. Final Rep., Contract No. AF 19(628)5946. Air Force Cambridge Research Laboratory (15 August 1967). Document AD 660 577 Nat. Tech. Information Service, Springfield, Virginia.

1533. Yim, W. M. (1969). Solid solutions in the pseudobinary (III–V)–(II–VI) systems and their optical energy gaps. *J. Appl. Phys.* **40**, 2617.

1534. Yim, W. M., Dismukes, J. P., and Kressel, H. (1970). Vapor growth of (II–VI)–(III–V) quarternary alloys and their properties. *RCA Rev.* **31**, 662.

1535. YING, C. F., and FARNSWORTH, H. E. (1952). Changes in work function of vacuum distilled gold films. *Phys. Rev.* **85**, 485.

1536. YINGLING, G. S. (1968). Studies on ohmic contact to shallow junction devices. *Trans. Electrochem. Soc. Meeting, October 1968*, Abstract No. 511.

1537. YOUNG, C. E. (1961). Extended curves of the space charge, electric field, and free carrier concentration at the surface of a semiconductor, and curves of the electrostatic potential inside a semiconductor. *J. Appl. Phys.* **32**, 329.

1538. YU, A. Y. C. (1970). The metal–semiconductor contact: On old devices with a new future. *IEEE Spectrum* **7**, March, p. 83.

1539. YU, A. Y. C., and SNOW, E. H. (1968). Surface effects on metal–silicon contacts. *J. Appl. Phys.* **39**, 3008.

1540. YU, A. Y. C., and MEAD, C. A. (1970). Characteristics of aluminum–silicon Schottky barrier diode. *Solid-State Electron.* **13**, 97.

1541. YUAN, H. T., and HOOPER, R. C. (1968). Ohmic contact to a diffused layer in silicon. *Trans. Electrochem. Soc. Meeting, October 1968*, Abstract No. 506.

1542. ZACHOS, T. H., and DYMENT, J. C. (1970). Resonant modes of GaAs junction lasers— III: Propagation characteristics of laser beams with rectangular symmetry. *IEEE J. Quantum Electron.* **QE–6**, 317.

1543. ZACHOS, T. H., and RIPPER, J. E. (1969). Resonant modes of GaAs junction lasers. *IEEE J. Quantum Electron.* **QE–5**, 29.

1544. ZANIO, K. (1970). Chemical diffusion in cadmium telluride. *J. Appl. Phys.* **41**, 1935.

1545. ZASED, V. S. (1969). Differential capacitance of an n–n heterojunction. *Sov. Phys.– Semicond.* **2**, 1538.

1546. ZAVADSKII, YU. I., and KORNILOV, B. V. (1970). Thermo-oscillistor: A new functional temperature sensitive device. *Sov. Phys.–Semicond.* **3**, 1211.

1547. ZEIDENBERGS, G., and ANDERSON, R. L. (1966). A proposed heterojunction field-effect transistor. *Proc. IEEE* **54**, 1960.

1548. ZEIDENBERGS, G., and ANDERSON, R. L. (1967). Si–GaP heterojunctions. *Solid-State Electron.* **10**, 113.

1549. ZETTLEMOYER, A. C. (1969). Chemistry and metallurgy of wetting solid surfaces. In "Ohmic Contacts to Semiconductors" (B. Schwartz, ed.). Electrochem. Soc., New York.

1550. ZETTLER, R. A., and COWLEY, A. M. (1969a). *P–n* junction–Schottky barrier hybrid diode. *IEEE Trans. Electron Devices* **ED–16**, 58.

1551. ZETTLER, R. A., and COWLEY, A. M. (1969b). Hybrid hot carrier diodes. *Hewlett-Packard J.* **20**, 13.

1552. ZOLOMY, I. (1970). Some remarks on the switching behaviour of n–n type heterodiodes. *Proc. Int. Conf. Phys. Chem. Semicond. Heterojunctions Layer Structures, Budapest, 1970* **1**, 377. Hung. Acad. Sciences, Budapest, Hungary.

1553. ZULEEG, R. (1963). A method for CdS thin-film deposition and film structure determination by x-ray and electron diffraction. *Electrochem. Soc. Spring Meeting, Pittsburgh, April 1963*, Abstract No. 95.

1554. ZULEEG, R. (1967). Thin Film Heterojunction Device. U.S. Patent 3,331,998.

1555. ZULEEG, R., and MULLER, R. S. (1964). Space-charge-limited currents and Schottky-emission currents in thin-film CdS diodes. *Solid-State Electron.* **7**, 575.

1556. ZULEEG, R., and SENKOVITS, E. J. (1963). A method for CdS thin film deposition and film structure determination by x-ray and electron diffraction. *Electrochem. Soc. Abstr. Spring Meeting, Pittsburgh, April 1963*, p. 110.

1557. ZWORYKIN, V. K., and RAMBERG, E. G. (1949). "Photoelectricity and Its Applications." Wiley, New York.

Author Index

The numbers without parentheses show the pages of the text on which the author is mentioned. Numbers in parentheses, however, refer to paper numbers in the bibliography.

A

Abdullaev, G. B., (1), (2)
Abe, T., (716)
Abrahams, M. S., (3)–(5), 97, 98, (732)
Abramov, B. G., (6)
Adams, M. J., (7), 149
Addiss, R. R., (1456)
Adirovich, E. I., (8), (9)
Advani, G. T., (10), 42
Agalarzade, P. S., (1248)
Agule, B. J., (1240)
Agusta, B., (11), (12)
Ahlstrom, E., (13), 194
Aitkhozhin, S. A., (14)
Akasaki, I., (85), (86), (89), 235
Akatsuka, M., (964)
Aladinskii, V. K., (15)
Albers, W., (16)
Aldrich, R. W., (817)
Alekseev, Yu. A., (17)
Aleskovsky, V. B., (959)
Alexander, B. H., (819)
Alexander, E. G., (1183)
Alferov, Zh. I., (18)–(48), 143–147
Alfrey, G. F., (49), 304
Alibozek, R. G., (240), 255
Allakhverdyan, R. G., (50)
Allen, F. G., (51), 52, (52), (53), 98, 122, 213, 235, (446)–(448)
Allen, H. A., (54), 300, 301
Allen, J. W., (55), 300
Alliston, H. W., (421)
Almasi, G. S., (56)
Altaf, N., (90)
Altschuler, B. L., (57), (778)
Alyanakyan, S. A., 235, (1125), (1129)
Amick, J. A., (58), 242, 243, (1394)
Amsterdam, M. F., (59)
Anand, Y., (60), 192
Anantha, N. G., (61)

Anderson, D. E., (982)
Anderson, J. S., (62)
Anderson, O. L., (63), 293
Anderson, P. A., (64)
Anderson, R. L., (11), 34, 35, 38–40, 45–47, 50, 53, (65)–(70), 95, 110, 118, 119, 125, 198, 236, (320), (846), (863), (1529), (1547), (1548)
Andreatch, P., (63)
Andreev, V. M., (34)–(36), (39), (40), (42), (45)–(47), (71), (72)
Andrews, J. M., 182, 185, 296, 297, (813)
Andreytchin, R., (667)
Andronik, I. K., (73)
Angelo, E. J., (1239)
Anner, G. E., (74), (75)
Anshon, A. V., (76), (472)
Anstead, R. J., (77)
Antell, G. R., (78), (368)
Antle, W. K., (79)
Antypas, G., (80), 216, (637)
Arata, H., (706)
Archer, R. J., (81)–(84), 164, 231, 290, 297, (1532)
Ariyoshi, H., (225)
Arizumi, T., (85)–(90), 234, 235, (606)
Arlett, R. H., 248, (1143), (1144)
Armantrout, G. A., (91)
Armstrong, A., (92)
Armstrong, H. L., (93), (1289)
Arnold, H., (860)
Arnold, S. R., 83, (94), 99
Arseni, K. A., (95)
Asakawa, I., (1244)
Atalla, M. M., (82), (96), (97), 164, 198, (1318)
Auber, F., (98)
August, R. R., (99)
Aven, M., 20, (100)–(113), 121, 122, 261, 297, 300, 301, 304, (515), (848), (855), (1350), (1513)

Subject Index

The numbers without parentheses show the pages of the text on which the subject matter is discussed. Numbers in parentheses, however, refer to the paper numbers in the bibliography relating to the subject.

A

Abrupt junction, 86, 97
Absorption coefficient, 201, 208
Accumulation region, 112
Ag–O–Cs, 201, 202
Al–Al$_2$O$_3$, (982)
  –CdS, 302, (971)
  –GaAs, 162, 187, 188, 279, (1040)
  –GaSb, 279
  –Mo–SiO$_2$–Si system, (1397)
  –n–Si, 182, 294, (1540)
AlAs, 8, 143, (380), (928), (1005), (1487)
  contacts, 297
  –GaAs, 9, (21), (35), (39), (42)–(45), (71), (72), 143, 279, (363)
Al$_x$Ga$_{1-x}$As, (19), (25), (36), (48), 143, 148, 154, (207), 224, 264, 279, (731), (733), (752), (827), (980), (1184), (1510)–(1512), (1531)
  contacts, 297
  –GaAs, (6), (7), 13, (19), (24), 33, (34), (44), (46), (47), 120, 142, (254), (467), (549), (551), (553), (563), (737)
Al$_x$Ga$_{1-x}$As$_{1-y}$P$_y$–GaAs$_{1-y}$P$_y$, (193)
Al$_x$Ga$_{1-x}$P, 128, 272, 279, (729)
  –GaP, (95), 120
Alkali photocathodes, 202, (1301)–(1303)
Alloying, 93, (140), (141), 226, 228, 267, 274, 277, 280, 289, 290
Alloy heterojunctions, (221), 274, (305), (307), (1122), (1123)
Alloyed junction transistors, 197
Alloys
  (II–VI)–(III–V) quarternary, (1165), (1533), (1534)
AlN, (820)
Al$_2$O$_3$, (170), 260, 271, (465), (529), (555), (685), (916)
  –GaAs, (865)
  –Ge, (720)

AlP, (1142)
  contacts, 297
  –GaP, 279
Alpha, 16, 60
AlSb, 8
  –GaSb, 9, 10, 279
Alumina polyphase heterojunction, (1073)
Amorphous substrate, (223), 250
Anderson model, 34, 114
Antireflection film, 132, 133
Anti-Stokes light converter, 30, (1441)
As, 228, 237, (764)
AsCl$_3$, 237, 245
Au, 69, (675)
  –GaAs, 179, 189, 298
  –Ge, 293, 298
  –Si, 178, 189, 295
Auger effect, 28, 216, 288, (747)
Auger electron spectroscopy, 212, (213), 288, (475), (542)–(544), (1375), (1412)
Auger transistor, 24, 28
Author's certificate, 143
Autodoping, 12, 228, 238, (764)
Avalanche, (24), (74), (104), (148), 194, 224, (256), (274), (751), (796), (840), (841), (891), (1065), (1092), (1353), (1532)

B

Back-scattered Laue X rays, 267, 285
BaF$_2$, 258
Balling, 292
Band diagrams, 3, 81, 98, 201, 289, (340)
Band-gap states, 42
Band minima, 208
Band-pass, 119
Band structure, (247), (370), (634), (741), (1315), (1445)